Advanced
Electrical Installation

T0074475

This new edition covers the City and Guilds 2365-03 course, updated in line with the 18th Edition of the Wiring Regulations. Written in an accessible style with a chapter dedicated to each unit of the syllabus, this book helps you to master each topic before moving on to the next. This new edition includes information on construction and demolition sites, fire proofing, energy efficiency and LED lights, as well as some updated diagrams. End of chapter revision questions help you to check your understanding and consolidate the key concepts learned in each chapter.

- Full colour diagrams and photographs explain difficult concepts
- Clear definitions of technical terms make the book a quick and easy reference
- Extensive online material helps both students and lecturers

The companion website contains videos, animations, worksheets and lesson plans, making it an invaluable resource to both students and lecturers alike. www.routledge.com/cw/linsley

Trevor Linsley was formerly Senior Lecturer in Electrical Engineering at Blackpool and the Fylde College of Technology. There he taught subjects at all levels from first year trainee to first year undergraduate courses; and was also head of the multidiscipline NVQ Assessment Centre and responsible for establishing and running the AM2 Electrical Skills Assessment Centre. He has had 28 books published and has also written many bespoke training packages for local SME electrical engineering companies in the North West of England.

To Joyce, Samantha and Victoria

Advanced
Electrical
Installation
Work 9th Edition

City and Guilds Edition

Trevor Linsley

Routledge
Taylor & Francis Group

LONDON AND NEW YORK

Ninth edition published 2020
by Routledge
2 Park Square, Milton Park, Abingdon, Oxon, OX14 4RN

and by Routledge
52 Vanderbilt Avenue, New York, NY 10017

Routledge is an imprint of the Taylor & Francis Group, an informa business

First edition published by ??
Eighth edition published by Routledge 2015

British Library Cataloguing-in-Publication Data
A catalogue record for this book is available from the British Library

Library of Congress Cataloging-in-Publication Data
A catalog record has been requested for this book

ISBN: 978-0-367-35976-8 (hbk)
ISBN: 978-0-367-35975-1 (pbk)
ISBN: 978-0-429-34297-4 (ebk)

Typeset in 10/13pt Helvetica Neue 55 Roman by
Servis Filmsetting Ltd, Stockport, Cheshire
Printed by CPI Group (UK) Ltd, Croydon CR0 4YY

Visit the companion website: www.routledge.com/linsley

Contents

Preface

The 9th Edition of *Advanced Electrical Installation Work* has been completely rewritten in six chapters to closely match the six units of the City and Guilds 2365-03 qualification. The technical content has been revised and updated to the requirements of the new 18th edition of the IET Wiring Regulations and BS 7671. Improved page design with new illustrations gives greater clarity to each topic.

This book of electrical installation theory and practice will be of value to the electrical trainee working towards:

- The City and Guilds 2365 Level 3 Diploma in Electrical Installations (Buildings and Structures);
- the EAL Diploma in Electrical Installation;
- The City and Guilds 2399 series of Environmental Technologies Qualification;
- The SCOTVEC and BTEC Electrical Utilization Units at Levels II and III.

Advanced Electrical Installation Work provides a sound basic knowledge of electrical theory and practice which other trades in the construction industry will find of value, particularly those involved in multi-skilled activities.

The book incorporates the requirements of the latest regulations, particularly:

- 18th Edition IET Wiring Regulations;
- *On Site Guide* BS 7671: 2018 Requirements for Electrical Installations;
- British Standards BS 7671: 2018;
- Part P of the Building Regulations, Electrical Safety in Dwellings 2006;
- Hazardous Waste Regulations 2005;
- Work at Height Regulations 2005.

Trevor Linsley
2019

Acknowledgements

I would like to acknowledge the assistance given by the following manufacturers and professional organizations in the preparation of this book:

- Crabtree Electrical Industries for technical information and data;
- RS Components Limited for technical information and photographs;
- Stocksigns Limited for technical information and photographs;
- Wylex Electrical Components for technical information and photographs;
- iStockphoto and Shutterstock for photographs;
- Martindale-electric.co.uk for technical information and photographs;
- Tenby electrical products for photographs;
- Legrand photographs visit legrand.co.uk for more information.
- Wago connectors for photographs.

I would like to thank the many college lecturers who responded to the questionnaire from Taylor & Francis the publishers regarding the proposed new edition of this book. Their recommendations have been taken into account in producing this improved 9th Edition.

I would also like to thank the editorial and production staff at Taylor & Francis the publishers for their enthusiasm and support and particularly my editors Gabriella Williams and Tony Moore. They were able to publish this 9th Edition within the very short time-scale created by the publication of the new 18th Edition of the IET Wiring Regulations.

A special thank you must also go to my colleagues Elliot Parkinson and John Gallagher, an electrical installation lecturer at Blackpool and The Fylde College, for writing the new section on preparing for the online assessment examination.

Finally, I would like to thank Joyce, Samantha and Victoria for their support and encouragement.

Health and safety in building services engineering

Unit 201/501 of the City and Guilds 2365-03 syllabus

Learning outcomes – when you have completed this chapter you should:

- know about health and safety legislation;
- know how to handle hazardous situations;
- know the electrical safety requirements when working in the building services industry;
- know the safety requirements when working with heat-producing equipment;
- know the safety requirements when using access equipment;
- know the safety requirements when working in confined spaces and excavations;
- be able to apply safe working practices to manual handling and using access equipment.

Advanced Electrical Installation Work. 978-0-367-35976-8
© 2019 Trevor Linsley. Published by Taylor & Francis. All rights reserved.

This chapter has free associated content, including animations and instructional videos, to support your learning.

When you see the logo, visit the companion website for more on this topic www.routledge.com/cw/linsley

The 9th Edition of *Advanced Electrical Installation Work* covers the topics in the City and Guilds Level 3 Diploma in Electrical Installations (Buildings and Structures).

The six chapters of this book closely match the learning outcomes in the six units of the City and Guilds Level 3 Diploma.

The first unit in both the Level 2 Diploma and the Level 3 Diploma covers the essential health and safety learning outcomes described in unit 201 of the City and Guilds syllabus. Therefore, Chapter 1 of this book covers unit 201 of the City and Guilds qualification.

However, if this is the beginning of your second year studying for the Diploma in Electrical Installation Work (Buildings and Structures), you may have already received a credit for unit 201. If you believe that you have already received a credit for unit 201, you **may not** be required to study the topics in Chapter 1 of this book and can go straight on to the other five chapters. You should check **your own position** with your trainer or assessor.

The City and Guilds unit 201 is covered in Chapter 1 of this book for those students who have not yet received a credit for unit 201.

Important note!

You may have already studied the material in this chapter – check with your trainer or assessor before going any further.

Safety regulations and laws

At the beginning of the nineteenth century children formed a large part of the working population of Great Britain. They started work early in their lives and they worked long hours for unscrupulous employers or masters.

The Health and Morals of Apprentices Act of 1802 was introduced by Robert Peel in an attempt at reducing apprentice working hours to 12 hours per day and improving the conditions of their employment. The Factories Act of 1833 restricted the working week for children aged 13–18 years to 69 hours in any working week.

With the introduction of the Factories Act of 1833, the first four full-time Factory Inspectors were appointed. They were allowed to employ a small number of assistants and were given the responsibility of inspecting factories throughout England, Scotland, Ireland and Wales. This small, overworked band of men were the forerunners of the modern HSE Inspectorate, enforcing the safety laws passed by Parliament. As the years progressed, new Acts of Parliament increased the powers of the Inspectorate and the growing strength of the trade unions meant that employers were increasingly being pressed to improve health, safety and welfare at work.

The most important recent piece of health and safety law was passed by Parliament in 1974 called the Health and Safety at Work Act. This Act gave added powers to the Inspectorate and is the basis of all modern statutory health and safety laws. This law not only increased the employer's liability for safety measures, but also put the responsibility for safety on employees too.

Health, safety and welfare legislation has increased the awareness of everyone to the risks involved in the workplace. All statutes within the Acts of Parliament must be obeyed and, therefore, we all need an understanding of the laws as they apply to the electrical industry.

Statutory laws

Acts of Parliament are made up of Statutes. Statutory Regulations have been passed by Parliament and have, therefore, become laws. Non-compliance with the laws of this land may lead to prosecution by the Courts and possible imprisonment for offenders.

We shall now look at some of the Statutory Regulations as they apply to the electrical industry.

The Health and Safety at Work Act 1974

Many governments have passed laws aimed at improving safety at work, but the most important recent legislation has been the Health and Safety at Work Act 1974. The purpose of the Act is to provide the legal framework for stimulating and encouraging high standards of health and safety at work; the Act puts the responsibility for safety at work on both workers and managers.

The employer has a duty to care for the health and safety of employees (Section 2 of the Act). To do this he or she must ensure that:

- the working conditions and standard of hygiene are appropriate;
- the plant, tools and equipment are properly maintained;
- the necessary safety equipment – such as personal protective equipment (PPE), dust and fume extractors and machine guards – is available and properly used;
- the workers are trained to use equipment and plant safely.

Figure 1.1 Both workers and managers are responsible for health and safety on site.

Employees have a duty to care for their own health and safety and that of others who may be affected by their actions (Section 7 of the Act). To do this they must:

- take reasonable care to avoid injury to themselves or others as a result of their work activity;
- cooperate with their employer, helping him or her to comply with the requirements of the Act;
- not interfere with or misuse anything provided to protect their health and safety.

Failure to comply with the Health and Safety at Work Act is a criminal offence and any infringement of the law can result in heavy fines, a prison sentence or both.

Enforcement of health and safety regulations

Laws and rules must be enforced if they are to be effective. The system of control under the Health and Safety at Work Act comes from the Health and Safety Executive (HSE) which is charged with enforcing the law. The HSE is divided into a number of specialist inspectorates or sections which operate from local offices throughout the United Kingdom. From the local offices the inspectors visit individual places of work.

The HSE inspectors have been given wide-ranging powers to assist them in the enforcement of the law. They can:

1 Enter premises unannounced and carry out investigations, take measurements or photographs.
2 Take statements from individuals.
3 Check the records and documents required by legislation.
4 Give information and advice to an employee or employer about safety in the workplace.
5 Demand the dismantling or destruction of any equipment, material or substance likely to cause immediate serious injury.
6 Issue an improvement notice which will require an employer to put right, within a specified period of time, a minor infringement of the legislation.
7 Issue a prohibition notice which will require an employer to stop immediately any activity likely to result in serious injury, and which will be enforced until the situation is corrected.
8 Prosecute all persons who fail to comply with their safety duties, including employers, employees, designers, manufacturers, suppliers and the self-employed.

Safety documentation

Under the Health and Safety at Work Act, the employer is responsible for ensuring that adequate instruction and information is given to employees to make them safety conscious. Part 1, Section 3 of the Act instructs all employers to prepare a written health and safety policy statement and to bring this to the notice of all employees. Figure 1.2 shows a typical Health and Safety Policy Statement of the type which will be available within your company. Your employer must let you know who your safety representatives are, and the new Health and Safety poster shown in Fig. 1.3 has a blank section into which the names and contact information of your specific representatives can be added. This is a large laminated poster, 595 × 415 mm, suitable for wall or notice-board display.

All workplaces employing five or more people had to display the type of poster shown in Fig. 1.3 after 30 June 2000.

To promote adequate health and safety measures the employer must consult with the employees' safety representatives. In companies which employ more than 20 people this is normally undertaken by forming a safety committee which is made up of a safety officer and employee representatives, usually nominated by a trade union. The safety officer is usually employed full-time in that role. Small companies might employ a safety supervisor who will have other duties within the company, or alternatively they could join a 'safety group'. The safety group then shares the cost of employing a safety adviser or safety officer, who visits each company in rotation. An employee who identifies a dangerous situation should initially report to his site safety representative. The safety representative should then bring the dangerous situation to the notice of the safety committee for action which will remove the danger. This may mean changing company policy or procedures or making modifications to equipment. All actions of the safety committee should be documented and recorded as evidence that the company takes its health and safety policy seriously.

FLASH-BANG ELECTRICAL

Statement of Health and Safety at Work Policy in accordance with the Health and Safety at Work Act 1974

Company objective

The promotion of health and safety measures is a mutual objective for the Company and for its employees at all levels. It is the intention that all the Company's affairs will be conducted in a manner which will not cause risk to the health and safety of its members, employees or the general public. For this purpose it is the Company policy that the responsibility for health and safety at work will be divided between all the employees and the Company in the manner outlined below.

Company's responsibilities

The Company will, as a responsible employer, make every endeavour to meet its legal obligations under the Health and Safety at Work Act to ensure the health and safety of its employees and the general public. Particular attention will be paid to the provision of the following:

1 Plant equipment and systems of work that are safe.
2 Safe arrangements for the use, handling, storage and transport of articles, materials and substances.
3 Sufficient information, instruction, training and supervision to enable all employees to contribute positively to their own safety and health at work and to avoid hazards.
4 A safe place of work, and safe access to it.
5 A healthy working environment.
6 Adequate welfare services.

Note: Reference should be made to the appropriate safety etc. manuals.

Employees' responsibilities

Each employee is responsible for ensuring that the work which he/she undertakes is conducted in a manner which is safe to himself or herself, other members of the general public, and for obeying the advice and instructions on safety and health matters issued by his/her superior. If any employee considers that a hazard to health and safety exists it is his/her responsibility to report the matter to his/her supervisor or through his/her Union Representative or such other person as may be subsequently defined.

Management and supervisors' responsibilities

Management and supervisors at all levels are expected to set an example in safe behaviour and maintain a constant and continuing interest in employee safety, in particular by:

1 acquiring the knowledge of health and safety regulations and codes of practice necessary to ensure the safety of employees in the workplace,
2 acquainting employees with these regulations on codes of practice and giving guidance on safety matters,
3 ensuring that employees act on instructions and advice given.

General Managers are ultimately responsible to the Company for the rectification or reporting of any safety hazard which is brought to their attention.

Joint consultations

Joint consultation on health and safety matters is important. The Company will agree with its staff, or their representatives, adequate arrangements for joint consultation on measures for promoting safety and health at work, and make and maintain satisfactory arrangements for the participation of their employees in the development and supervision of such measures. Trade Union representatives will initially be regarded as undertaking the role of Safety Representatives envisaged in the Health and Safety at Work Act. These representatives share a responsibility with management to ensure the health and safety of their members and are responsible for drawing the attention of management to any shortcomings in the Company's health and safety arrangements. The Company will in so far as is reasonably practicable provide representatives with facilities and training in order that they may carry out this task.

Review

A review, addition or modification of this statement may be made at any time and may be supplemented as appropriate by further statements relating to the work of particular departments and in accordance with any new regulations or codes of practice.

This policy statement will be brought to the attention of all employees.

Figure 1.2 Typical Health and Safety Policy Statement.

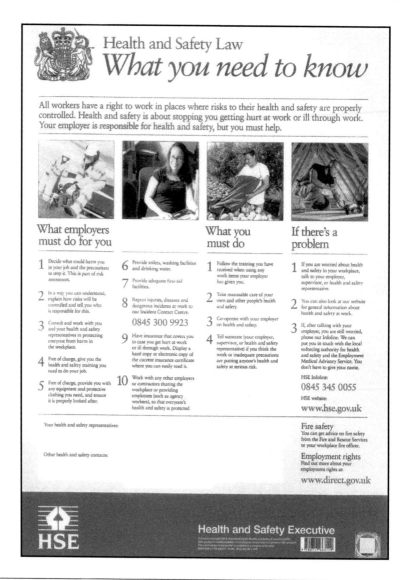

Figure 1.3 Health and Safety law poster. *Source*: HSE © Crown copyright material is reproduced with the permission of the Controller of HMSO and Her Majesty's Stationery Office, Norwich.

The Electricity Safety, Quality and Continuity Regulations 2002 (formerly Electricity Supply Regulations 1989)

The Electricity Safety, Quality and Continuity Regulations 2002 are issued by the Department of Trade and Industry. They are statutory regulations which are enforceable by the laws of the land. They are designed to ensure a proper and safe supply of electrical energy up to the consumer's terminals.

These regulations impose requirements upon the regional electricity companies regarding the installation and use of electric lines and equipment. The regulations are administered by the Engineering Inspectorate of the Electricity Division of the Department of Energy and will not normally concern the electrical contractor, except that it is these regulations which lay down the earthing requirement of the electrical supply at the meter position.

The regional electricity companies must declare the supply voltage and maintain its value between prescribed limits or tolerances.

The government agreed on 1 January 1995 that the electricity supplies in the United Kingdom would be harmonized with those of the rest of Europe. Thus the voltages used previously in low-voltage supply systems of 415 V and 240 V have become 400 V for three-phase supplies and 230 V for single-phase supplies. The permitted tolerances to the nominal voltage have also been changed from ±6% to +10% and −6%. This gives a voltage range of 216–253 V for a nominal voltage of 230 V and 376–440 V for a nominal supply voltage of 400 V.

The next proposed change is for the tolerance levels to be adjusted to ±10% of the declared nominal voltage (IET Regulation, Appendix 2:14).

The frequency is maintained at an average value of 50 Hz over 24 hours so that electric clocks remain accurate.

Regulation 29 gives the area boards the power to refuse to connect a supply to an installation which in their opinion is not constructed, installed and protected to an appropriately high standard. This regulation would only be enforced if the installation did not meet the requirements of the IET Regulations for Electrical Installations.

The Electricity at Work Regulations 1989 (EWR)

This legislation came into force in 1990 and replaced earlier regulations such as the Electricity (Factories Act) Special Regulations 1944. The regulations are made under the Health and Safety at Work Act 1974, and enforced by the Health and Safety Executive. The purpose of the regulations is to 'require precautions to be taken against the risk of death or personal injury from electricity in work activities'.

Section 4 of the EWR tells us that 'all systems must be constructed so as to prevent danger …, and be properly maintained. … Every work activity shall be carried out in a manner which does not give rise to danger. … In the case of work of an electrical nature, it is preferable that the conductors be made dead before work commences.'

The EWR do not tell us specifically how to carry out our work activities and ensure compliance, but if proceedings were brought against an individual for breaking the EWR, the only acceptable defence would be 'to prove that all reasonable steps were taken and all diligence exercised to avoid the offence' (Regulation 29).

An electrical contractor could reasonably be expected to have 'exercised all diligence' if the installation was wired according to the IET Wiring Regulations (see p. 13). However, electrical contractors must become more 'legally aware' following the conviction of an electrician for manslaughter at Maidstone Crown Court in 1989. The court accepted that an electrician had caused the death of another man as a result of his shoddy work in wiring up a central heating system. He received a nine-month suspended prison sentence. This case has set an important legal precedent, and in future any tradesman or professional who causes death through negligence or poor workmanship risks prosecution and possible imprisonment.

Duty of care

The Health and Safety at Work Act and the Electricity at Work Regulations make numerous references to employer and employees having a '**duty of care**' for the health and safety of others in the work environment. In this context the Electricity at Work Regulations refer to a person as a '**duty holder**'. This phrase recognizes the level of responsibility which electricians are expected to take on as a part of their job in order to control electrical safety in the work environment.

Everyone has a duty of care, but not everyone is a duty holder. The regulations recognize the amount of control that an individual might exercise over the whole electrical installation. The person who exercises 'control over the whole systems, equipment and conductors' and is the electrical company's representative on-site is the *duty holder*. He might be a supervisor or manager, but he will have a duty of care on behalf of his employer for the electrical, health, safety and environmental issues on that site.

Duties referred to in the regulations may have the qualifying terms '**reasonably practicable**' or '**absolute**'. If the requirement of the regulation is absolute, then that regulation must be met regardless of cost or any other consideration. If the regulation is to be met 'so far as is reasonably practicable', then risks, cost, time, trouble and difficulty can be considered.

Often there is a cost-effective way to reduce a particular risk and prevent an accident from occurring. For example, placing a fireguard in front of the fire at home when there are young children in the family is a reasonably practicable way of reducing the risk of a child being burned.

If a regulation is not qualified with 'so far as is reasonably practicable', then it must be assumed that the regulation is absolute. In the context of the Electricity at Work Regulations, where the risk is very often death by electrocution, the level of duty to prevent danger more often approaches that of an absolute duty of care.

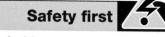

Safety first

Duty holder

This person has the responsibility to control electrical safety in the work environment.

The Management of Health and Safety at Work Regulations 1999

The Health and Safety at Work Act 1974 places responsibilities on employers to have robust health and safety systems and procedures in the workplace. Directors and managers of any company who employ more than five employees can be held personally responsible for failures to control health and safety.

The Management of Health and Safety at Work Regulations 1999 tell us that employers must systematically examine the workplace, the work activity and the management of safety in the establishment through a process of 'risk assessments'. A record of all significant risk assessment findings must be kept in a safe place and be available to an HSE Inspector if required. Information based on these findings must be communicated to relevant staff and, if changes in work behaviour patterns are recommended in the interests of safety, they must be put in place. The process of risk assessment is considered in detail later in this chapter.

Risks, which may require a formal assessment in the electrical industry, might be:

- working at heights;
- using electrical power tools;
- falling objects;
- working in confined places;
- electrocution and personal injury;
- working with 'live' equipment;
- using hire equipment;
- manual handling – pushing – pulling – lifting;

- site conditions – falling objects – dust – weather – water – accidents and injuries.

And any other risks which are particular to a specific type of workplace or work activity.

The Control of Substances Hazardous to Health Regulations 2002 (COSHH)

The original COSHH Regulations were published in 1988 and came into force in October 1989. They were re-enacted in 1994 with modifications and improvements, and the latest modifications and additions came into force in 2002.

The COSHH Regulations control people's exposure to hazardous substances in the workplace. Regulation 6 requires employers to assess the risks to health from working with hazardous substances, to train employees in techniques which will reduce the risk and provide personal protective equipment (PPE) so that employees will not endanger themselves or others through exposure to hazardous substances. Employees should also know what cleaning, storage and disposal procedures are required and what emergency procedures to follow. The necessary information must be available to anyone using hazardous substances as well as to visiting HSE Inspectors.

Hazardous substances include:

1 any substance which gives off fumes causing headaches or respiratory irritation;
2 man-made fibres which might cause skin or eye irritation (e.g. loft insulation);
3 acids causing skin burns and breathing irritation (e.g. car batteries, which contain dilute sulphuric acid);
4 solvents causing skin and respiratory irritation (strong solvents are used to cement together PVC conduit fittings and tube);
5 fumes and gases causing asphyxiation (burning PVC gives off toxic fumes);
6 cement and wood dust causing breathing problems and eye irritation;
7 exposure to asbestos – although the supply and use of the most hazardous asbestos material is now prohibited, huge amounts were installed between 1950 and 1980 in the construction industry and much of it is still in place today. In their latest amendments, the COSHH Regulations focus on giving advice and guidance to builders and contractors on the safe use and control of asbestos products. These can be found in Guidance Notes EH 71 or visit www.hse.gov.uk/hiddenkiller.

Figure 1.4

Where PPE is provided by an employer, employees have a duty to use it to safeguard themselves as shown in Fig. 1.4.

Provision and Use of Work Equipment Regulations 1998

These regulations tidy up a number of existing requirements already in place under other regulations such as the Health and Safety at Work Act 1974, the Factories Act 1961 and the Offices, Shops and Railway Premises Act 1963.

The Provision and Use of Work Equipment Regulations 1998 place a general duty on employers to ensure minimum requirements of plant and equipment.

Figure 1.5 All workers on site must wear head protection.

If an employer has purchased good-quality plant and equipment which is well maintained, there is little else to do. Some older equipment may require modifications to bring it into line with modern standards of dust extraction, fume extraction or noise, but no assessments are required by the regulations other than those generally required by the Management Regulations 1999 discussed previously.

The Construction (Health, Safety and Welfare) Regulations 1996

An electrical contractor is a part of the construction team, usually as a subcontractor, and therefore the regulations particularly aimed at the construction industry also influence the daily work procedures and environment of an electrician. The most important recent piece of legislation is the Construction Regulations.

The temporary nature of construction sites makes them one of the most dangerous places to work. These regulations are made under the Health and Safety at Work Act 1974 and are designed specifically to promote safety at work in the construction industry. Construction work is defined as any building or civil engineering work, including construction, assembly, alterations, conversions, repairs, upkeep, maintenance or dismantling of a structure.

The general provision sets out minimum standards to promote a good level of safety on-site. Schedules specify the requirements for guardrails, working platforms, ladders, emergency procedures, lighting and welfare facilities. Welfare facilities set out minimum provisions for site accommodation: washing facilities, sanitary conveniences and protective clothing. There is now a duty for all those working on construction sites to wear head protection, and this includes electricians working on-site as subcontractors.

Personal Protective Equipment (PPE) at Work Regulations 1998

PPE is defined as all equipment designed to be worn, or held, to protect against a risk to health and safety. This includes most types of protective clothing, and equipment such as eye, foot and head protection, safety harnesses, life-jackets and high-visibility clothing.

Under the Health and Safety at Work Act, employers must provide free of charge any PPE and employees must make full and proper use of it. Safety signs such as those shown at Fig. 1.6 are useful reminders of the type of PPE to be used in a particular area. The vulnerable parts of the body which may need protection are the head, eyes, ears, lungs, torso, hands and feet; in addition, protection from falls may need to be considered. Objects falling from a height present the major hazard against which head protection is provided. Other hazards include striking the head against projections and hair becoming entangled in machinery. Typical methods of protection include helmets, light-duty scalp protectors called 'bump caps' and hairnets.

The eyes are very vulnerable to liquid splashes, flying particles and light emissions such as ultraviolet light, electric arcs and lasers. Types of eye protectors include safety spectacles, safety goggles and face shields. Screen-based workstations are being used increasingly in industrial and commercial

---

Definition

PPE is defined as all equipment designed to be worn, or held, to protect against a risk to health and safety.

Safety first

PPE

Always wear or use the PPE (personal protective equipment) provided by your employer for your safety.

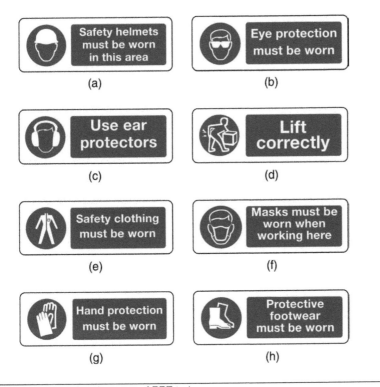

(a) Safety helmets must be worn in this area

(b) Eye protection must be worn

(c) Use ear protectors

(d) Lift correctly

(e) Safety clothing must be worn

(f) Masks must be worn when working here

(g) Hand protection must be worn

(h) Protective footwear must be worn

Figure 1.6 Safety signs showing type of PPE to be worn.

locations by all types of personnel. Working with VDUs (visual display units) can cause eye strain and fatigue.

Noise is accepted as a problem in most industries and surprisingly there has been very little control legislation. The Health and Safety Executive have published a 'Code of Practice' and 'Guidance Notes' HSG 56 for reducing the exposure of employed persons to noise. A continuous exposure limit of below 85 dB for an eight-hour working day is recommended by the Code.

Noise may be defined as any disagreeable or undesirable sound or sounds, generally of a random nature, which do not have clearly defined frequencies. The usual basis for measuring noise or sound level is the decibel scale. Whether noise of a particular level is harmful or not also depends on the length of exposure to it. This is the basis of the widely accepted limit of 85 dB of continuous exposure to noise for eight hours per day.

A peak sound pressure of above 200 pascals or about 120 dB is considered unacceptable and 130 dB is the threshold of pain for humans. If a person has to shout to be understood at 2 metres, the background noise is about 85 dB. If the distance is only 1 metre, the noise level is about 90 dB. Continuous noise at work causes deafness, makes people irritable, affects concentration, causes fatigue and accident proneness, and may mask sounds which need to be heard in order to work efficiently and safely.

It may be possible to engineer out some of the noise, for example, by placing a generator in a separate sound-proofed building. Alternatively, it may be possible to provide job rotation, to rearrange work locations or provide acoustic refuges.

Where individuals must be subjected to some noise at work, it may be reduced by ear protectors. These may be disposable ear plugs, reusable ear plugs or ear muffs. The chosen ear protector must be suited to the user and suitable for the type of noise, and individual personnel should be trained in its correct use.

Breathing reasonably clean air is the right of every individual, particularly at work. Some industrial processes produce dust which may present a potentially serious hazard. The lung disease asbestosis is caused by the inhalation of asbestos dust or particles and the coal dust disease pneumoconiosis, suffered by many coal-miners, has made people aware of the dangers of breathing in contaminated air.

Some people may prove to be allergic to quite innocent products such as flour dust in the food industry or wood dust in the construction industry. The main effect of inhaling dust is a measurable impairment of lung function. This can be avoided by wearing an appropriate mask, respirator or breathing apparatus as recommended by the company's health and safety policy and indicated by local safety signs.

A worker's body may need protection against heat or cold, bad weather, chemical or metal splash, impact or penetration and contaminated dust. Alternatively, there may be a risk of the worker's own clothes causing contamination of the product, as in the food industry. Appropriate clothing will be recommended in the company's health and safety policy. Ordinary working clothes and clothing provided for food hygiene purposes are not included in the Personal Protective Equipment at Work Regulations.

Hands and feet may need protection from abrasion, temperature extremes, cuts and punctures, impact or skin infection. Gloves or gauntlets provide protection from most industrial processes, but should not be worn when operating machinery because they may become entangled in it. Care in selecting the appropriate protective device is required; for example, barrier creams provide only a limited protection against infection.

Safety first

Safety signs

Always follow the instructions given in the safety signs where you are working – it will help to keep you safe.

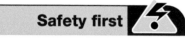

Try this

PPE

* Make a list of any PPE which you have used at work.
* What was this PPE protecting you from?

Boots or shoes with in-built toe-caps can give protection against impact or falling objects and, when fitted with a mild steel sole plate, can also provide protection from sharp objects penetrating through the sole. Special slip-resistant soles can also be provided for employees working in wet areas.

Whatever the hazard to health and safety at work, the employer must be able to demonstrate that he or she has carried out a risk analysis, made recommendations which will reduce that risk and communicated these recommendations to the workforce. Where there is a need for PPE to protect against personal injury and to create a safe working environment, the employer must provide that equipment and any necessary training which might be required and the employee must make full and proper use of such equipment and training.

Definition

Statutory laws and regulations are written in a legal framework.

Non-statutory regulations

Statutory laws and regulations are written in a legal framework; some don't actually tell us how to comply with the laws at an everyday level.

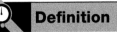

Non-statutory regulations and codes of practice interpret the statutory regulations, telling us how we can comply with the law.

They have been written for every specific section of industry, commerce and situation, to enable everyone to comply with or obey the written laws.

When the Electricity at Work Regulations (EWR) tell us to 'ensure that all systems are constructed so as to prevent danger' they do not tell us how to actually do this in a specific situation. However, the IET Regulations tell us precisely how to carry out our electrical work safely in order to meet the statutory requirements of the EWR. In Part 1 of the IET Regulations, at 114, it states: 'the Regulations are non-statutory. They may, however, be used in a court of law in evidence to claim compliance with a statutory requirement.' If your electrical installation work meets the requirements of the IET Regulations, you will also meet the requirements of EWR.

Over the years, non-statutory regulations and codes of practice have built upon previous good practice and responded to changes by bringing out new editions of the various regulations and codes of practice to meet the changing needs of industry and commerce.

We will now look at one non-statutory regulation, what is sometimes called 'the electrician's bible', the most important set of regulations for anyone working in the electrical industry, the BS 7671: 2008 Requirements for Electrical Installations, IET Wiring Regulations 17th Edition.

The IET Wiring Regulations 18th edition requirements for electrical installations to BS 7671: 2018

The Institution of Engineering and Technology Requirements for Electrical Installations (the IET Regulations) are non-statutory regulations. They relate principally to the design, selection, erection, inspection and testing of electrical installations, whether permanent or temporary, in and about buildings generally and to agricultural and horticultural premises, construction sites and caravans and their sites. Paragraph 7 of the introduction to the EWR says: 'the IET Wiring Regulations is a code of practice which is widely recognized and accepted in the United Kingdom and compliance with them is likely to achieve compliance with all relevant aspects of the Electricity at Work Regulations.' The IET Wiring Regulations are the national standard in the United Kingdom and apply to installations operating at a voltage up to 1000 V a.c. They do not apply to electrical installations in mines and quarries, where special regulations apply because of the adverse conditions experienced there.

The current edition of the IET Wiring Regulations is the 18th Edition 2018. The main reason for incorporating the IET Wiring Regulations into British Standard BS 7671: 2018 was to create harmonization with European Standards.

The IET Regulations take account of the technical intent of the CENELEC European Standards, which in turn are based on the IEC International Standards.

The purpose in harmonizing British and European Standards is to help develop a single European market economy so that there are no trade barriers to electrical goods and services across the European Economic Area.

To assist electricians in their understanding of the regulations a number of guidance notes have been published. The guidance notes which I will frequently

Figure 1.7 This kind of thing can happen if you're not up to date with IET wiring regulations.

make reference to in this book are those contained in the *On Site Guide.* Eight other guidance notes booklets are also currently available. These are:

- *Selection and Erection*;
- *Isolation and Switching*;
- *Inspection and Testing*;
- *Protection against Fire*;
- *Protection against Electric Shock*;
- *Protection against Overcurrent*;
- *Special Locations*;
- *Earthing and Bonding*.

These guidance notes are intended to be read in conjunction with the regulations.

The IET Wiring Regulations are the electrician's bible and provide the authoritative framework of information for anyone working in the electrical industry.

Health and safety responsibilities

We have now looked at statutory and non-statutory regulations which influence working conditions in the electrical industry today. So, who has *responsibility* for these workplace Health and Safety Regulations?

In 1970, a Royal Commission was set up to look at the health and safety of employees at work. The findings concluded that the main cause of accidents at work was apathy on the part of *both* employers and employees.

The Health and Safety at Work Act 1974 was passed as a result of recommendations made by the Royal Commission and, therefore, the Act puts legal responsibility for safety at work on *both* the employer and employee.

In general terms, the employer must put adequate health and safety systems in place at work and the employee must use all safety systems and procedures responsibly.

In specific terms the employer must:

- provide a Health and Safety Policy Statement if there are five or more employees such as that shown in Fig. 1.2;
- display a current employers liability insurance certificate as required by the Employers Liability (Compulsory Insurance) Act 1969;
- report certain injuries, diseases and dangerous occurrences to the enforcing authority (HSE area office – see Appendix B for address);
- provide adequate first aid facilities (see Tables 1.1 and 1.2);
- provide PPE;
- provide information, training and supervision to ensure staffs' health and safety;
- provide adequate welfare facilities;
- put in place adequate precautions against fire, provide a means of escape and means of fighting fire;
- ensure plant and machinery are safe and that safe systems of operation are in place;
- ensure articles and substances are moved, stored and used safely;
- make the workplace safe and without risk to health by keeping dust, fumes and noise under control.

Table 1.1 Suggested numbers of first aid personnel

Category of risk	Numbers employed at any location	Suggested number of first aid personnel
Lower risk e.g. shops and offices, libraries	Fewer than 50 50–100 More than 100	At least one appointed person At least one first aider One additional first aider for every 100 employed
Medium risk e.g. light engineering and assembly work, food processing, warehousing	Fewer than 20 20–100 More than 100	At least one appointed person At least one first aider for every 50 employed (or part thereof) One additional first aider for every 100 employed
Higher risk e.g. most construction, slaughterhouses, chemical manufacture, extensive work with dangerous machinery or sharp instruments	Fewer than five 5–50 More than 50	At least one appointed person At least one first aider One additional first aider for every 50 employed

Table 1.2 Contents of first aid boxes

Item	Number of employees				
	1–5	6–10	11–50	51–100	101–150
Guidance card on general first aid	1	1	1	1	1
Individually wrapped sterile adhesive dressings	10	20	40	40	40
Sterile eye pads, with attachment (Standard Dressing No. 16 BPC)	1	2	4	6	8
Triangular bandages	1	2	4	6	8
Sterile covering for serious wounds (where applicable)	1	2	4	6	8
Safety-pins	6	6	12	12	12
Medium-sized sterile unmedicated dressings (Standard Dressings No. 9 and No. 14 and the Ambulance Dressing No. 1)	3	6	8	10	12
Large sterile unmedicated dressings (Standard Dressings No. 9 and No. 14 and the Ambulance Dressing No. 1)	1	2	4	6	10
Extra-large sterile unmedicated dressings (Ambulance Dressing No. 3)	1	2	4	6	8
Where tap water is not available, sterile water or sterile normal saline in disposable containers (each holding a minimum of 300 ml) must be kept near the first aid box. The following minimum quantities should be kept:					
Number of employees	1–10	11–50	51–100	101–150	
Quantity of sterile water	1 × 300 ml	3 × 300 ml	6 × 300 ml	6 × 300 ml	

In specific terms the employee must:

- take reasonable care of his/her own health and safety and that of others who may be affected by what they do;
- cooperate with his/her employer on health and safety issues by not interfering with or misusing anything provided for health, safety and welfare in the working environment;
- report any health and safety problem in the workplace to, in the first place, a supervisor, manager or employer.

Categories of safety signs

The rules and regulations of the working environment are communicated to employees by written instructions, signs and symbols. All signs in the working environment are intended to inform. They should give warning of possible dangers and must be obeyed. At first there were many different safety signs, but British Standard BS 5499 Part 1 and the Health and Safety (Signs and Signals) Regulations 1996 have introduced a standard system which gives health and safety information with the minimum use of words. The purpose of the regulations is to establish an internationally understood system of safety signs and colours which draw attention to equipment and situations that do, or could, affect health and safety. Text-only safety signs became illegal from 24 December 1998. From that date, all safety signs have had to contain a pictogram or symbol such as those shown in Fig. 1.9. Signs fall into four categories: prohibited activities; warnings; mandatory instructions, and safe conditions.

Prohibition signs

These are *must not do* signs. These are circular white signs with a red border and red cross-bar, and are given in Fig. 1.9. They indicate an activity *which must not* be done.

Warning signs

These give safety information. These are triangular yellow signs with a black border and symbol, and are given in Fig. 1.10. They *give warning* of a hazard or danger.

Mandatory signs

These are *must do* signs. These are circular blue signs with a white symbol, and are given in Fig. 1.11. They *give instructions* which must be obeyed.

Advisory or safe condition signs

These give safety information. These are square or rectangular green signs with a white symbol, and are given in Fig. 1.12. They *give information* about safety provision.

Figure 1.8 Text-only safety signs do not comply.

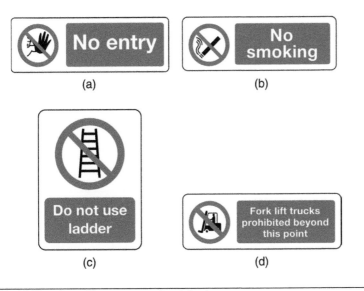

Figure 1.9 Prohibition signs. These are *must not* do signs.

Figure 1.10 Warning signs. These give safety information.

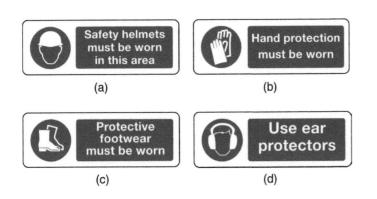

Figure 1.11 Mandatory signs. These are *must do* signs.

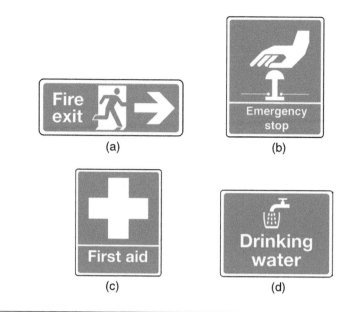

Figure 1.12 Advisory or safe condition signs. These also give safety information.

Accidents at work

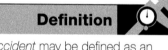

Despite new legislation, improved information, education and training, accidents at work do still happen. An **accident** may be defined as an uncontrolled event causing injury or damage to an individual or property. An accident can nearly always be avoided if correct procedures and methods of working are followed. Any accident which results in an absence from work for more than three days or causes a major injury or death is notifiable to the HSE. There are more than 40,000 accidents reported to the HSE each year which occur as a result of some building-related activity. To avoid having an accident you should:

1 follow all safety procedures (e.g. fit safety signs when isolating supplies and screen off work areas from the general public);
2 not misuse or interfere with equipment provided for health and safety;
3 dress appropriately and use PPE when appropriate;
4 behave appropriately and with care;
5 avoid over-enthusiasm and foolishness;
6 stay alert and avoid fatigue;
7 not use alcohol or drugs at work;
8 work within your level of competence;
9 attend safety courses and read safety literature;
10 take a positive decision to act and work safely.

If you observe a hazardous situation at work, first make the hazard safe, using an appropriate method, or screen it off, but only if you can do so without putting yourself or others at risk, then report the situation to your safety representative or supervisor.

Definition

Fire is a chemical reaction which will continue if fuel, oxygen and heat are present.

Fire control

Fire is a chemical reaction which will continue if fuel, oxygen and heat are present. To eliminate a fire *one* of these components must be removed. This is often expressed by means of the fire triangle shown in Fig. 1.13; all three corners of the triangle must be present for a fire to burn.

Fuel

Fuel is found in the construction industry in many forms: petrol and paraffin for portable generators and heaters; bottled gas for heating and soldering. Most solvents are flammable. Rubbish also represents a source of fuel: offcuts of wood, roofing felt, rags, empty solvent cans and discarded packaging will all provide fuel for a fire.

To eliminate fuel as a source of fire, all flammable liquids and gases should be stored correctly, usually in an outside locked store. The working environment should be kept clean by placing rags in a metal bin with a lid. Combustible waste material should be removed from the work site or burned outside under controlled conditions by a competent person.

Oxygen

Oxygen is all around us in the air we breathe, but can be eliminated from a small fire by smothering with a fire blanket, sand or foam. Closing doors and windows, but not locking them will limit the amount of oxygen available to a fire in a building and help to prevent it from spreading.

Most substances will burn if they are at a high enough temperature and have a supply of oxygen. The minimum temperature at which a substance will burn is called the 'minimum ignition temperature' and for most materials this is considerably higher than the surrounding temperature. However, a danger does exist from portable heaters, blowtorches and hot-airguns which provide heat and can cause a fire by raising the temperature of materials placed in their path above the minimum ignition temperature. A safe distance must be maintained between heat sources and all flammable materials.

Heat

Heat can be removed from a fire by dousing with water, but water must not be used on burning liquids since the water will spread the liquid and the fire. Some fire extinguishers have a cooling action which removes heat from the fire.

Fires in industry damage property and materials, injure people and sometimes cause loss of life. Everyone should make an effort to prevent fires, but those which do break out should be extinguished as quickly as possible.

In the event of fire you should:

- raise the alarm;
- turn off machinery, gas and electricity supplies in the area of the fire;
- close doors and windows but without locking or bolting them;
- remove combustible materials and fuels away from the path of the fire, if the fire is small, and if this can be done safely;
- attack small fires with the correct extinguisher.

Only attack the fire if you can do so without endangering your own safety in any way. Always leave your own exit from the danger zone clear. Those not involved in fighting the fire should walk to a safe area or assembly point.

COMPANION WEBSITE @ Visit the companion website for more on this topic.

Figure 1.13 The fire triangle.

Figure 1.14 Once started, fires can spread rapidly if the conditions are favourable.

Safety first

Fire

If you discover a fire:
- raise the alarm;
- attack small fires with an extinguisher;
- BUT only if you can do so without risk to your own safety.

Fires are divided into four classes or categories:

- Class A are wood, paper and textile fires;
- Class B are liquid fires such as paint, petrol and oil;
- Class C are fires involving gas or spilled liquefied gas;
- Class D are very special types of fire involving burning metal.

Electrical fires do not have a special category because, once started, they can be identified as one of the four above types.

Fire extinguishers are for dealing with small fires, and different types of fire must be attacked with a different type of extinguisher. Using the wrong type of extinguisher could make matters worse. For example, water must not be used on a liquid or electrical fire. The normal procedure when dealing with electrical fires is to cut off the electrical supply and use an extinguisher which is appropriate to whatever is burning. Figure 1.15 shows the correct type of extinguisher to be used on the various categories of fire. The colour coding shown is in accordance with BS EN3: 1996.

Definition

Fire extinguishers remove heat from a fire and are a first response for small fires.

 Visit the companion website for more on this topic.

Type of fire extinguisher / Type of fire	(i) Water — Signal red flash on red	(ii) Foam — Pale cream flash on red	(iii) Carbon dioxide gas — Black flash on red	(iv) Dry powder — French blue flash on red	(v) Vapourizing foam — Emerald green flash on red
Class A Paper, wood and fabric	✓ Yes	✓ Yes	✗ No	✓ Yes	✓ Yes
Class B Flammable liquids	✗ No	✓ Yes	✓ Yes	✓ Yes	✓ Yes
Class C Flammable gases	✗ No	✗ No	✓ Yes	✓ Yes	✓ Yes
Electrical fires	✗ No	✗ No	✓ Yes	✓ Yes	✓ Yes
Motor vehicle protection	✗ No	✓ Yes	✗ No	✓ Yes	✓ Yes

Figure 1.15 Fire extinguishers and their applications (colour codes to BS EN3: 1996). The base colour of all fire extinguishers is red, with a different-coloured flash to indicate the type.

Heat-producing equipment

Electrical installers usually make their connections 'cold', terminating cables in joint boxes or using crimping tools for larger cable connections. However, mechanical service trades on site, such as plumbers and heating and ventilation engineers, make many of their connections by 'hot' working.

Hot working is potentially more hazardous than cold working because of the additional risk of fire. Operatives carrying out hot working must follow strict procedures to reduce the risk of fire. These safety procedures must be written down in a method statement and the work procedure strictly followed.

The fuel source of the hot working might be:

* bottled gas – butane in blue containers or propane in red containers are often used with a blowtorch or boiling ring;
* oxy-acetylene gas – this is always used with a torch which mixes the two gases for cutting or welding steel. This is not work which is normally carried out by an electrician.

When hot working is carried out, operatives must be trained and competent to use the equipment. They must also follow all safety procedures to reduce the risk of fire. For example, a fire extinguisher must be placed in the immediate work area and hot working must be completed at least one hour before the operator leaves the site.

Bottles, torches and flexible hoses must be in good condition and tested. If the bottled gas is stored on-site it must be placed in a secured outdoor location open to the elements to reduce the risk of a gas buildup from a possibly faulty or leaking on-off valve.

See also Appendix F (B2.11.1 and B2.11.2) of the Electricians Guide to the Building Regulations.

Evacuation procedures

When the fire alarm sounds you must leave the building immediately by any one of the escape routes indicated. **Exit routes** are usually indicated by a green and white 'running man' symbol. Evacuation should be orderly; do not run but walk purposefully to your designated assembly point.

The purpose of an **assembly point** is to get you away from danger to a place of safety where you will not be in the way of the emergency services. It also allows for people to be accounted for and to make sure that no one is left in the building. You must not re-enter the building until a person in authority gives permission to do so.

An evacuation in a real emergency can be a frightening experience, especially if you do not really know what to do, so take time to familiarize yourself with the fire safety procedures where you are working before an emergency occurs.

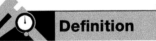

Definition

Exit routes are usually indicated by a green and white 'running man' symbol. Evacuation should be orderly; do not run but walk purposefully to your designated assembly point.

Definition

The purpose of an *assembly point* is to get you away from danger to a place of safety where you will not be in the way of the emergency services.

Health and safety risks, precautions and procedures

Earlier in this chapter, we looked at some of the health and safety rules and regulations. In particular, we now know that the Health and Safety at Work Act is the most important piece of recent legislation, because it places responsibilities for safety at work on both employers and employees. This responsibility is enforceable by law. We know what the regulations say about the control of substances that might be hazardous to our health at work, because we briefly looked at the COSHH Regulations 2002 earlier in this chapter. We also know that if there is a risk to health and safety at work our employer must provide personal protective equipment (PPE) free of charge, for us to use so that we are safe at work. The law is in place, we all apply the principles of health and safety at work and we always wear the appropriate PPE, so what are the risks? Well, getting injured at work is not a pleasant subject to think about but each year about 300 people in Great Britain lose their lives at work. In addition, there are about 158,000 non-fatal injuries reported to the Health and Safety Executive (HSE) each year and an estimated 2.2 million people suffer ill health caused, or made worse by work. It is a mistake to believe that these things only happen in dangerous occupations such as deep-sea diving, mining and quarrying, fishing industry, tunnelling and fire-fighting, or that they only happen in exceptional circumstances such as would never happen in your workplace. This is not the case. Some basic thinking and acting beforehand could have prevented most of these accident statistics from happening.

The most common categories of risk and causes of accidents at work are:

- slips, trips and falls;
- manual handling, that is, moving objects by hand;
- using equipment, machinery or tools;
- storage of goods and materials which then become unstable;
- fire;
- electricity;
- mechanical handling.

Precautions taken to control risks:

- eliminate the cause;
- substitute a procedure or product with less risk;
- enclose the dangerous situation;
- put guards around the hazard;
- use safe systems of work;
- supervise, train and give information to staff;
- if the hazard cannot be removed or minimized then provide PPE.

Let us now look at the application of some of the procedures that make the workplace a safer place to work, but first I want to explain what I mean when I use the words hazard and risk.

Hazard and risk

A **hazard** is something with the 'potential' to cause harm, for example, chemicals, electricity or working above ground.

A **risk** is the 'likelihood' of harm actually being done.

Competent persons are often referred to in the Electricity at Work Regulations and the Health and Safety at Work Regulations, but who is 'competent'? For the purposes of the Act, a competent person is anyone who has the necessary technical skills, training and expertise to safely carry out the particular activity. Therefore, a competent person dealing with a hazardous situation reduces the risk. The 3rd Amendment to the 17th Edition of the IET Wiring Regulations has chosen not to use the word 'competent person' in the IET Regulations, replacing it with the more specific definitions of skilled (electrically) or instructed (electrically) persons.

Definition

A *hazard* is something with the 'potential' to cause harm, for example, chemicals, electricity or working above ground.

A *risk* is the 'likelihood' of harm actually being done.

A *competent person* is anyone who has the necessary technical skills, training and expertise to safely carry out the particular activity.

Think about your workplace and each stage of what you do, then think about what might go wrong. Some simple activities may be hazardous. Here are some typical activities where accidents might happen.

Typical activity	Potential hazard
Receiving materials	Lifting and carrying
Stacking and storing	Falling materials
Movement of people	Slips, trips and falls
Building maintenance	Working at heights or in confined spaces
Movement of vehicles	Collisions

How high are the risks? Think about what might be the worst result; is it a broken finger, or someone suffering permanent lung damage or being killed? How likely is it to happen? How often is that type of work carried out and how close do people get to the hazard? How likely is it that something will go wrong?

How many people might be injured if things go wrong? Might this also include people who do not work for your company?

Employers of more than five people must document the risks at work and the process is known as **hazard risk assessment**.

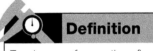

Definition

Employers of more than five people must document the risks at work and the process is known as *hazard risk assessment*.

Hazard risk assessment: the procedure

The Management of Health and Safety at Work Regulations 1999 tells us that employers must systematically examine the workplace, the work activity and the management of safety in the establishment through a process of risk assessments. A record of all significant risk assessment findings must be kept in a safe place and be made available to an HSE Inspector if required. Information based on the risk assessment findings must be communicated to relevant staff and, if changes in work behaviour patterns are recommended in the interests of safety, they must be put in place.

So risk assessment must form a part of any employer's robust policy of health and safety. However, an employer only needs to 'formally' assess the significant risks. He or she is not expected to assess the trivial and minor types of household risks. Staff are expected to read and to act upon these formal risk assessments, and they are unlikely to do so enthusiastically if the file is full of trivia. An assessment of risk is nothing more than a careful examination of what, in your work, could cause harm to people. It is a record that shows whether sufficient precautions have been taken to prevent harm.

The HSE recommends five steps to any risk assessment.

Step 1

Look at what might reasonably be expected to cause harm. Ignore the trivial and concentrate only on significant hazards that could result in serious harm or injury. Manufacturers' data sheets or instructions can also help you spot hazards and put risks in their true perspective.

Step 2

Decide who might be harmed and how. Think about people who might not be in the workplace all the time – cleaners, visitors, contractors or maintenance personnel. Include members of the public or people who share the workplace. Is there a chance that they could be injured by activities taking place in the workplace?

Step 3

Evaluate what is the risk arising from an identified hazard. Is it adequately controlled or should more be done? Even after precautions have been put in place, some risk may remain. What you have to decide, for each significant hazard, is whether this remaining risk is low, medium or high. First of all, ask yourself if you have done all the things that the law says you have got to do. For example, there are legal requirements on the prevention of access to dangerous machinery. Then ask yourself whether generally accepted industry standards are in place, but do not stop there – think for yourself, because the law also says that you must do what is reasonably practicable to keep the workplace safe. Your real aim is to make all risks small by adding precautions, if necessary.

If you find that something needs to be done, ask yourself:

- Can I get rid of this hazard altogether?
- If not, how can I control the risk so that harm is unlikely?

Only use PPE when there is nothing else that you can reasonably do.

If the work that you do varies a lot, or if there is movement between one site and another, select those hazards which you can reasonably foresee, the ones that apply to most jobs and assess the risks for them. After that, if you spot any unusual hazards when you get on-site, take what action seems necessary.

Step 4

Record your findings and say what you are going to do about risks that are not adequately controlled. If there are fewer than five employees you do not need to write anything down but if there are five or more employees the significant findings of the risk assessment must be recorded. This means writing down the more significant hazards and assessing if they are adequately controlled and recording your most important conclusions. Most employers have a standard risk assessment form which they use, such as that shown in Fig. 1.14, but any format is suitable. The important thing is to make a record.

There is no need to show how the assessment was made, provided you can show that:

1 a proper check was made
2 you asked those who might be affected
3 you dealt with all obvious and significant hazards
4 the precautions are reasonable and the remaining risk is low
5 you informed your employees about your findings.

Risk assessments need to be *suitable* and *sufficient*, not perfect. The two main points are:

1 Are the precautions reasonable?
2 Is there a record to show that a proper check was made?

File away the written assessment in a dedicated file for future reference or use. It can help if an HSE Inspector questions the company's precautions or if the company becomes involved in any legal action. It shows that the company has done what the law requires.

Step 5

Review the assessments from time to time and revise them if necessary.

Key fact

Definition

- A hazard is something that might cause harm.
- A risk is the chance that harm will be done.

Definition

Risk assessments need to be *suitable* and *sufficient*, not perfect.

HAZARD RISK ASSESSMENT	FLASH-BANG ELECTRICAL CO.
For Company name or site: _____ Address: _____ _____	Assessment undertaken by: _____ Signed: _____ Date: _____

STEP 5 Assessment review date: _____

STEP 1 List the hazards here	STEP 2 Decide who might be harmed
_____ _____ _____ _____ _____ _____	_____ _____ _____ _____ _____ _____
STEP 3 Evaluate (what is) the risk – is it adequately controlled? State risk level as low, medium or high	STEP 4 Further action – what else is required to control any risk identified as medium or high?
_____ _____ _____ _____ _____	_____ _____ _____ _____ _____

Figure 1.16 Hazard risk assessment standard form.

Method statement

The Construction, Design and Management Regulations and Approved Codes of Practice define a method statement as a written document laying out the work procedure and sequence of operations to ensure health and safety.

If the method statement is written as a result of a risk assessment carried out for a task or operation, then following the prescribed method will reduce the risk.

The safe isolation procedure described in Fig. 1.28 is a method statement. Following this method meets the requirements of the Electricity at Work Regulations, the IET Regulations, and reduces the risk of electric shock to the operative and other people who might be affected by his actions.

Completing a risk assessment

When completing a risk assessment such as that shown in Fig. 1.16, do not be over-complicated. In most firms in the commercial, service and light industrial sector, the hazards are few and simple. Checking them is common sense but necessary.

Background information	
Company details	Flash Bang Electrical Company Contact street Tel Fax email
Site address	Contact name Address Contact No.
Activity – risk	To access and work safely in the attic space during the replacement of a fluorescent light fitting.

Implementation and control of risk

Hazardous task – risk	Method of control
Access roof space	Access to the roof space will be via a suitably secured stepladder of the correct height for the task. All relevant PPE will be worn.
Access working area	Walk boards will be used to walk across ceiling joists. Electric lead lights shall illuminate access and work areas.
Replacement of existing light fitting	Safe isolation procedure to be carried out.
Removal of fluorescent light fitting	The fluorescent light fitting will be carefully lowered through the loft hatch. Use a secured strap and rope system if necessary.
Safe deposit of fluorescent light fitting	Fluorescent light fitting to be sealed in one piece and by prior arrangement, deposited at the council tip by the customer.

Site control

Inspection and testing of equipment	All equipment such as stepladder, tooling and electrical equipment shall be regularly inspected before commencement of work. All test equipment to be calibrated annually.
Customer awareness	The customer will be made aware of all potential dangers throughout the contract. They will also be made aware if any electrical isolation is required.

Figure 1.17 Typical method statement.

Step 1

List only hazards which you could reasonably expect to result in significant harm under the conditions prevailing in your workplace. Use the following examples as a guide:

* slipping or tripping hazards (e.g. from poorly maintained or partly installed floors and stairs);
* fire (e.g. from flammable materials you might be using, such as solvents);
* chemicals (e.g. from battery acid);
* moving parts of machinery (e.g. blades);
* rotating parts of hand tools (e.g. drills);
* accidental discharge of cartridge-operated tools;
* high-pressure air from airlines (e.g. air-powered tools);
* pressure systems (e.g. steam boilers);
* vehicles (e.g. fork lift trucks);
* electricity (e.g. faulty tools and equipment);
* dust (e.g. from grinding operations or thermal insulation);
* fumes (e.g. from welding);
* manual handling (e.g. lifting, moving or supporting loads);
* noise levels too high (e.g. machinery);
* poor lighting levels (e.g. working in temporary or enclosed spaces);
* low temperatures (e.g. working outdoors or in refrigeration plant);
* high temperatures (e.g. working in boiler rooms or furnaces).

Step 2

Decide who might be harmed; do not list individuals by name. Just think about groups of people doing similar work or who might be affected by your work:

* office staff;
* electricians;
* maintenance personnel;
* other contractors on-site;
* operators of equipment;
* cleaners;
* members of the public.

Pay particular attention to those who may be more vulnerable, such as:

* staff with disabilities;
* visitors;
* young or inexperienced staff;
* people working in isolation or enclosed spaces.

Step 3

Calculate what is the risk – is it adequately controlled? Have you already taken precautions to protect against the hazards which you have listed in Step 1? For example:

* have you provided adequate information to staff?
* have you provided training or instruction?

Do the precautions already taken

* meet the legal standards required?
* comply with recognized industrial practice?

Safety first

Safety procedures

* Hazard risk assessment is *an essential part* of any health and safety management system.
* The aim of the planning process is to minimize risk.
* HSE Publication HSG (65).

- represent good practice?
- reduce the risk as far as is reasonably practicable?

If you can answer 'yes' to the above points then the risks are adequately controlled, but you need to state the precautions you have put in place. You can refer to company procedures, company rules, company practices, etc. in giving this information. For example, if we consider there might be a risk of electric shock from using electrical power tools, then the risk of a shock will be *less* if the company policy is to portable appliance test (PAT) all power tools each year and to fit a label to the tool showing that it has been tested for electrical safety. If the stated company procedure is to use battery drills whenever possible, or 110 V drills when this is not possible, and to *never* use 230 V drills, then this again will reduce the risk. If a policy such as this is written down in the company safety policy statement, then you can simply refer to the appropriate section of the safety policy statement and the level of risk will be low. (Note: PAT testing is described in Advanced Electrical Installation Work.)

Step 4

Further action – what more could be done to reduce those risks which were found to be inadequately controlled?

You will need to give priority to those risks that affect large numbers of people or which could result in serious harm. Senior managers should apply the principles below when taking action, if possible in the following order:

1 Remove the risk completely.
2 Try a less risky option.
3 Prevent access to the hazard (e.g. by guarding).
4 Organize work differently in order to reduce exposure to the hazard.
5 Issue PPE.
6 Provide welfare facilities (e.g. washing facilities for removal of contamination and first aid).

Any hazard identified by a risk assessment as *high risk* must be brought to the attention of the person responsible for health and safety within the company. Ideally, in Step 4 of the risk assessment you should be writing: 'No further action is required. The risks are under control and identified as low risk.'

The assessor may use as many standard hazard risk assessment forms, such as that shown in Fig. 1.16, as the assessment requires. Upon completion they should be stapled together or placed in a plastic wallet and stored in the dedicated file.

You might like to carry out a risk assessment on a situation you are familiar with at work, using the standard form of Fig. 1.16, or your employer's standard forms.

Safe manual handling

Manual handling is lifting, transporting or supporting loads by hand or by bodily force. The load might be any heavy object, a printer, a VDU, a box of tools or a stepladder. Whatever the heavy object is, it must be moved thoughtfully and carefully, using appropriate lifting techniques if personal pain and injury are to be avoided. *Many people hurt their back, arms and feet, and over one-third of all three-day reported injuries submitted to the HSE each year are the result of manual handling.*

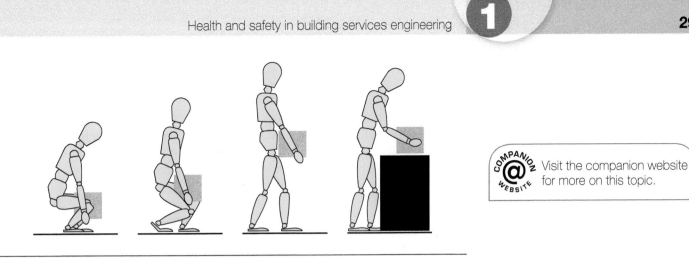

Figure 1.18 Correct manual lifting and carrying procedure.

When lifting heavy loads, correct lifting procedures must be adopted to avoid back injuries. Figure 1.18 demonstrates the technique. Do not lift objects from the floor with the back bent and the legs straight as this causes excessive stress on the spine. Always lift with the back straight and the legs bent so that the powerful leg muscles do the lifting work. Bend at the hips and knees to get down to the level of the object being lifted, positioning the body as close to the object as possible. Grasp the object firmly and, keeping the back straight and the head erect, use the leg muscles to raise in a smooth movement. Carry the load close to the body. When putting the object down, keep the back straight and bend at the hips and knees, reversing the lifting procedure. A bad lifting technique will result in sprains, strains and pains. *There have been too many injuries over the years resulting from bad manual handling techniques. The problem has become so serious that the HSE has introduced new legislation* under the Health and Safety at Work Act 1974, the Manual Handling Operations Regulations 1992. Publications such as *Getting to Grips with Manual Handling* can be obtained from HSE Books; the address and Infoline are given in Appendix B.

Where a job involves considerable manual handling, employers must now train employees in the correct lifting procedures and provide the appropriate equipment necessary to promote the safe manual handling of loads:

Consider some 'good practice' when lifting loads:

- Do not lift the load manually if it is more appropriate to use a mechanical aid. Only lift or carry what you can easily manage.
- Always use a trolley, wheelbarrow or truck such as that shown in Fig. 1.19 when these are available.
- Plan ahead to avoid unnecessary or repeated movement of loads.
- Take account of the centre of gravity of the load when lifting – the weight acts through the centre of gravity.
- Never leave a suspended load unsupervised.
- Always lift and lower loads gently.
- Clear obstacles out of the lifting area.
- Use the manual lifting techniques described above and avoid sudden or jerky movements.
- Use gloves when manual handling to avoid injury from rough or sharp edges.
- Take special care when moving loads wrapped in grease or bubble-wrap.
- Never move a load over other people or walk under a suspended load.

COMPANION @ WEBSITE Visit the companion website for more on this topic.

Figure 1.19 Always use a mechanical aid to transport a load when available.

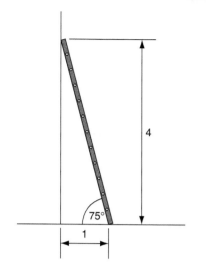

Figure 1.20 A correctly erected ladder.

Working at height regulations

Working above ground level creates added dangers and slows down the work rate of the electrician. New Work at Height Regulations came into force on 6 April 2005. Every precaution should be taken to ensure that the working platform is appropriate for the purpose and in good condition.

Ladders

The term ladder is generally taken to include stepladders and trestles. The use of ladders for working above ground level is only acceptable for access and work of short duration (Work at Height Regulations 2005).

It is advisable to inspect the ladder before climbing it. It should be straight and firm. All rungs and tie rods should be in position and there should be no cracks in the stiles. The ladder should not be painted since the paint may be hiding defects.

Extension ladders should be erected in the closed position and extended one section at a time. Each section should overlap by at least the number of rungs indicated below:

• ladder up to 4.8 m length – two rungs overlap;
• ladder up to 6.0 m length – three rungs overlap;
• ladder over 6.0 m length – four rungs overlap.

The angle of the ladder to the building should be in the proportion 4 up to 1 out or 75° as shown in Fig. 1.20. The ladder should be lashed at the top and bottom when possible to prevent unwanted movement and placed on firm and level ground. Footing is only considered effective for ladders smaller than 6 m and manufactured securing devices should always be considered. When ladders provide access to a roof or working platform the ladder must extend at least 1.05 m or five rungs above the landing place.

Short ladders may be carried by one person resting the ladder on the shoulder, but longer ladders should be carried by two people, one at each end, to avoid accidents when turning corners.

Long ladders or extension ladders should be erected by two people as shown in Fig. 1.21. One person stands on or 'foots' the ladder, while the other person lifts and walks under the ladder towards the wall. When the ladder is upright it can be positioned in the correct place, at the correct angle and secured before being climbed.

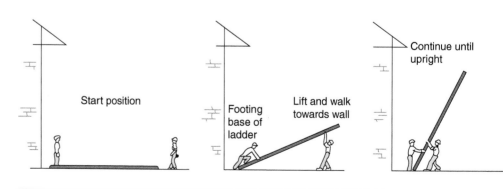

Figure 1.21 Correct procedure for erecting long or extension ladder.

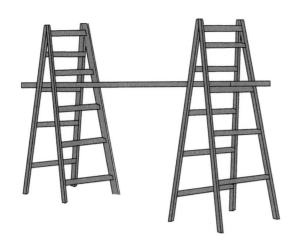

Figure 1.22 A trestle scaffold.

Trestle scaffold

Figure 1.22 shows a trestle scaffold. Two pairs of trestles spanned by scaffolding boards provide a simple working platform. The platform must be at least two boards or 450 mm wide. At least one-third of the trestle must be above the working platform. If the platform is more than 2 m above the ground, toeboards and guardrails must be fitted, and a separate ladder provided for access. The boards which form the working platform should be of equal length and not overhang the trestles by more than four times their own thickness. The maximum span of boards between trestles is:

- 1.3 m for boards 40 mm thick;
- 2.5 m for boards 50 mm thick.

Trestles which are higher than 3.6 m must be tied to the building to give them stability. Where anyone can fall more than 4.5 m from the working platform, trestles may not be used.

Mobile scaffold towers

Mobile scaffold towers may be constructed of basic scaffold components or made from light alloy tube. The tower is built up by slotting the sections together until the required height is reached. A scaffold tower is shown in Fig. 1.23.

If the working platform is above 2 m from the ground it must be close boarded and fitted with guardrails and toeboards. When the platform is being used, all four wheels must be locked. The platform must not be moved unless it is clear of tools, equipment and workers, and should be pushed at the base of the tower and not at the top.

The stability of the tower depends upon the ratio of the base width to tower height. A ratio of base to height of 1:3 gives good stability. Outriggers can be used to increase stability by effectively increasing the base width. If outriggers are used then they must be fitted diagonally across all four corners of the tower and not on one side only. The tower must not be built more than 12 m high unless it has been specially designed for that purpose. Any tower higher than 9 m should be secured to the structure of the building to increase stability.

Access to the working platform of a scaffold tower should be by a ladder securely fastened vertically to the inside of the tower. Ladders must never be leaned against a tower since this might push the tower over.

Safety first

Scaffold
Scaffold or mobile towers are always safer than ladders for working above ground.

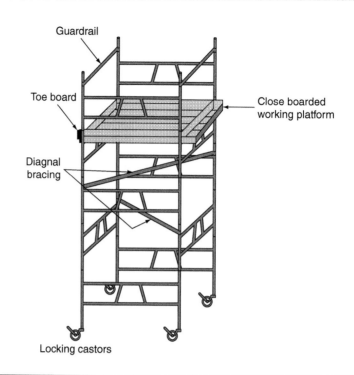

Figure 1.23 A mobile scaffold tower.

Labels on Figure 1.23: Guardrail, Toe board, Close boarded working platform, Diagnal bracing, Locking castors

Figure 1.24 An example of safe ladder use.

Working in excavations and confined spaces

Working alone

Some working situations are so potentially hazardous that not only must PPE be worn but you must also never work alone and safe working procedures must be in place before your work begins to reduce the risk.

It is unsafe to work in isolation in the following situations:

- when working above ground;
- when working below ground;
- when working in confined spaces;
- when working in excavations;
- when working close to unguarded machinery;
- when a fire risk exists;
- when working close to toxic or corrosive substances.

Working above ground

We have already looked at this topic as it applies to electrical personnel. The new Work at Height Regulations 2005 tell us that a person is at height if that person could be injured by falling from it. The regulations require that:

- we should avoid working at height if at all possible;
- no work should be done at height which can be done on the ground (e.g. equipment can be assembled on the ground then taken up to height, perhaps for fixing);
- ensure the work at height is properly planned;
- take account of any risk assessments carried out under Regulation 3 of the Management of Health and Safety at Work Regulations.

Working in excavations or below ground

Working below ground might be working in a cellar or an unventilated basement with only one entrance/exit. There is a risk that this entrance/exit might become blocked by materials, fumes or fire. When working in trenches or excavations there is always the risk of the sides collapsing if they are not adequately supported by temporary steel sheets and adequate side bracing. There is also the risk of falling objects, so always:

- wear a hard hat;
- never go into an unsupported excavation;
- erect barriers around the excavation;
- provide good ladder access;
- ensure the work is properly planned;
- take account of the risk assessment before starting work.

Working in confined spaces

When working in confined spaces there is always the risk that you may become trapped or overcome by a lack of oxygen, or by gas, fumes, heat or an accumulation of dust. Examples of confined spaces are:

- storage tanks and silos on farms;
- enclosed sewers and pumping stations;
- furnaces;
- ductwork.

In my experience, electricians spend a lot of time on their knees in confined spaces because many electrical cable systems run out of sight away from public areas of a building.

The Confined Spaces Regulations 1997 require that:

- a risk assessment is carried out before work commences;
- if there is a serious risk of injury in entering the confined space then the work should be done on the outside of the vessel;
- a safe working procedure, such as a 'permit-to-work procedure', is followed and adequate emergency arrangements put in place before work commences.

Working near unguarded machinery

There is an obvious risk in working close to unguarded machinery, and indeed, most machinery will be guarded, but in some production processes and with overhead travelling cranes, this is not always possible. To reduce the risks associated with these hazards:

- have the machinery stopped during your work activity if possible;
- put temporary barriers in place;
- make sure the machine operator knows that you are working on the equipment;
- identify the location of emergency stop buttons;
- take account of the risk assessment before work commences.

A risk of fire

When working in locations containing stored flammable materials such as petrol, paraffin, diesel or bottled gas, there is always the risk of fire. To minimize the risk:

- take account of the risk assessment before work commences;
- keep the area well ventilated;

- locate the fire extinguishers;
- secure your exit from the area;
- locate the nearest fire alarm point;
- follow a safe working procedure and put adequate emergency arrangements in place before work commences.

Permit-to-work system

Definition

The *permit-to-work procedure* is a type of 'safe system to work' procedure used in specialized and potentially dangerous plant process situations.

The **permit-to-work procedure** is a type of 'safe system to work' procedure used in specialized and potentially dangerous plant process situations. The procedure was developed for the chemical industry, but the principle is equally applicable to the management of complex risk in other industries or situations. For example:

- working on part of an assembly line process where goods move through a complex, continuous process from one machine to another (e.g. the food industry);
- repairs to railway tracks, tippers and conveyors;
- working in confined spaces (e.g. vats and storage containers);
- working on or near overhead crane tracks;
- working underground or in deep trenches;
- working on pipelines;
- working near live equipment or unguarded machinery;
- roof work;
- working in hazardous atmospheres (e.g. the petroleum industry);
- working near or with corrosive or toxic substances.

All the above situations are high-risk working situations that should be avoided unless you have received special training and will probably require the completion of a permit-to-work. Permits-to-work must adhere to the following eight principles:

1 Wherever possible the hazard should be eliminated so that the work can be done safely without a permit-to-work.
2 The site manager has overall responsibility for the permit-to-work even though he or she may delegate the responsibility for its issue.
3 The permit must be recognized as the master instruction, which, until it is cancelled, overrides all other instructions.
4 The permit applies to everyone on-site, other trades and subcontractors.
5 The permit must give detailed information; for example: (i) which piece of plant has been isolated and the steps by which this has been achieved; (ii) what work is to be carried out; (iii) the time at which the permit comes into effect.
6 The permit remains in force until the work is completed and is cancelled by the person who issued it.
7 No other work is authorized. If the planned work must be changed, the existing permit must be cancelled and a new one issued.
8 Responsibility for the plant must be clearly defined at all stages because the equipment that is taken out of service is released to those who are to carry out the work.

The people doing the work, the people to whom the permit is given, take on the responsibility of following and maintaining the safeguards set out in the permit, which will define what is to be done (no other work is permitted) and the time-scale in which it is to be carried out.

The permit-to-work system must help communication between everyone involved in the process or type of work. Employers must train staff in the use of such permits and, ideally, training should be designed by the company issuing the permit, so that sufficient emphasis can be given to particular hazards present and the precautions which will be required to be taken. For further details see Permit to Work at www.hse.gov.uk.

Secure electrical isolation

Electric shock occurs when a person becomes part of the electrical circuit. The level or intensity of the shock will depend upon many factors, such as age, fitness and the circumstances in which the shock is received. The lethal level is approximately 50mA, above which muscles contract, the heart flutters and breathing stops. A shock above the 50mA level is therefore fatal unless the person is quickly separated from the supply. Below 50mA only an unpleasant tingling sensation may be experienced or you may be thrown across a room or shocked enough to fall from a roof or ladder, but the resulting fall may lead to serious injury.

To prevent people from receiving an electric shock accidentally, all circuits contain protective devices. All exposed metal is earthed; fuses and miniature circuit-breakers (MCBs) are designed to trip under fault conditions, and residual current devices (RCDs) are designed to trip below the fatal level as described in Chapter 4.

Construction workers and particularly electricians do receive electric shocks, usually as a result of carelessness or unforeseen circumstances. As an electrician working on electrical equipment you must always make sure that the equipment is switched off or electrically isolated before commencing work. Every circuit must be provided with a means of isolation (IET Regulation 132.15). When working on portable equipment or desk top units it is often simply a matter of unplugging the equipment from the adjacent supply. Larger pieces of equipment and electrical machines may require isolating at the local isolator switch before work commences. To deter anyone from re-connecting the supply while work is being carried out on equipment, a sign 'Danger – Electrician at Work' should be displayed on the isolator and the isolation 'secured' with a small padlock or the fuses removed so that no one can re-connect while work is being carried out on that piece of equipment. The Electricity at Work Regulations 1989 are very specific at Regulation 12(1) that we must ensure the disconnection and separation of electrical equipment from every source of supply and that this disconnection and separation is secure. Where a test instrument or voltage indicator is used to prove the supply dead, Regulation 4(3) of the Electricity at Work Regulations 1989 recommends that the following procedure is adopted.

1 First, connect the test device such as that shown in Fig. 1.25 to the supply which is to be isolated. The test device should indicate mains voltage.

2 Next, isolate the supply and observe that the test device now reads zero volts.

3 Then connect the same test device to a known live supply or proving unit such as that shown in Fig. 1.26 to 'prove' that the tester is still working correctly.

4 Finally, secure the isolation and place warning signs; only then should work commence.

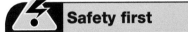

Safety first

Isolation
- never work 'live';
- isolate;
- secure the isolation;
- prove the supply 'dead' before starting work.

Definition

Electrical isolation – we must ensure the disconnection and separation of electrical equipment from every source of supply and that this disconnection and separation is secure.

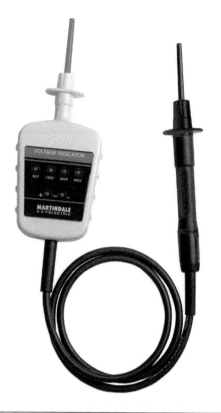

Figure 1.25 Typical voltage indicator.

Figure 1.26 Voltage proving unit.

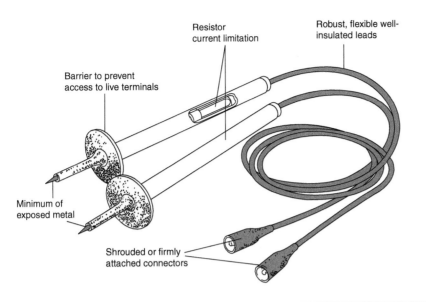

Barrier to prevent
access to live terminals

Resistor
current limitation

Robust, flexible well-
insulated leads

Minimum of
exposed metal

Shrouded or firmly
attached connectors

Figure 1.27 Recommended type of test probe and leads.

The test device being used by the electrician must incorporate safe test leads which comply with the Health and Safety Executive Guidance Note 38 on electrical test equipment. These leads should incorporate barriers to prevent the user from touching live terminals when testing and incorporating a protective resistor and be well insulated and robust, such as those shown in Fig. 1.27.

To isolate a piece of equipment or individual circuit successfully, competently, safely and in accordance with all the relevant regulations, we must follow a procedure such as that given by the flow diagram in Fig. 1.29. Start at the top and work down the flow diagram.

When the heavy outlined amber boxes are reached, pause and ask yourself whether everything is satisfactory up until this point. If the answer is 'yes', move on. If the answer is 'no', go back as indicated by the diagram.

Live testing

The Electricity at Work Regulations 1989 at Regulation 4(3) tell us that it is preferable that supplies be made dead before work commences. However, it does acknowledge that some work, such as fault finding and testing, may require the electrical equipment to remain energized. Therefore, if the fault finding and testing can only be successfully carried out live, then the person carrying out the fault diagnosis must:

- be trained so that they understand the equipment and the potential hazards of working live and can, therefore, be deemed 'competent' to carry out that activity;
- only use approved test equipment;
- set up appropriate warning notices and barriers so that the work activity does not create a situation dangerous to others.

While live testing may be required by workers in the electrical industries in order to find the fault, live repair work must not be carried out. The individual circuit or piece of equipment must first be isolated before work commences in order to comply with the Electricity at Work Regulations 1989.

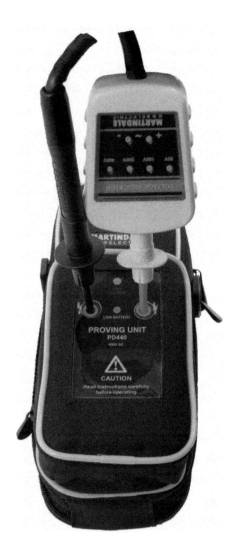

Figure 1.28 Voltage indicator being proved.

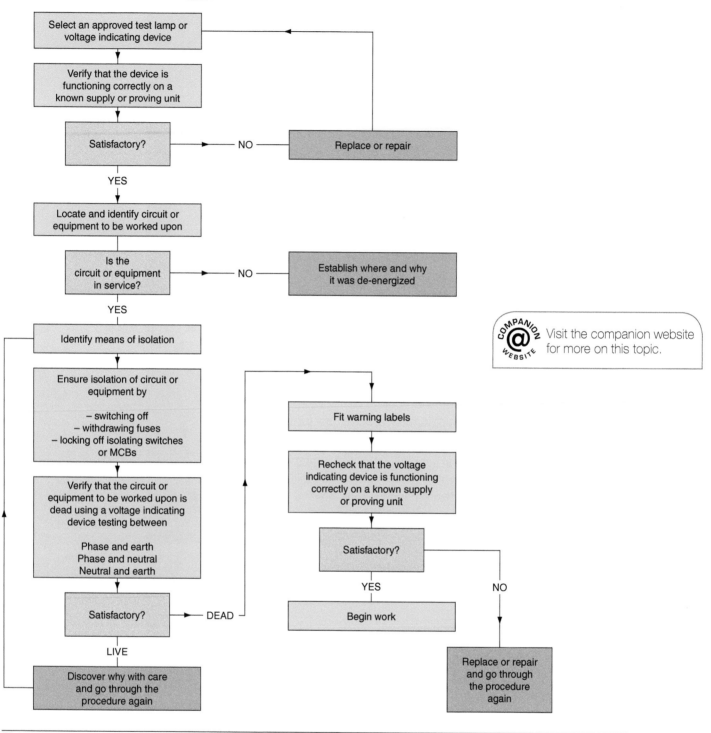

Figure 1.29 Flowchart for a secure isolation procedure.

On-site electrical supplies and tools

Temporary electrical supplies on construction sites can save many person-hours of labour by providing energy for fixed and portable tools and lighting. However, as stated previously in this chapter, construction sites are dangerous places and the temporary electrical supplies must be safe. IET Regulation 110.1 tells us that *all* the regulations apply to temporary electrical installations such as construction sites. The frequency of inspection of construction sites is increased

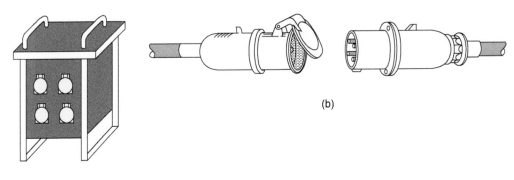

(b)

(a)

Figure 1.30 110 volts distribution unit and cable connector suitable for construction site electrical supplies: (a) reduced-voltage distribution unit incorporating industrial sockets to BS 4343; (b) industrial plug and connector.

to every three months because of the inherent dangers. Regulation 704.313.4 recommends the following voltages for distributing to plant and equipment on construction sites:

400V – fixed plant such as cranes

230V – site offices and fixed floodlighting robustly installed

110V – portable tools and hand lamps

SELV – portable lamps used in damp or confined places.

Portable tools must be fed from a 110V socket outlet unit (see Fig. 1.29(a)) incorporating splash-proof sockets and plugs with a keyway which prevents a tool from one voltage from being connected to the socket outlet of a different voltage.

Socket outlet and plugs are also colour coded for voltage identification: 25V violet, 50V white, 110V yellow, 230V blue and 400V red, as shown in Fig. 1.29(b).

Electric shock

Electric shock occurs when a person becomes part of the electrical circuit, as shown in Fig. 1.31. The lethal level is approximately 50mA, above which muscles contract, the heart flutters and breathing stops. A shock above the 50mA level is therefore fatal unless the person is quickly separated from the supply. Below 50mA only an unpleasant tingling sensation may be experienced or you may be thrown across a room or fall from a roof or ladder, but the resulting fall may lead to serious injury.

To prevent people from receiving an electric shock accidentally, all circuits must contain protective devices.

Construction workers and particularly electricians do receive electric shocks, usually as a result of carelessness or unforeseen circumstances. When this happens it is necessary to act quickly to prevent the electric shock from becoming fatal. Actions to be taken upon finding a workmate receiving an electric shock are as follows:

- switch off the supply if possible;
- alternatively, remove the person from the supply *without touching him*, e.g. push him off with a piece of wood, or pull him off with a scarf, dry towel or coat;

> **Definition**
>
> *Electric shock* occurs when a person becomes part of the electrical circuit.

Visit the companion website for more on this topic.

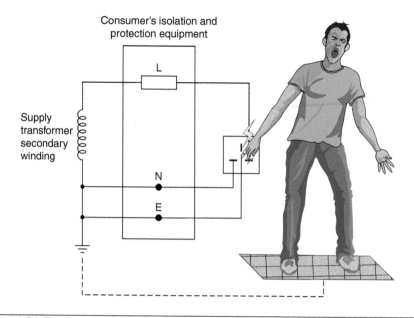

Consumer's isolation and protection equipment

Supply transformer secondary winding

Figure 1.31 Touching live and earth or live and neutral makes a person part of the electrical circuit and can lead to an electric shock.

- if breathing or heart has stopped, immediately call professional help by dialling 999 and asking for the ambulance service. Give precise directions to the scene of the accident. The casualty stands the best chance of survival if the emergency services can get a rapid response paramedic team quickly to the scene. They have extensive training and will have specialist equipment with them;
- only then should you apply resuscitation or cardiac massage until the patient recovers, or help arrives;
- treat for shock.

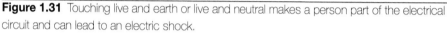

First aid

Despite all the safety precautions taken on construction sites to prevent injury to the workforce, accidents do happen and *you* may be the only other person able to take action to assist a workmate. This section is not intended to replace a first aid course but to give learners the knowledge to understand the types of injuries they may come across in the workplace. If you are not a qualified first aider limit your help to obvious common-sense assistance and call for help, *but* do remember that if a workmate's heart or breathing has stopped as a result of an accident they have only minutes to live unless you act quickly. The Health and Safety (First Aid) Regulations 1981 and relevant approved codes of practice and guidance notes place a duty of care on all employers to provide *adequate* first aid facilities appropriate to the type of work being undertaken. *Adequate* facilities will relate to a number of factors such as:

- How many employees are employed?
- What type of work is being carried out?
- Are there any special or unusual hazards?
- Are employees working in scattered and/or isolated locations?
- Is there shift work or 'out of hours' work being undertaken?

- Is the workplace remote from emergency medical services?
- Are there inexperienced workers on-site?
- What were the risks of injury and ill health identified by the company's hazard risk assessment?

The regulations state that:

Employers are under a duty to provide such numbers of suitable persons as is adequate and appropriate in the circumstances for rendering first aid to his employees if they are injured or become ill at work. For this purpose a person shall not be suitable unless he or she has undergone such training and has such qualifications as the Health and Safety Executive may approve.

This is typical of the way in which the Health and Safety Regulations are written. The regulations and codes of practice do not specify numbers, but set out guidelines in respect of the number of first aiders needed, dependent upon the type of company, the hazards present and the number of people employed.

Let us now consider the questions 'what is first aid?' and 'who might become a first aider?' The regulations give the following definitions of first aid. '*First aid* is the treatment of minor injuries which would otherwise receive no treatment or do not need treatment by a doctor or nurse'; *or* 'In cases where a person will require help from a doctor or nurse, first aid is treatment for the purpose of preserving life and minimizing the consequences of an injury or illness until such help is obtained'. A more generally accepted definition of first aid might be as follows: **first aid** is the initial assistance or treatment given to a casualty for any injury or sudden illness before the arrival of an ambulance, doctor or other medically qualified person.

Now having defined first aid, who might become a first aider? A **first aider** is someone who has undergone a training course to administer first aid at work and holds a current first aid certificate. The training course and certification must be approved by the HSE. The aims of a first aider are to preserve life, to limit the worsening of the injury or illness and to promote recovery.

A first aider may also undertake the duties of an appointed person. An **appointed person** is someone who is nominated to take charge when someone is injured or becomes ill, including calling an ambulance if required. The appointed person will also look after the first aid equipment, including restocking the first aid box.

Appointed persons should not attempt to give first aid for which they have not been trained, but should limit their help to obvious common-sense assistance and summon professional assistance as required. Suggested numbers of first aid personnel are given in Table 1.1. The actual number of first aid personnel must take into account any special circumstances such as remoteness from medical services, the use of several separate buildings and the company's hazard risk assessment. First aid personnel must be available at all times when people are at work, taking into account shift-working patterns and providing cover for sickness absences.

Every company must have at least one first aid kit under the regulations. The size and contents of the kit will depend upon the nature of the risks involved in the particular working environment and the number of employees. Table 1.2 gives a list of the contents of any first aid box to comply with the HSE Regulations.

There now follows a description of some first aid procedures which should be practised under expert guidance before they are required in an emergency.

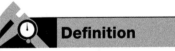

Definition

First aid is the initial assistance or treatment given to a casualty for any injury or sudden illness before the arrival of an ambulance, doctor or other medically qualified person.

Definition

A *first aider* is someone who has undergone a training course to administer first aid at work and holds a current first aid certificate.

Definition

An *appointed person* is someone who is nominated to take charge when someone is injured or becomes ill, including calling an ambulance if required. The appointed person will also look after the first aid equipment, including restocking the first aid box.

Asphyxiation

Asphyxiation is a condition caused by lack of air in the lungs leading to suffocation. Suffocation may cause discomfort by making breathing difficult or it may kill by stopping the breathing. There is a risk of asphyxiation to workers when:

- working in confined spaces;
- working in poorly ventilated spaces;
- working in paint stores and spray booths;
- working in the petrochemical industry;
- working in any environment in which toxic fumes and gases are present.

Under the Management of Health and Safety at Work Regulations a risk assessment must be made if the environment may be considered hazardous to health. Safety procedures, including respiratory protective equipment, must be in place before work commences.

The treatment for fume inhalation or asphyxia is to get the patient into fresh air but only if you can do this without putting yourself at risk. If the patient is unconscious, proceed with resuscitation as described below.

Bleeding

If the wound is dirty, rinse it under clean running water. Clean the skin around the wound and apply a plaster, pulling the skin together.

If the bleeding is severe, apply direct pressure to reduce the bleeding and raise the limb if possible. Apply a sterile dressing or pad and bandage firmly before obtaining professional advice.

To avoid possible contact with hepatitis or the HIV virus when dealing with open wounds, first aiders should avoid contact with fresh blood by wearing plastic or rubber protective gloves, or by allowing the casualty to apply pressure to the bleeding wound.

Burns

Remove heat from the burn to relieve the pain by placing the injured part under clean, cold water. Do not remove burnt clothing sticking to the skin. Do not apply lotions or ointments. Do not break blisters or attempt to remove loose skin. Cover the injured area with a clean, dry dressing.

Broken bones

Make the casualty as comfortable as possible by supporting the broken limb either by hand or with padding. Do not move the casualty unless by remaining in that position they are likely to suffer further injury. Obtain professional help as soon as possible.

Contact with chemicals

Wash the affected area very thoroughly with clean, cold water. Remove any contaminated clothing. Cover the affected area with a clean, sterile dressing and seek expert advice. It is a wise precaution to treat all chemical substances as possibly harmful; even commonly used substances can be dangerous if contamination is from concentrated solutions. When handling dangerous substances, it is also good practice to have a neutralizing agent to hand.

Disposal of dangerous substances must not be into the main drains since this can give rise to an environmental hazard, but should be undertaken in accordance with local authority regulations.

Exposure to toxic fumes

Get the casualty into fresh air quickly and encourage deep breathing if he/she is conscious. Resuscitate if breathing has stopped. Obtain expert medical advice as fumes may cause irritation of the lungs.

Sprains and bruising

A cold compress can help to relieve swelling and pain. Soak a towel or cloth in cold water, squeeze it out and place it on the injured part. Renew the compress every few minutes.

Resuscitation – breathing stopped

Remove any restrictions from the face and any vomit, loose or false teeth from the mouth. Loosen tight clothing around the neck, chest and waist. To ensure a good airway, lay the casualty on their back and support the shoulders on some padding. Tilt the head backwards and open the mouth. If the casualty is faintly breathing, lifting the tongue and clearing the airway may be all that is necessary to restore normal breathing.

Heart stopped beating – CPR (Cardio Pulmonary Resuscitation)

This sometimes happens following a severe electric shock. If the casualty's lips are blue, the pupils of their eyes widely dilated and the pulse in their neck cannot be felt, then they may have gone into cardiac arrest. Act quickly and lay the casualty on their back. Kneel down beside them and place the heel of one hand in the centre of their chest. Cover this hand with your other hand and interlace

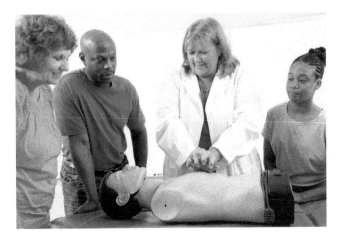

Figure 1.32 Following a few simple rules can save lives.

the fingers. Straighten your arms and press down on their chest sharply with the heel of your hands and then release the pressure. Continue to do this 15 times at the rate of one push per second. 'Push hard and fast, where a gold medallion would rest on the chest, to the rhythm of Ha, Ha, Ha, Ha, staying alive, staying alive'. That is the music from *Saturday Night Fever*, as demonstrated by Vinnie Jones in a recent TV advertisement. Check the casualty's pulse. If none is felt, give a further 15 chest compressions. Continue this procedure until the heartbeat is restored. Pay close attention to the condition of the casualty while giving heart massage. When a pulse is restored the blueness around the mouth will quickly go away and you should stop the heart massage. Treat the casualty for shock, place them in the recovery position and obtain professional help.

Shock

Everyone suffers from shock following an accident. The severity of the shock depends upon the nature and extent of the injury. In cases of severe shock the casualty will become pale and their skin will become clammy from sweating. They may feel faint, have blurred vision, feel sick and complain of thirst. Reassure the casualty that everything that needs to be done is being done. Loosen tight clothing and keep them warm and dry until help arrives. *Do not* move them unnecessarily or give them anything to drink.

Accident reports

Every accident must be reported to an employer and minor accidents reported to a supervisor, safety officer or first aider and the details of the accident and treatment given suitably documented. A first aid log-book or accident book such as that shown in Fig. 1.33 containing first aid treatment record sheets could be used to effectively document accidents which occur in the workplace and the treatment given. Failure to do so may influence the payment of compensation at a later date if an injury leads to permanent disability. To comply with the Data Protection Regulations, from 31 December 2003 all first aid treatment log-books or accident report books must contain perforated sheets which can be removed after completion and filed away for personal security.

If the accident results in death, serious injury or an injury that leads to an absence from work of more than three days, your employer must report the accident to the local office of the HSE. The quickest way to do this is to call the Incident Contact Centre on 0845 300 9923. They will require the following information:

* the name of the person injured;
* a summary of what happened;
* a summary of events prior to the accident;
* information about the injury or loss sustained;
* details of witnesses;
* date and time of accident;
* name of the person reporting the incident.

The Incident Contact Centre will forward a copy of every report they complete to the employer for them to check and hold on record. However, good practice would recommend an employer or their representative to make an extensive report of any serious accident that occurs in the workplace. In addition to recording the above information, the employer or their representative should:

Figure 1.33 First aid log-book/accident book with data protection-compliant removable sheets.

- sketch diagrams of how the accident occurred, where objects were before and after the accident, where the victim fell, etc.;
- take photographs or video that show how things were after the accident, for example, broken step-ladders, damaged equipment, etc.;
- collect statements from witnesses. Ask them to write down what they saw;
- record the circumstances surrounding the accident. Was the injured person working alone – in the dark – in some other adverse situation or condition – was PPE being worn – was PPE recommended in that area?

The above steps should be taken immediately after the accident has occurred and after the victim has been sent for medical attention. The area should be made safe and the senior management informed so that any actions to prevent a similar occurrence can be put in place. Taking photographs and obtaining witnesses' statements immediately after an accident happens means that evidence may still be around and memories still sharp.

RIDDOR

RIDDOR stands for Reporting of Injuries, Diseases and Dangerous Occurrences Regulation 1995, which is sometimes referred to as RIDDOR 95, or just RIDDOR for short. The HSE requires employers to report some work-related accidents or diseases so that they can identify where and how risks arise, investigate serious accidents and publish statistics and data to help reduce accidents at work.

What needs reporting? Every work-related death, major injury, dangerous occurrence, disease or any injury which results in an absence from work of over three days.

Where an employee or member of the public is killed as a result of an accident at work the employer or his representative must report the accident to the Environmental Health Department of the Local Authority by telephone that day and give brief details. Within 10 days this must be followed up by a completed accident report form (Form No. F2508). Major injuries sustained as a result of an accident at work include amputations, loss of sight (temporary or permanent), fractures to the body other than to fingers, thumbs or toes and any other serious injury. Once again, the Environmental Health Department of the Local Authority must be notified by telephone on the day that the serious injury occurs and the telephone call followed up by a completed Form F2508 within 10 days. Dangerous occurrences are listed in the regulations and include the collapse of a lift, an explosion or injury caused by an explosion, the collapse of a scaffold over 5 m high, the collision of a train with any vehicle, the unintended collapse of a building and the failure of fairground equipment.

Depending upon the seriousness of the event, it may be necessary to immediately report the incident to the Local Authority. However, the incident must be reported within 10 days by completing Form F2508. If a doctor notifies an employer that an employee is suffering from a work-related disease then form F2508A must be completed and sent to the Local Authority. Reportable diseases include certain poisonings, skin diseases, lung disease, infections and occupational cancer. The full list is given within the pad of report forms.

An accident at work resulting in an over-three-day injury (that is, an employee being absent from work for over three days as a result of an accident at work) requires that accident report form F2508 be sent to the Local Authority within 10 days.

An over-three-day injury is one which is not major but results in the injured person being away from work for more than three days not including the day the injury occurred.

Who are the reports sent to? They are sent to the Environmental Health Department of the Local Authority or the area HSE offices (see Appendix B for area office addresses). Accident report forms F2508 can also be obtained from them or by ringing the HSE Infoline, or by ringing the Incident Contact Centre on 0845 300 9923.

For most businesses, a reportable accident, dangerous occurrence or disease is a very rare event. However, if a report is made, the company must keep a record of the occurrence for three years after the date on which the incident happened. The easiest way to do this would probably be to file a photocopy of the completed accident report form F2508, but a record may be kept in any form which is convenient.

Dangerous occurrences and hazardous malfunctions

Definition

Dangerous occurrence – a 'near miss' that could easily have led to serious injury or loss of life. Near-miss accidents occur much more frequently than injury accidents and are, therefore, a good indicator of hazard, which is why the HSE collects this data.

A **dangerous occurrence** is a 'near miss' that could easily have led to serious injury or loss of life. Dangerous occurrences are defined in the Reporting of Injuries, Diseases and Dangerous Occurrences Regulations (RIDDOR) 1995. Near-miss accidents occur much more frequently than injury accidents and are, therefore, a good indicator of hazard, which is why the HSE collects this data. In January 2008 a BA passenger aeroplane lost power to both engines as it prepared to land at Heathrow airport. The pilots glided the plane into a crash landing on the grass just short of the runway. This is one example of a dangerous occurrence which could so easily have been a disaster.

Consider another example: on a wet and windy night a large section of scaffold around a town centre building collapses. Fortunately this happens at about midnight when no one is around because of the time and the bad weather. However, if it had occurred at midday, workers would have been using the scaffold and the streets would have been crowded with shoppers. This would be classified as a dangerous occurrence and must be reported to the HSE, who will investigate the cause and, using their wide range of powers, would either:

- stop all work;
- demand the dismantling of the structure;
- issue an Improvement Notice;
- issue a Prohibition Notice;
- prosecute those who have failed in their health and safety duties.

Other reportable dangerous occurrences are:

- the collapse of a lift;
- plant coming into contact with overhead power lines;
- any unexpected collapse which projects material beyond the site boundary;
- the overturning of a road tanker;
- a collision between a car and a train.

Definition

Hazardous malfunction – if a piece of equipment was to fail in its function (that is, fail to do what it is supposed to do) and, as a result of this failure have the potential to cause harm, then this would be defined as a hazardous malfunction.

Hazardous malfunction – if a piece of equipment was to fail in its function (that is, fail to do what it is supposed to do) and, as a result of this failure have the potential to cause harm, then this would be defined as a hazardous malfunction. Consider an example: if a 'materials lift' on a construction site was to collapse when the supply to its motor failed, this would be a hazardous malfunction. All the regulations concerning work equipment state that it must be:

- suitable for its intended use;
- safe in use;
- maintained in a safe condition;
- used only by instructed persons;
- provided with suitable safety measures, protective devices and warning signs.

Disposing of waste

We have said many times in this book so far that having a good attitude to health and safety, working conscientiously and neatly, keeping passageways clear and regularly tidying up the workplace is the sign of a good and competent craftsman. But what do you do with the rubbish that the working environment produces? Well, all the packaging material for electrical fittings and accessories usually goes into either your employer's skip or the skip on-site designated for that purpose. All the offcuts of conduit, trunking and tray also go into the skip. In fact, most of the general site debris will probably go into the skip and the waste disposal company will take the skip contents to a designated local council land-fill area for safe disposal.

The part coils of cable and any other reusable left-over lengths of conduit, trunking or tray will be taken back to your employer's stores area. Here it will be stored for future use and the returned quantities deducted from the costs allocated to that job.

What goes into the skip for normal disposal into a land-fill site is usually a matter of common sense. However, some substances require special consideration and disposal. We will now look at asbestos and large quantities of used fluorescent tubes.

Control of asbestos at work regulations

In October 2010 the HSE launched a national campaign to raise awareness among electricians and other trades of the risk to their health of coming into contact with asbestos. It is called the 'Hidden Killer Campaign' because asbestos is a killer. It can cause asbestosis, lung cancer and mesothelioma which affect the lining of the lungs and surrounding digestive system.

The British Lung Foundation has calculated that every week on average six electricians, four plumbers and eight joiners die in the UK as a result of exposure to asbestos, making it the greatest cause of work-related deaths. Chief Executive Dr Penny Woods said that 'twice as many people die in Britain each year from asbestos-related illnesses than those who die from accidents on the roads'. For more information about asbestos hazards, visit www.hse.uk/hiddenkiller, www.hse.gov.uk/asbestos and the British Lung Foundation www.take 5and stayalive.com.

Asbestos is a mineral found in many rock formations. When separated it becomes a fluffy, fibrous material with many uses. It was used extensively in the construction industry during the 1960s and 1970s for roofing material, ceiling and floor tiles, fire-resistant board for doors and partitions, for thermal insulation and commercial and industrial pipe lagging.

There are three main types of asbestos:

- chysotile which is white and accounts for about 90% of the asbestos in use today;
- amosite which is brown;
- crocidolite which is blue.

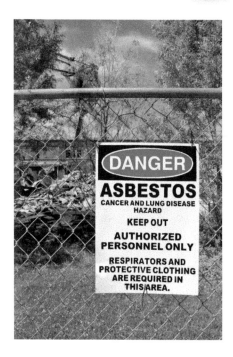

Figure 1.34 Hazardous areas should be clearly signposted.

Asbestos cannot be identified by colour alone and a laboratory analysis is required to establish its type. Blue and brown are the two most dangerous forms of asbestos and have been banned from use since 1985. White asbestos was banned from use in 1999.

In the buildings where it was installed some 40 years ago, when left alone, it did not represent a health hazard, but those buildings are increasingly becoming in need of renovation and modernization. It is in the dismantling, drilling and breaking up of these asbestos materials that the health hazard increases. Asbestos is a serious health hazard if the dust is inhaled. The tiny asbestos particles find their way into delicate lung tissue and remain embedded for life, causing constant irritation and eventually, serious lung disease.

Asbestos materials may be encountered by electricians in decorative finishes such as artex ceiling finishes, plaster and floor tiles. It is also found in control gear such as flash guards and matting in fuse carriers and distribution fuse boards, and in insulation materials in vessels, containers, pipework, ceiling ducts and wall and floor partitions.

Working with asbestos materials is not a job for anyone in the electrical industry. If asbestos is present in situations or buildings where you are expected to work, it should be removed by a specialist contractor before your work commences. Specialist contractors, who will wear fully protective suits and use breathing apparatus, are the only people who can safely and responsibly carry out the removal of asbestos. They will wrap the asbestos in thick plastic bags and store the bags temporarily in a covered and locked skip. This material is then disposed of in a special land-fill site with other toxic industrial waste materials and the site monitored by the local authority for the foreseeable future.

There is a lot of work for electrical contractors in my part of the country, updating and improving the lighting in government buildings and schools. This work often involves removing the old fluorescent fittings hanging on chains or fixed to beams, and installing a suspended ceiling and an appropriate number of recessed modular fluorescent fittings. So what do we do with the old fittings? Well, the fittings are made of sheet steel, a couple of plastic lamp holders, a little cable, a starter and ballast. All of these materials can go into the ordinary skip. However, the fluorescent tubes contain a little mercury and fluorescent powder with toxic elements, which cannot be disposed of in the normal land-fill sites.

Hazardous Waste Regulations 2005

New Hazardous Waste Regulations were introduced in July 2005 and under these regulations lamps and tubes are classified as hazardous. While each lamp contains only a small amount of mercury, vast numbers of lamps and tubes are disposed of in the United Kingdom every year, resulting in a significant environmental threat.

The environmentally responsible way to dispose of fluorescent lamps and tubes is to recycle them.

The process usually goes like this:

- your employer arranges for the local electrical wholesaler to deliver a plastic used lamp waste container of an appropriate size for the job;
- expired lamps and tubes are placed whole into the container, which often has a grating inside to prevent the tubes from breaking when being transported;
- when the container is full of used lamps and tubes, you telephone the electrical wholesaler and ask them to pick up the filled container and deliver it to one of the specialist recycling centres;

Safety first

Waste
- clean up before you leave the job;
- put waste in the correct skip;
- recycle used lamps and tubes;
- get rid of all waste responsibly.

- your electrical company will receive a 'Duty of Care Note' and full recycling documents which should be filed safely as proof that the hazardous waste was recycled safely;
- the charge is approximately 50p for each 1800 mm tube and this cost is passed on to the customer through the final account.

Safe working procedures

The principles which were laid down in the many Acts of Parliament and the regulations that we have already looked at in this chapter control our working environment. They make our workplace safer, but despite all this legislation, workers continue to be injured and killed at work or die as a result of a work-related injury. The number of deaths has consistently averaged about 200 each year for the past eight years. These figures only relate to employees. If you include the self-employed and members of the public killed in work-related accidents, the numbers almost double.

In addition to the deaths, about 28,000 people have major accidents at work and about 130,000 people each year receive minor work-related injuries which keep them off work for more than three days.

It is a mistake to believe that these things only happen in dangerous occupations such as deep-sea diving, mining and quarrying, fishing industry, tunnelling and fire-fighting, or that they only happen in exceptional circumstances that would never occur in your workplace. This is not the case. Some basic thinking and acting beforehand could have prevented most of these accident statistics from happening.

Causes of accidents

Most accidents are caused by either human error or environmental conditions. **Human errors** include behaving badly or foolishly, being careless and not paying attention to what you should be doing at work, doing things that you are not competent to do or have not been trained to do. You should not work when tired or fatigued and should never work when you have been drinking alcohol or taking drugs.

Environmental conditions include unguarded or faulty machinery, damaged or faulty tools and equipment, poorly illuminated or ventilated workplaces and untidy, dirty or overcrowded workplaces.

The most common causes of accidents

These are:

- slips, trips and falls;
- manual handling, that is, moving objects by hand;
- using equipment, machinery or tools;
- storage of goods and materials which then become unstable;
- fire;
- electricity;
- mechanical handling.

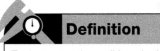

Definition

Human errors include behaving badly or foolishly, being careless and not paying attention to what you should be doing at work.

Definition

Environmental conditions include unguarded or faulty machinery.

Safety first

Safety procedures
- Hazard risk assessment *is an essential part* of any health and safety management system.
- The aim of the planning process is to minimize risk.

HSE publication HSG(65)

Accident prevention measures

To control the risk of an accident we usually:

* eliminate the cause;
* substitute a procedure or product with less risk;
* enclose the dangerous situation;
* put guards around the hazard;
* use safe systems of work;
* supervise, train and give information to staff;
* if the hazard cannot be removed or minimized, provide PPE.

In other chapters of this book we will look at the application of procedures that make the workplace a safer place to work.

Site visitors

A construction site will have many people working on a building project throughout the construction period. The groundwork people lay the foundations, the steel work is erected, the bricklayers build the walls, the carpenters put on the roof, and only when the building is waterproof do the electricians and mechanical service trades begin to install the electrical and mechanical systems.

If there was an emergency and the site had to be evacuated, how would you know who, or how many people were on-site? You can see that there has to be a procedure for logging people in and out, so that the main contractor can identify who is on-site at any one time.

If the architect pops in to make a quick inspection or resolve a problem, he must first 'sign in'. If the managing director of your electrical company drops by to see how work is progressing, he must first 'sign in'. When they leave, they must 'sign out'. How else would you know who is on-site at any one time?

A formal site visitor procedure must be in place to meet the requirements of the health and safety regulations and to maintain site security.

The site visitor procedure must also meet the needs of the building project. For example, a high-profile building project, one which pushes the architectural boundaries such as the new Wembley Football Stadium in London, would have attracted many more visitors during its construction period than, say, a food distribution warehouse for a supermarket being constructed close to a motorway junction.

The site visitor procedure is only required to be 'fit for purpose', but whatever system is put in place, here are some suggestions for consideration:

* you will want your visitors to be safe when they are on-site and so you might insist that they wear a hard hat and a high-visibility jacket;
* in some cases, you might want visitors to be escorted for their own safety while on-site. Of course this would not apply to professional visitors such as the architect or quantity surveyor;
* the procedure should identify who the visitor is and who they wish to visit;
* you will want to know the time of their arrival and the time of their departure.

Many of these requirements can be met with a simple log-book divided into columns with headings such as date and time in, visitor's name, company name, name of the person to be visited and time out or time of leaving.

Figure 1.35 Slips, trips and falls are one of the most common causes of accidents.

Check your understanding

When you have completed the questions, check out the answers at the back of the book.

Note: more than one multiple-choice answer may be correct.

1 For any fire to continue to burn, three components must be present. These are:
  a. fuel, wood and cardboard
  b. petrol, oxygen and bottled gas
  c. flames, fuel and heat
  d. fuel, oxygen and heat.

2 A water fire extinguisher is suitable for use on small fires of burning:
  a. wood, paper and fabric
  b. flammable liquids
  c. flammable gas
  d. all of the above.

3 A foam fire extinguisher is suitable for use on small fires of burning:
  a. wood, paper and fabric
  b. flammable liquids
  c. flammable gas
  d. all of the above.

4 A carbon dioxide gas fire extinguisher is suitable for use on small fires of burning:
  a. wood, paper and fabric
  b. flammable liquids
  c. flammable gas
  d. all of the above.

5 A dry powder fire extinguisher is suitable for use on small fires of burning:
  a. wood, paper and fabric
  b. flammable liquids
  c. flammable gas
  d. all of the above.

6 A vaporizing foam fire extinguisher is suitable for use on small fires of burning:
  a. wood, paper and fabric
  b. flammable liquids
  c. flammable gas
  d. all of the above.

7   You should only attack a fire with a fire extinguisher if:
   a.   it is burning brightly
   b.   you can save someone's property
   c.   you can save someone's life
   d.   you can do so without putting yourself at risk.

8   A fire extinguisher should only be used to fight:
   a.   car fires
   b.   electrical fires
   c.   small fires
   d.   big fires.

9   A Statutory Regulation:
   a.   is the law of the land
   b.   must be obeyed
   c.   tells us how to comply with the law
   d.   is a code of practice.

10   A Non-statutory Regulation:
   a.   is the law of the land
   b.   must be obeyed
   c.   tells us how to comply with the law
   d.   is a code of practice.

11   Under the Health and Safety at Work Act an employer is responsible for:
   a.   maintaining plant and equipment
   b.   providing PPE
   c.   wearing PPE
   d.   taking reasonable care to avoid injury.

12   Under the Health and Safety at Work Act an employee is responsible for:
   a.   maintaining plant and equipment
   b.   providing PPE
   c.   wearing PPE
   d.   taking reasonable care to avoid injury.

13   The IET Wiring Regulations:
   a.   are statutory regulations
   b.   are non-statutory regulations
   c.   are codes of good practice
   d.   must always be complied with.

14   Before beginning work on a 'live' circuit or piece of equipment you should:
   a.   only work 'live' if your supervisor is with you
   b.   only work 'live' if you feel that you are 'competent' to do so
   c.   isolate the circuit or equipment before work commences
   d.   secure the isolation before work commences.

15   The initial assistance or treatment given to a casualty for any injury or sudden illness before the arrival of an ambulance or medically qualified person is one definition of:

    a.   an appointed person

    b.   a first aider

    c.   first aid

    d.   an adequate first aid facility.

16   Someone who has undergone a training course to administer medical aid at work and holds a current qualification is one definition of:

    a.   a doctor

    b.   a nurse

    c.   a first aider

    d.   a supervisor.

17   Two of the most common categories of risk and causes of accidents at work are:

    a.   slips, trips and falls

    b.   put guards around the hazard

    c.   manual handling

    d.   use safe systems of work.

18   Two of the most common precautions taken to control risks are:

    a.   slips, trips and falls

    b.   put guards around the hazard

    c.   manual handling

    d.   use safe systems of work.

19   Something which has the potential to cause harm is one definition of:

    a.   health and safety

    b.   risk

    c.   competent person

    d.   hazard.

20   The chances of harm actually being done is one definition of:

    a.   electricity

    b.   risk

    c.   health and safety

    d.   hazard.

21   A competent person dealing with a hazardous situation:

    a.   must wear appropriate PPE

    b.   display a health and safety poster

    c.   reduces the risk

    d.   increases the risk.

22   Employers of companies employing more than five people must:

    a.   become a member of the NICEIC

    b.   provide PPE if appropriate

    c.   carry out a hazard risk assessment

    d.   display a health and safety poster.

23   There are five parts to a hazard risk assessment procedure. Identify one from the list below:

    a.   wear appropriate PPE

    b.   notify the HSE that you intend to carry out a risk assessment

    c.   list the hazards and who might be harmed

    d.   substitute a procedure with less risk.

24   Lifting, transporting or supporting heavy objects by hand or bodily force is one definition of:

    a.   working at height

    b.   a mobile scaffold tower

    c.   a sack truck

    d.   manual handling.

25   When working above ground for long periods of time the most appropriate piece of equipment to use would be:

    a.   a ladder

    b.   a trestle scaffold

    c.   a mobile scaffold tower

    d.   a pair of sky hooks.

26   The most appropriate piece of equipment to use for gaining access to a permanent scaffold would be:

    a.   a ladder

    b.   a trestle scaffold

    c.   a mobile scaffold tower

    d.   a pair of sky hooks.

27   The Electricity at Work Regulations tell us that 'we must ensure the disconnection and separation of electrical equipment from every source of supply and the separation must be secure'. A procedure to comply with this regulation is called:

    a.   work at height

    b.   a hazard risk assessment

    c.   a safe isolation procedure

    d.   a workstation risk assessment.

28   The Electricity at Work Regulations absolutely forbid the following work activity:

    a.   working at height

    b.   testing live electrical systems

    c.   live repair work on electrical circuits

    d.   working without the appropriate PPE.

29   'Good housekeeping' at work is about:

    a.   cleaning up and putting waste in the skip

    b.   working safely

    c.   making the tea and collecting everyone's lunch

    d.   putting tools and equipment away after use.

30   Use bullet points to describe a safe isolation procedure of a 'live' electrical circuit.

31   How does the law enforce the regulations of the Health and Safety at Work Act?

32   List the responsibilities under the Health and Safety at Work Act of:
a.   an employer to his employees
b.   an employee to his employer.

33   Safety signs are used in the working environment to give information and warning. Sketch and colour one sign from each of the four categories of signs and state the message given by that sign.

34   State the name of two important Statutory Regulations and one Non-statutory Regulation relevant to the electrical industry.

35   Define what is meant by PPE.

36   State five pieces of PPE which a trainee could be expected to wear at work and the protection given by each piece.

37   Describe the action to be taken upon finding a workmate apparently dead on the floor and connected to an electrical supply.

38   State how the Data Protection Act has changed the way in which we record accident and first aid information at work.

39   List five common categories of risk.

40   List five common precautions which might be taken to control risk.

41   Use bullet points to list the main stages involved in lifting a heavy box from the floor, carrying it across a room and placing it on a worktop, using a safe manual handling technique.

42   Describe a safe manual handling technique for moving a heavy electric motor out of the stores, across a yard and into the back of a van for delivery to site.

43   Use bullet points to list a step-by-step safe electrical isolation procedure for isolating a circuit in a three-phase distribution fuse board.

44   Use bullet points to list each stage in the erection and securing of a long extension ladder. Identify all actions which would make the ladder safe to use.

45   Describe how you would use a mobile scaffold tower to re-lamp all the light fittings in a supermarket. Use bullet points to give a step-by-step account of re-lamping the first two fittings.

46   What is a proving unit used for?

47   The HSE Guidance Note GS 38 tells us about suitable test probe leads. Use a sketch to identify the main recommendations.

48 State how you would deal with the following materials when you are cleaning up at the end of the job:

- pieces of conduit and tray
- cardboard packaging material
- empty cable rolls
- half-full cable rolls
- bending machines for conduit and tray
- your own box of tools
- your employer's power tools
- 100 old fluorescent light fittings
- 200 used fluorescent tubes.

Understand the fundamental principles and requirements of environmental technology systems

Advanced Electrical Installation Work. 978-0-367-35976-8

Unit 301 of the City and Guilds 2365-03 syllabus

Learning outcomes – when you have completed this chapter you should:

- know the regulations which safeguard the environment;
- know the working principles of micro-renewable energy systems;
- know the working principles of water conservation technologies;
- know the regulations which relate to the installation of microgeneration technology;
- know how the building location and features affect the potential to install microgeneration technologies;
- state the advantages and disadvantages associated with microgeneration technologies.

This chapter has free associated content, including animations and instructional videos, to support your learning.

When you see the logo, visit the website below to access this material: www.routledge.com/cw/linsley

Environmental laws and regulations

The **environment** describes the world in which we live, work and play; it relates to our neighbourhood and surroundings and the situation in which we find ourselves.

Environmental laws protect the environment in which we live by setting standards for the control of pollution to land, air and water.

If a wrong is identified in the area in which we now think of as 'environmental' it can be of two kinds:

1 An offence in common law which means damage to property, nuisance or negligence leading to a claim for damages.
2 A statutory offence against one of the laws dealing with the protection of the environment. These offences are nearly always 'crimes' and punished by fines or imprisonment rather than by compensating any individual.

The legislation dealing with the environment has evolved for each part – air, water, land noise, radioactive substances. Where organizations' activities impact upon the environmental laws they are increasingly adopting environmental management systems which comply with ISO 14001. Let us now look at some of the regulations and try to see the present picture at the beginning of the new millennium.

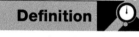

Definition

The *environment* describes the world in which we live, work and play; it relates to our neighbourhood and surroundings and the situation in which we find ourselves.

Environmental Protection Act 1990

In the context of environmental law, the Environmental Protection Act 1990 was a major piece of legislation. The main sections of the Act are:

Part 1	Integrated pollution control by HM Inspectorate of Pollution, and air pollution control by Local Authorities
Part 2	Wastes on land
Part 3	Statutory nuisances and clean air
Part 4	Litter
Part 5	Radioactive Substances Act 1960
Part 6	Genetically modified organisms
Part 7	Nature conservation
Part 8	Miscellaneous, including contaminated land.

The Royal Commission of 1976 identified that a reduction of pollutant to one medium, air, water or land, then led to an increase of pollutant in another. It therefore stressed the need to take an integrated approach to pollution control. The processes subject to an integrated pollution control are:

• Air emissions.
• Processes which give rise to significant quantities of special waste; that is, waste defined in law in terms of its toxicity or flammability.

- Processes giving rise to emissions to sewers or 'Red List' substances. These are 23 substances including mercury, cadmium and many pesticides which are subject to discharge consent to the satisfaction of the Environment Agency.

The Inspectorate is empowered to set conditions to ensure that the best practicable environmental option (BPEO) is employed to control pollution. This is the cornerstone of the Environmental Protection Act.

Pollution Prevention and Control Regulations 2000

The system of Pollution Prevention and Control replaced that of Integrated Pollution Control established by the Environmental Protection Act 1990, thus bringing environmental law into the new millennium and implementing the European Directive (EC/96/61) on integrated pollution prevention and control. The new system was fully implemented in 2007.

Pollution Prevention and Control is a regime for controlling pollution from certain industrial activities. This regime introduces the concept of Best Available Technique (BAT) for reducing and preventing pollution to an acceptable level.

Industrial activities are graded according to their potential to pollute the environment:

- A(1) installations are regulated by the Environment Agency;
- A(2) installations are regulated by the Local Authorities;
- Part B installations are also regulated by the Local Authority.

All three systems require the operators of certain industrial installations to obtain a permit to operate. Once an operator has submitted a permit application, the regulator then decides whether to issue a permit. If one is issued it will include conditions aimed at reducing and preventing pollution to acceptable levels. A(1) installations are generally perceived as having the greatest potential to pollute the environment. A(2) installations and Part B installations would have the least potential to pollute.

The industries affected by these regulations are those dealing with petrol vapour recovery, incineration of waste, mercury emissions from crematoria, animal rendering, non-ferrous foundry processes, surface treating of metals and plastic materials by powder coating, galvanizing of metals and the manufacture of certain specified composite wood-based boards.

Clean Air Act 1993

We are all entitled to breathe clean air but until quite recently the only method of heating houses and workshops was by burning coal, wood or peat in open fires. The smoke from these fires created air pollution and the atmosphere in large towns and cities was of poor quality. On many occasions in the 1950s the burning of coal in London was banned because the city was grinding to a halt owing to the combined effect of smoke and fog, called smog. Smog was a very dense fog in which you could barely see more than a metre in front of you and which created serious breathing difficulties. In the new millennium we are no longer dependent upon coal and wood to heat our buildings; smokeless coal has been created and the gaseous products of combustion are now diluted and dispersed by new chimney design regulations. Using well-engineered combustion equipment together with the efficient arrestment of small particles in commercial chimneys of sufficient height, air pollution has been much reduced. This is what the **Clean Air Act** set out to achieve and it has been largely successful.

Definition

The *Clean Air Act* applies to all small and medium-sized companies operating furnaces, boilers or incinerators.

The Clean Air Act applies to all small and medium-sized companies operating furnaces, boilers or incinerators. Compliance with the Act does not require an application for authorization and so companies must make sure that they do not commit an offence. In general the emission of dark smoke from any chimney is unacceptable. The emission of dark smoke from any industrial premises is also unacceptable. This might be caused by, for example, the burning of old tyres or old cable.

In England, Scotland and Wales it is not necessary for the Local Authority to have witnessed the emission of dark smoke before taking legal action. Simply the evidence of burned materials, which potentially give rise to dark smoke when burned, is sufficient. In this way the law aims to stop people creating dark smoke under the cover of darkness.

A **public nuisance** is 'an act unwarranted by law or an omission to discharge a legal duty which materially affects the life, health, property, morals or reasonable comfort or convenience of Her Majesty's subjects'. This is a criminal offence and Local Authorities can prosecute, defend or appear in proceedings that affect the inhabitants of their area.

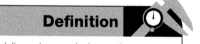
Controlled Waste Regulations 1998

Under these regulations we have a 'Duty of Care to handle, recover and dispose of all waste responsibly'. This means that all waste must be handled, recovered and disposed of by individuals or businesses that are authorized to do so under a system of signed Waste Transfer Notes.

The Environmental Protection (Duty of Care) Regulations 1991 state that as a business you have a duty to ensure that any waste you produce is handled safely and in accordance with the law. This is the Duty of Care and applies to anyone who produces, keeps, carries, treats or disposes of waste from business or industry.

You are responsible for the waste that you produce, even after you have passed it on to another party such as a skip hire company, a scrap metal merchant, recycling company or local council. The Duty of Care has no time limit and extends until the waste has either been finally and properly disposed of or fully recovered.

So what does this mean for your company?

*   Make sure that waste is only transferred to an authorized company.
*   Make sure that waste being transferred is accompanied by the appropriate paperwork showing what was taken, where it was to be taken and by whom.
*   Segregate the different types of waste that your work creates.
*   Label waste skips and waste containers so that it is clear to everyone what type of waste goes into that skip.
*   Minimize the waste that you produce and do not leave waste behind for someone else to clear away. Remember there is no time limit on your Duty of Care for waste.

Occupiers of domestic properties are exempt from the Duty of Care for the household waste that they produce. However, they do have a Duty of Care for the waste produced by, for example, a tradesperson working at a domestic property.

Special waste is covered by the Special Waste Regulations 1996 and is waste that is potentially hazardous or dangerous and which may, therefore, require

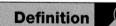

special precautions during handling, storage, treatment or disposal. Examples of special waste are asbestos, lead-acid batteries, used engine oil, solvent-based paint, solvents, chemical waste and pesticides. The disposal of special waste must be carried out by a competent person, with special equipment and a licence.

Hazardous Waste Regulations 2005

New Hazardous Waste Regulations were introduced in July 2005 and under these regulations electric discharge lamps and tubes such as fluorescent, sodium, metal halide and mercury vapour are classified as hazardous waste. While each lamp only contains a very small amount of mercury, vast numbers are used and disposed of each year, resulting in a significant environmental threat. The environmentally responsible way to dispose of lamps and tubes is to recycle them and this process is now available through electrical wholesalers.

Electrical companies produce relatively small amounts of waste and even smaller amounts of special waste. Most companies buy in the expertise of specialist waste companies these days and build these costs into the contract.

Packaging (Essential Requirements) Regulations 2003

The new **Packaging Regulations** were introduced on 25 August 2003 bringing the UK into harmony with Europe. The regulations deal with the essential requirements of packaging for the storage and transportation of goods. There are two essential elements to the regulations:

1 The packaging shall be designed and manufactured so that the volume and weight is to the minimum amount required to maintain the necessary level of safety for the packaged product.
2 The packaging shall be designed and manufactured in such a way that the packaging used is either reusable or recyclable.

The regulations are enforced by the Weights and Measures Authority in Great Britain, the Department of Enterprise Trade and Investment in Northern Ireland and the Procurator Fiscal or Lord Advocate in Scotland.

> **Definition**
>
> The new *Packaging Regulations* were introduced on 25 August 2003 bringing the UK into harmony with Europe. The regulations deal with the essential requirements of packaging for the storage and transportation of goods.

Waste Electrical and Electronic Equipment EU Directive 2007

The Waste Electrical and Electronic Equipment (WEEE) Regulations will ensure that Britain complies with its EU obligation to recycle waste from electrical products. The regulations came into effect in July 2007 and from that date any company which makes, distributes or trades in electrical or electronic goods such as household appliances, sports equipment and even torches and toothbrushes has to make arrangements for recycling these goods at the end of their useful life. Batteries will be covered separately by yet another forthcoming EU directive.

Some sectors are better prepared for the new regulations than others. Mobile phone operators, O2, Orange, Virgin and Vodafone, along with retailers such as Currys and Dixons, have already joined together to recycle their mobile phones collectively. In Holland the price of a new car now includes a charge for the recycling costs.

Further information is available on the DTI and DEFRA website under WEEE.

Radioactive Substances Act 1993

These regulations apply to the very low-ionizing radiation sources used by specialized industrial contractors. The radioactive source may be sealed or unsealed. Unsealed sources are added to a liquid in order to trace the direction or rate of flow of that liquid. Sealed radioactive sources are used in radiography for the non-destructive testing of materials or in liquid level and density gauges.

This type of work is subject to the Ionizing Radiations Regulations 1999 (IRR), which impose comprehensive duties on employers to protect people at work against exposure to ionizing radiation. These regulations are enforced by the HSE, while the Radioactive Substances Act (RSA) is enforced by the Environmental Agency.

The RSA 1993 regulates the keeping, use, accumulation and disposal of radioactive waste, while the IRR 1999 regulate the working and storage conditions when using radioactive sources. The requirements of RSA 1993 are in addition to and separate from IRR 1999 for any industry using radioactive sources. These regulations also apply to offshore installations and to work in connection with pipelines.

Dangerous Substances and Preparations and Chemicals Regulations 2000

Chemical substances that are classified as carcinogenic, mutagenic or toxic, or preparations which contain those substances, constitute a risk to the general public because they may cause cancer, genetic disorders and birth defects, respectively.

These regulations were introduced to prohibit the supply of these dangerous substances to the general public, to protect consumers from contracting fatal diseases through their use.

The regulations require that new labels be attached to the containers of these substances which identify the potential dangers and indicate that they are restricted to professional users only.

The regulations implement Commission Directive 99/43/EC, known as the 17th Amendment, which brings the whole of Europe to an agreement that these substances must not be sold to the general public, this being the only way of offering the highest level of protection for consumers.

The regulations are enforced by the Local Authority Trading Standards Department.

Noise regulations

Before 1960 noise nuisance could only be dealt with by common law as a breach of the peace under various Acts or local by-laws. In contrast, today there are many statutes, Government circulars, British Standards and EU Directives dealing with noise matters. Environmental noise problems have been around for many years. During the eighteenth century, in the vicinity of some London hospitals, straw was put on the roads to deaden the sound of horses' hooves and the wheels of carriages. Today we have come a long way from this self-regulatory situation.

In the context of the Environmental Protection Act 1990, noise or vibration is a **statutory nuisance** if it is prejudicial to health or is a nuisance. However,

Definition

'A *statutory nuisance* must materially interfere with the enjoyment of one's dwelling. It is more than just irritating or annoying and does not take account of the undue sensitivity of the receiver.'

nuisance is not defined and has exercised the minds of lawyers, magistrates and judges since the concept of nuisance was first introduced in the 1936 Public Health Act. There is a wealth of case law but a good working definition might be 'a statutory nuisance must materially interfere with the enjoyment of one's dwelling. It is more than just irritating or annoying and does not take account of the undue sensitivity of the receiver.'

The line that separates nuisance from no nuisance is very fine and non-specific. Next door's intruder alarm going off at 3 a.m. for an hour or more is clearly a statutory nuisance, whereas one going off a long way from your home would not be a nuisance. Similarly, an all-night party with speakers in the garden would be a nuisance, whereas an occasional party finishing at, say, midnight would not be a statutory nuisance.

At Stafford Crown Court on 1 November 2004, Alton Towers, one of the country's most popular Theme Parks, was ordered by a judge to reduce noise levels from its 'white knuckle' rides. In the first judgment of its kind, the judge told the Park's owners that neighbouring residents must not be interrupted by noise from rides such as Nemesis, Air, Corkscrew, Oblivion, or from loudspeakers or fireworks.

The owners of Alton Towers, Tussauds Theme Parks Ltd, were fined the maximum sum of £5,000 and served with a Noise Abatement Order for being guilty of breaching the 1990 Environmental Protection Act. Mr Richard Buxton, for the prosecution, said that the £5,000 fine reflected the judge's view that Alton Towers had made little or no effort to reduce the noise nuisance.

Many nuisance complaints under the Act are domestic and are difficult to assess and investigate. Barking dogs, stereos turned up too loud, washing machines running at night to use 'low-cost' electricity, television and DIY activities are all difficult to assess precisely as statutory nuisance. Similarly, sources of commercial noise complaints are also varied and commonly include deliveries of goods during the night, general factory noises, refrigeration units and noise from public houses and clubs.

Industrial noise can be complex and complaints difficult to resolve both legally and technically. Industrial noise assessment is aided by BS 4142 but no guidance exists for other noise nuisance. The Local Authority has a duty to take reasonable steps to investigate all complaints and to take appropriate action.

The Noise and Statutory Nuisance Act 1993

This Act extended the statutory nuisance provision of the Environmental Protection Act 1990 to cover noise from vehicles, machinery or equipment in the streets. The definition of equipment includes musical instruments but the most common use of this power is to deal with car alarms and house intruder alarms being activated for no apparent reason and which then continue to cause a nuisance for more than one hour.

Home intruder alarms that have been sounding for one hour can result in a 'Notice' being served on the occupier of the property, even if he or she is absent from the property at the time of the offence. Sections 7–9 of the Act make provision for incorporating the 'Code of Practice relating to Audible Intruder Alarms' into the statute. The two key points of the Code are the installation of a 20-minute cut-off of the external sounder and the notification to the police and Local Authority of two key holders who can silence the alarm.

Noise Act 1996

This Act clarifies the powers which may be taken against work which is in default under the nuisance provision of the Environmental Protection Act 1990. It provides a mechanism for permanent deprivation, return of seized equipment and charges for storage.

The Act also includes an *adoptive* provision making night-time noise between 23:00 and 07:00 hours a criminal offence if the noise exceeds a certain level to be prescribed by the Secretary of State. If a Notice is not complied with, a fixed penalty may be paid instead of going to court.

Noise at Work Regulations 1989

The Noise at Work Regulations, unlike the previous vague or limited provisions, apply to all workplaces and require employers to carry out assessments of the noise levels within their premises and to take appropriate action where necessary. The 1989 Regulations came into force on 1 January 1990 implementing in the United Kingdom the EC Directive 86/188/EEC 'The Protection of Workers from Noise'.

Three action levels are defined by the regulations:

1 The first action level is a daily personal noise exposure of 85 dB, expressed as 85 dB(A).
2 The second action level is a daily personal noise exposure of 90 dB(A).
3 The third defined level is a peak action level of 140 dB(A) or 200 Pa of pressure which is likely to be linked to the use of cartridge operated tools, shooting guns or similar loud explosive noises. This action level is likely to be most important where workers are subjected to a small number of loud impulses during an otherwise quiet day.

The Noise at Work Regulations are intended to reduce hearing damage caused by loud noise. So, what is a loud noise? If you cannot hear what someone is saying when they are 2 m away from you or if they have to shout to make themselves heard, then the noise level is probably above 85 dB and should be measured by a competent person.

At the first action level an employee must be provided with ear protection (earmuffs or earplugs) on request. At the second action level the employer must reduce, so far as is reasonably practicable, other than by providing ear protection, the exposure to noise of that employee.

Hearing damage is cumulative; it builds up, leading eventually to a loss of hearing ability. Young people, in particular, should get into the routine of avoiding noise exposure before their hearing is permanently damaged. The damage can also take the form of permanent tinnitus (ringing noise in the ears) and an inability to distinguish words of similar sound such as bit and tip.

Vibration is also associated with noise. Direct vibration through vibrating floors or from vibrating tools can lead to damage to the bones of the feet or hands. A condition known as 'vibration white finger' is caused by an impaired blood supply to the fingers, associated with vibrating hand tools.

Employers and employees should not rely too heavily on ear protectors. In practice, they reduce noise exposure far less than is often claimed, because they may be uncomfortable or inconvenient to wear. To be effective, ear protectors need to be worn all the time when in noisy places. If left off for even a short time, the best protectors cannot reduce noise exposure effectively.

Safety first

The Noise at Work Regulations are intended to reduce hearing damage caused by loud noise.

Figure 2.1 Ear protectors protect workers from noise at work.

Protection against noise is best achieved by controlling it at source. Wearing ear protection must be a last resort. Employers should:

- Design machinery and processes to reduce noise and vibration (mounting machines on shock-absorbing materials can dampen out vibration).
- When buying new equipment, where possible, choose quiet machines. Ask the supplier to specify noise levels at the operator's working position.
- Enclose noisy machines in sound-absorbing panels.
- Fit silencers on exhaust systems.
- Install motor drives in a separate room away from the operator.
- Inform workers of the noise hazard and get them to wear ear protection.
- Reduce a worker's exposure to noise by job rotation or provide a noise refuge.

New regulations introduced in 2006 reduced the first action level to 80 dB(A) and the second level to 85 dB(A) with a peak action level of 98 dB(A) or 140 Pa of pressure. Every employer must make a 'noise' assessment and provide workers with information about the risks to hearing if the noise level approaches the first action level. He must do all that is reasonably practicable to control the noise exposure of his employees and clearly mark ear protection zones. Employees must wear personal ear protection while in such a zone.

Water Regulations 1999

Water supply and installations in England and Wales are controlled by the Water Regulations known as the Water Supply (Water Fittings) Regulations 1999 which came into force on 1 July 1999. Separate arrangements apply to Scotland and Ireland.

The **water** that finds its way to our taps is derived from rainfall, and the treatment of that water depends upon where it is sourced from and the impurities it contains. Water sourced from springs and wells is naturally purified and needs little disinfection. The quality of the water from reservoirs and rivers, called raw water, will determine the level of treatment. This usually involves several stages of treatment including settling, filtering and a final 'polishing' with carbon grains to remove minute traces of impurities and to improve the water taste. Water suppliers store water either in its raw state in impounding reservoirs or lakes, or as treated wholesome water in service reservoirs.

After being treated, water is distributed from the water supplier to individual consumers through a network of pipes known as 'mains'. The mains belong to the water suppliers and it is their responsibility to maintain them in a way that will conserve this important resource. The local mains provide the final leg of the journey to our homes.

The water at our taps is of the very highest quality and it seems a little irresponsible to flush it down the drains without giving some consideration to water conservation.

Building Regulations – Part P 2006 and the self-certification scheme

The Building Regulations lay down the design and build standards for construction work in buildings in a series of Approved Documents. The scope of each Approved Document is given below:

Part A structure

Part B fire safety

Part C site preparation and resistance to moisture

Part D toxic substances

Part E resistance to the passage of sound

Part F ventilation

Part G hygiene

Part H drainage and waste disposal

Part J combustion appliances and fuel storage systems

Part K protection from falling, collision and impact

Part L conservation of fuel and power

Part M access and facilities for disabled people

Part N glazing – safety in relation to impact, opening and cleaning

Part P electrical safety.

The Building Regulations are one of the most important pieces of legislation controlling the installation of environmental technology systems.

Part P of the Building Regulations was published on 22 July 2004, bringing domestic electrical installations in England and Wales under building regulations control. This means that anyone carrying out domestic electrical installation work from 1 January 2005 must comply with Part P of the Building Regulations. An amended document was published in an attempt at greater clarity and this came into effect on 6 April 2006.

If the electrical installation meets the requirements of the IET Regulations BS 7671, then it will also meet the requirements of Part P of the Building

Regulations, so no change there. What is going to change under Part P is this new concept of 'notification' to carry out electrical work.

Notifiable electrical work

Any work to be undertaken by a firm or individual who is *not* registered under an 'approved competent person scheme' must be notified to the Local Authority Building Control Body before work commences. That is, work that involves:

- the provision of at least one new circuit;
- work carried out in kitchens;
- work carried out in bathrooms;
- work carried out in special locations such as swimming pools and hot-air saunas.

Upon completion of the work, the Local Authority Building Control Body will test and inspect the electrical work for compliance with Part P of the Building Regulations.

Non-notifiable electrical work

This is work carried out by a person or firm registered under an authorized Competent Persons Self-Certification Scheme or electrical installation work that does not include the provision of a new circuit. This includes work such as:

- replacing accessories such as socket outlets, control switches and ceiling roses;
- replacing a like-for-like cable for a single circuit which has become damaged by, for example, impact, fire or rodent;
- refixing or replacing the enclosure of an existing installation component provided that the circuit's protective measures are unaffected;
- providing mechanical protection to existing fixed installations;
- adding lighting points (light fittings and switches) to an existing circuit, provided that the work is not in a kitchen, bathroom or special location;
- installing or upgrading the main or supplementary equipotential bonding provided that the work is not in a kitchen, bathroom or special location.

All replacement work is non-notifiable even when carried out in kitchens, bathrooms and special locations, but certain work carried out in kitchens, bathrooms and special locations may be notifiable, even when carried out by an authorized competent person. The IET have published a guide called the *Electricians' Guide to the Building Regulations* which brings clarity to this subject. In specific cases the Local Authority Building Control Officer or an approved inspector will be able to confirm whether Building Regulations apply.

Failure to comply with the Building Regulations is a criminal offence and Local Authorities have the power to require the removal or alteration of work that does not comply with these requirements.

Electrical work carried out by DIY homeowners will still be permitted after the introduction of Part P. Those carrying out notifiable DIY work must first submit a building notice to the Local Authority before the work begins. The work must then be carried out to the standards set by the IET Wiring Regulations BS 7671 and a building control fee paid for such work to be inspected and tested by the Local Authority.

Competent Persons Scheme

The Competent Persons Self-Certification Scheme is aimed at those who carry out electrical installation work as the primary activity of their business. The government has approved schemes to be operated by BRE Certification Ltd, British Standards Institution, ELECSA Ltd, NICEIC Certification Services Ltd and Napit Certification Services Ltd. All the different bodies will operate the scheme to the same criteria and will be monitored by the Department for Communities and Local Government, formally called the Office of the Deputy Prime Minister. Installers of environmental technology systems must also be registered under the scheme.

Those individuals or firms wishing to join the Competent Persons Scheme will need to demonstrate their competence, if necessary, by first undergoing training. The work of members will then be inspected at least once each year. There will be an initial registration and assessment fee and then an annual membership and inspection fee.

Microgeneration certification scheme

The official definition of microgeneration is in the Energy Act 2004, Section 82. Microgeneration is the generation of energy from renewable and from low- or zero-carbon technologies. Microgeneration has the potential to contribute to the UK's challenging 2020 renewable and greenhouse targets by generating some energy from secure and reliable small-scale installations of up to 45 kW producing heat or 50 kW producing electricity. Typical microgeneration technology systems are listed below.

Technologies generating electricity:

* solar photovoltaic (PV);
* wind turbines;
* micro hydro.

Technologies generating heat:

* solar water heating;
* heat pumps (using a ground, air or water source);
* biomass heating.

Cogeneration technologies:

* combined heat and power (CHP);
* fuel cells.

The microgeneration certification scheme (MCS) is endorsed by the Department of Energy and Climate Change to assess the competence of companies which wish to register as installers of microgeneration technology systems. Only systems installed by MCS-accredited companies will be eligible for government grants and payments from local energy suppliers through the 'feed-in tariff' schemes. Companies wishing to join the microgeneration certification scheme must demonstrate the competence of their office and installation systems, if necessary by first undergoing training in the microgeneration technology for which accreditation is sought. The work of MCS-accredited companies will be inspected at least once each year. There will be an initial registration and assessment fee followed by an annual membership and inspection fee.

The MCS scheme is very similar to the Part P Competent Persons Scheme to which many electrical contractors now belong. The aim of both schemes is to provide greater protection for customers by registering installers who have demonstrated their technical competence to carry out work in accordance with the relevant standards.

The EHO (Environmental Health Officer)

The responsibilities of the EHOs are concerned with reducing risks and eliminating the dangers to human health associated with the living and working environment. They are responsible for monitoring and ensuring the maintenance of standards of environmental and public health, including food safety, workplace health and safety, housing, noise, odour, industrial waste, pollution control and communicable diseases in accordance with the law. Although they have statutory powers with which to enforce the relevant regulations, the majority of their work involves advising and educating in order to implement public health policies.

The majority of EHOs are employed by Local Authorities, which are the agencies concerned with the protection of public health. Increasingly, however, officers are being employed by the private sector, particularly those concerned with food, such as large hotel chains, airlines and shipping companies.

Your Local Authority EHOs would typically have the responsibility of enforcing the environmental laws discussed above. Their typical work activities are to:

- ensure compliance with the Health and Safety at Work Act 1974, the Food Safety Act 1990 and the Environmental Protection Act 1990;
- carry out health and safety investigations, food hygiene inspections and food standards inspections;
- investigate public health complaints such as illegal dumping of rubbish, noise complaints and inspect contaminated land;
- investigate complaints from employees about their workplace and carry out accident investigations;
- investigate food-poisoning outbreaks;
- obtain food samples for analysis where food is manufactured, processed or sold;
- visit housing and factory accommodation to deal with specific incidents such as vermin infestation and blocked drains;
- test recreational water, such as swimming pool water and private water supplies in rural areas;
- inspect and license pet shops, animal boarding kennels, riding stables and zoos;
- monitor air pollution in heavy traffic areas and remove abandoned vehicles;
- work in both an advisory capacity and as enforcers of the law, educating managers of premises on issues which affect the safety of staff and members of the public.

In carrying out these duties, officers have the right to enter any workplace without giving notice, although notice may be given if they think it appropriate. They may also talk to employees, take photographs and samples and serve an Improvement Notice, detailing the work which must be carried out if they feel that there is a risk to health and safety that needs to be dealt with.

Enforcement Law Inspectors

If the laws relating to work, the environment and people are to be effective, they must be able to be enforced. The system of control under the Health and Safety at Work Act comes from the HSE or the Local Authority. Local Authorities are responsible for retail and service outlets such as shops, garages, offices, hotels, public houses and clubs. The HSE are responsible for all other work premises including the Local Authorities themselves. Both groups of inspectors have the same powers. They are allowed to:

* enter premises, accompanied by a police officer if necessary;
* examine, investigate and require the premises to be left undisturbed;
* take samples and photographs as necessary, dismantle and remove equipment;
* require the production of books or documents and information;
* seize, destroy or render harmless any substance or article;
* issue enforcement notices and initiate prosecutions.

There are two types of enforcement notices: an 'improvement notice' and a 'prohibition notice'.

An **improvement notice** identifies a contravention of the law and specifies a date by which the situation is to be put right. An appeal may be made to an Employment Tribunal within 21 days.

A **prohibition notice** is used to stop an activity which the inspector feels may lead to serious injury. The notice will identify which legal requirement is being contravened and the notice takes effect as soon as it is issued. An appeal may be made to the Employment Tribunal but the notice remains in place and work is stopped during the appeal process.

Cases may be heard in the Magistrates' or Crown Courts.

Magistrates' Court (Summary Offences) for health and safety offences: employers may be fined up to £20,000 and employees or individuals up to £5,000. For failure to comply with an enforcement notice or a court order, anyone may be imprisoned for up to six months.

Crown Court (Indictable Offences) for failure to comply with an enforcement notice or a court order: fines are unlimited in the Crown Court and may result in imprisonment for up to two years.

Actions available to an inspector upon inspection of premises:

* Take no action – the law is being upheld.
* Give verbal advice – minor contraventions of the law identified.
* Give written advice – omissions have been identified and a follow-up visit will be required to ensure that they have been corrected.
* Serve an improvement notice – a contravention of the law has or is taking place and the situation must be remedied by a given date. A follow-up visit will be required to ensure that the matter has been corrected.
* Serve a prohibition notice – an activity has been identified which may lead to serious injury. The law has been broken and the activity must stop immediately.
* Prosecute – the law has been broken and the employer prosecuted.

On any visit one or more of the above actions may be taken by the inspector.

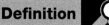

Definition

An *improvement notice* identifies a contravention of the law and specifies a date by which the situation is to be put right.

Definition

A *prohibition notice* is used to stop an activity which the inspector feels may lead to serious injury.

Environmental technology systems and renewable energy

Energy is vital to the modern industrial economy in the UK and Europe. We also need energy in almost every aspect of our lives, to heat and light our homes and offices, to enable us to travel on business or for pleasure, and to power our business and industrial machines.

In the past the UK has benefited from its fossil fuel resources of coal, oil and gas but respectable scientific sources indicate that the fossil fuel era is drawing to a close. Popular estimates suggest that gas and oil will reach peak production in the year 2060 with British coal reserves lasting only a little longer. Therefore we must look to different ways of generating electricity so that:

- the remaining fossil fuel is conserved;
- our CO_2 emissions are reduced to avoid the consequences of climate change;
- we ensure that our energy supplies are secure, and not dependent upon supplies from other countries.

Following the introduction of the Climate Change Act in 2008 the UK and other Member States agreed an EU-wide target of 20% renewable energy by the year 2020 and 60% by 2050. Meeting these targets will mean basing much of the new energy infrastructure around renewable energy, particularly offshore wind power.

The 'Energy Hierarchy' states that organizations and individuals should address energy issues in the following order so as to achieve the agreed targets:

1 Reduce the need for energy – reducing energy demand is cost saving, reduces greenhouse gas emissions and contributes to the security of supply. Reducing the energy loss from buildings by better insulation and switching off equipment when not in use is one way of achieving this target.

2 Use energy more efficiently – use energy-efficient lamps and 'A'-rated equipment.

3 Use renewable energy – renewable energy refers to the energy that occurs naturally and repeatedly in the environment. This energy may come from wind waves or water, the sun, or heat from the ground or air.

4 Any continuing use of fossil fuels should use clean and efficient technology. Power stations generating electricity from coal and oil (fossil fuel) release a lot of CO_2 in the generating process. New-build power stations must now be fitted with carbon capture filters to reduce the bad environmental effects.

Funding for environmental technology systems

Renewable energy is no less reliable than energy generated from more traditional sources. Using renewable energy does not mean that you have to change your lifestyle or your domestic appliances. There has never been a better time to consider generating energy from renewable technology than now because grants and funding are available to help individuals and companies.

The Low Carbon Building Programme implemented by the Department of Energy and Climate Change (DECC) provides grants towards the installation of renewable technologies and is available to householders, public non-profit-making organizations and commercial organizations across the UK.

The government's 'Feed-in Tariff' pays a tax-free sum which is guaranteed for 25 years. It is called 'Clean energy cash back' and has been introduced to promote the uptake of small-scale renewable and low-carbon electricity generation technologies. The customer receives a generation tariff from the electricity supplier, whether or not any electricity generated is exported to the national grid, and an additional export tariff when electricity is transported to the electricity grid through a smart meter.

Key fact

Renewable energy is no less reliable than energy generated from more traditional sources.

Key fact

The government's 'Feed-in Tariff' pays a tax-free sum which is guaranteed for 25 years.

From April 2010, clean energy generators were paid 41.3p for each kWh of electricity generated. Surplus energy fed back into the national grid earns an extra 3p per unit. However, the scheme has been very popular so the feedback tariffs were reduced to 21p, and from 1 August 2012 the fee fell to only 16p for each kWh. If you add to this the electricity bill savings, a normal householder could still make some savings. Savings vary according to energy use and type of system used. The Energy Saving Trust at www.energysavingtrust.org.uk, British Gas at www.britishgas.co.uk and Ofgem at www.ofgem.gov.uk/fits provide an online calculator to determine the cost, size of system and CO_2 savings for PV systems.

Microgeneration technologies

Microgeneration is defined in The Energy Act 2004 Section 82 as the generation of heat energy up to 45 kW and electricity up to 50 kW.

Today, microgeneration systems generate relatively small amounts of energy at the site of a domestic or commercial building. However, it is estimated that by 2050, 30 to 40% of the UK's electricity demand could be met by installing microgeneration equipment to all types of building.

In the USA, the EU and the UK buildings consume more than 70% of the nation's electricity and contribute almost 40% of the polluting CO_2 greenhouse gases. Any reductions which can be made to these figures will be good for the planet, and hence the great interest today in the microgeneration systems. Microgeneration technologies include small wind turbines, solar photovoltaic (PV) systems, small-scale hydro and micro-CHP (combined heat and power) systems. Microgenerators that produce electricity may be used as stand-alone systems, or may be run in parallel with the low-voltage distribution network; that is, the a.c. mains supply.

The April 2008 amendments to the Town and Country Planning Act 1990 brought in 'permitted development' to allow the installation of microgeneration systems within the boundary of domestic premises without obtaining planning permission. However, size limitations have been set to reduce the impact upon neighbours. For example, solar panels attached to a building must not protrude more than 200 mm from the roof slope, and stand-alone panels must be no higher than four metres above ground level and no nearer than five metres from the property boundary. See the Electrical Safety Council site for advice on connecting microgeneration systems at www.esc.org.uk/bestpracticeguides.html.

Smart electricity meters

Smart electricity meters are designed to be used in conjunction with micro-generators. Electricity generated by the consumer's microgenerator can be sold back to the energy supplier using the 'smart' two-way meter.

From 2012 the Department for Energy and Climate Change began introducing smart meters into consumers' homes, and this is expected to run until 2020, with the aim being to help consumers reduce their energy bills.

When introducing the proposal Edward Davey, the Energy and Climate Change Secretary, said, 'the meters which most of us have in our homes were designed for a different age, before climate change. Now we need to get smarter with our energy. This is a big project affecting 26 million homes and several million businesses. The project will lead to extra work for electrical contractors through

Key fact

Microgeneration technologies include small wind turbines, solar photovoltaic (PV) systems, small-scale hydro and micro-CHP (combined heat and power) systems.

Key fact

Smart electricity meters are designed to be used in conjunction with micro-generators.

Figure 2.2 Smart electricity meter.

installing the meters on behalf of the utility companies and implementing more energy-efficient devices once customers can see how much energy they are using.'

Already available is the Real Time Display (RTD) wireless monitor which enables consumers to see exactly how many units of electricity they are using through an easy to read portable display unit. By seeing the immediate impact in pence per hour of replacing existing lamps with low-energy ones or switching off unnecessary devices throughout the home or office, consumers are naturally motivated to consider saving energy. RTD monitors use a clip-on sensor on the meter tails and includes desktop software for PC and USB links.

Let us now look at some of these microgeneration technologies.

Figure 2.3 Solar photovoltaic (PV)

Microgeneration technologies

Electricity-producing technologies

1 Solar photovoltaic (PV) microgenerators
2 Wind microgenerators
3 Hydro microgenerators

Heat-producing technology

4 Solar thermal (hot water) microgenerators
5 Ground source heat pump microgenerators
6 Air source heat pump microgenerators
7 Water source heat pump microgenerators
8 Biomass microgenerators

Co-generation technologies

9 Combined heat and power (CHP) microgenerators

Water conservation technologies

10 Rainwater harvesting

11 Grey water recycling

1. Solar photovoltaic (PV) microgeneration

Definition

A *solar photovoltaic* (PV) system is a collection of PV cells known as a PV string, that forms a PV array and collectively are called a PV generator which turns sunlight directly into electricity.

A **solar photovoltaic** (PV) system is a collection of PV cells known as a PV string, that forms a PV array and collectively are called a PV generator which turns sunlight directly into electricity. PV systems may be 'stand-alone' power supplies or be designed to operate in parallel with the public low-voltage distribution network; that is, the a.c. mains supply.

Stand-alone PV systems are typically a small PV panel of maybe 300mm by 300mm tilted to face the southern sky, where it receives the maximum amount of sunlight. They typically generate 12 to 15 volts and are used to charge battery supplies on boats, weather stations, road signs and any situation where electronic equipment is used in remote areas away from a reliable electrical supply.

The developing nations are beginning to see stand-alone PV systems as the way forward for electrification of rural areas beyond the National Grid rather than continuing with expensive diesel generators and polluting kerosine lamps.

The cost of PV generators is falling. The period 2009 to 2010 saw PV cells fall by 30% and with new 'thin-film' cells being developed, the cost is expected to continue downward. In the rural areas of the developing nations they see PV systems linked to batteries bringing information technology, radio and television to community schools. This will give knowledge and information to the next generation which will help these countries to develop a better economy, a better way of life and to have a voice in the developed world.

Stand-alone systems are not connected to the electricity supply system and are therefore exempt from much of BS 7671, the IET Regulations. However, Regulation 134.1.1, 'good workmanship by competent persons and proper materials shall be used in all electrical installations,' will apply to any work done by an electrician who must also pay careful attention to the manufacturer's installation instructions.

PV systems designed to operate in parallel with the public low-voltage distribution network are the type of microgenerator used on commercial and domestic buildings. The PV cells operate in exactly the same way as the stand-alone system described above, but will cover a much greater area. The PV cells are available in square panels which are clipped together and laid over the existing roof tiles as shown in Fig. 2.4, or the PV cells may be manufactured to look just like the existing roof tiles which are integrated into the existing roof.

A solar PV system for a domestic three-bedroom house will require approximately 15 to 20 square metres generating 2 to 3 kilowatts of power and the cost at the time of going to press of the PV cells alone will be in the region of £10,000 to £12,000, although grants are available. On the positive side, a PV system for a three-bedroom house will save around 1,200 kg of CO_2 per year.

These bigger microgeneration systems are designed to be connected to the power supply system and the installation must therefore comply with Section 712 of BS 7671: 2008. Section 712 contains the requirements for protective measures comprising automatic disconnection of the supply wiring systems, isolation, switching and control, earthing arrangements and labelling. In addition,

Figure 2.4 PV system in a domestic situation.

the installation must meet the requirements of the Electricity Safety Quality and Continuity Regulations 2006. This is a mandatory requirement. However, where the output does not exceed 16A per line, they are exempt from some of the requirements, provided that:

- the equipment is type tested and approved by a recognized body;
- the installation complies with the requirements of BS 7671, the IET Regulations;
- the PV equipment must disconnect from the distributor's network in the event of a network fault;
- the distributor must be advised of the installation before or at the time of commissioning.

Installations of less than 16A per phase but up to 5 kilowatt peak (kWp) will also be required to meet the requirements of the Energy Network Association's Engineering Recommendation G83/1 for small-scale embedded generators in parallel with public low-voltage distribution networks. Installations generating more than 16A must meet the requirements of G59/1 which requires approval from the distributor before any work commences. The installer must be MCS-accredited if the client is to receive the feedback tariff from the energy supplier.

2. Wind energy microgenerators

Modern large-scale wind machines are very different from the traditional windmill of the last century which gave no more power than a small car engine. Very large structures are needed to extract worthwhile amounts of energy from the wind. Modern large-scale wind generators are taller than electricity pylons, with a three-blade aeroplane-type propeller to catch the wind and turn the generator. If a wind turbine was laid down on the ground, it would be longer and wider than a football pitch. They are usually sited together in groups in what has become known as 'wind energy farms,' as shown in Fig. 2.5.

Each modern grid-connected wind turbine generates about 600 kW of electricity. A wind energy farm of 20 generators will therefore generate 12 MW, a useful contribution to the national grid, using a naturally occurring, renewable, non-polluting source of energy. The Department of Energy and Climate Change

Figure 2.5 Offshore wind farm energy generators.

considers wind energy to be the most promising of the renewable energy sources.

In 2010 there were 253 wind energy farms in operation in the UK with 12 operating offshore. The 3,000 turbines on these farms have the capacity to generate 4,600 MW of electricity, enough for 2.5 million homes. There are a further 500 wind energy farms planned or in construction. However, because of the unpredictable nature of the wind and inefficiencies in the generation process, the amount of power produced will be less than the installed capacity.

The Countryside Commission, the government's adviser on land use, has calculated that to achieve a target of generating 10% of the total electricity supply by wind power will require 40,000 generators of the present size. At the time of writing, we are generating only about 8% of the total electricity supply from wind power and all hopes are pinned on large offshore wind farms to achieve the government and EU targets. In march 2019 the UK Government announced that they would double the present capacity of off shore wind farms around the UK **in order to meet a new target** of generating 30% of the total electricity demand from renewable energy by the year 2030.

Wind power is an endless renewable source of energy, is safer than nuclear power and provides none of the polluting emissions associated with fossil fuel. If there was such a thing as a morally pure form of energy, then wind energy would be it. However, wind farms are, by necessity, sited in some of the most beautiful landscapes in the UK. Building wind energy farms in these areas of outstanding natural beauty has outraged conservationists. Prince Charles has reluctantly joined the debate saying that he was in favour of renewable energy sources but believed that 'wind farms are an horrendous blot on the landscape'. He believes that if they are to be built at all they should be constructed well out to sea.

The next generation of wind farms will mostly be built offshore, where there is more space and more wind, but the proposed size of these turbines creates considerable engineering problems. From the sea bed foundations to the top of the turbine blade will be up to a staggering 250 metres, three times the height

Definition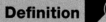

Wind power is an endless renewable source of energy, is safer than nuclear power and provides none of the polluting emissions associated with fossil fuel.

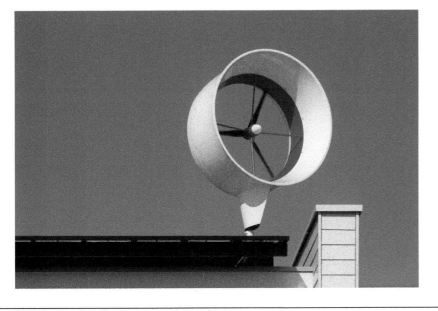

Figure 2.6 Small wind generator on a domestic property.

of the Statue of Liberty. Each offshore turbine, generating between 5 and 7 MW, will weigh between 200 and 300 tonnes. When you put large wind forces onto that structure you will create a very big cantilever effect which creates a major engineering challenge.

The 100 turbine 'Thanet' wind farm just off the Kent coast generates enough power to supply 200,000 homes. The Thanet projects cost £780 million pounds to build and was the world's largest, when it opened in 2010. The turbines are up to 380 feet high and cover an area as large as 4,000 football pitches. The Thanet project did not retain its title as the world's largest wind farm for long though, because the 'Greater Gabbard' wind farm, off the north-east coast with 140 turbines, opened in 2013. These projects bring Britain's total wind energy capacity above 5,000 megawatts for the first time and more are being built. The UK now has the largest installed wind base in the world and, for a small island like the UK, wind energy makes sense.

The Department of Energy and Climate Change has calculated that 10,000 wind turbines could provide the energy equivalent of eight million tonnes of coal per year and reduce CO_2 emissions. While this is a worthwhile saving of fossil fuel, opponents point out the obvious disadvantages of wind machines, among them the need to maintain the energy supply during periods of calm, which means that wind machines can only ever supplement a more conventional electricity supply.

Small wind microgenerators can be used to make a useful contribution to a domestic property or a commercial building. They can be stand-alone about the size of a tall street lamp. A 12 m-high turbine costs about £24,000 and, with a good wind, will generate 10,000 kWh per year enough for three small domestic homes. However, if you live in a village, town or city you are unlikely to obtain the the Local Authority Building and Planning permissions to install a wind generator because your neighbours will object.

Small wind generators of the type shown in Fig. 2.6 typically generate between 1.5A and 15A in wind speeds of 10 mph to 40 mph.

Key fact

The Department of Energy and Climate Change has calculated that 10,000 wind turbines could provide the energy equivalent of eight million tonnes of coal per year and reduce CO_2 emissions.

Figure 2.7 Tidal flow water turbines.

Hydroelectric power generation

The UK is a small island surrounded by water. Surely we could harness some of the energy contained in tides, waves, lakes and rivers? Many different schemes have been considered in the past 20 years and a dozen or more experimental schemes are currently being tested.

Water power makes a useful contribution to the energy needs of Scotland but the possibility of building similar hydroelectric schemes in England is unlikely chiefly due to the topographical nature of the country.

The Severn Estuary has a tidal range of 15 m, the largest in Europe, and a reasonable shape for building a dam across the estuary. This would allow the basin to fill with water as the tide rises, and then allow the impounded water to flow out through electricity-generating turbines as the tide falls. However, such a tidal barrier might have disastrous ecological consequences upon the many wildfowl and wading bird species by the submerging of the mudflats which now provide winter shelter for these birds. Therefore, the value of the power which might be produced must be balanced against the possible ecological consequences.

France has successfully operated a 240 MW tidal power station at Rance in Brittany for the past 25 years.

Marine Current Turbines Ltd are carrying out research and development on submerged turbines which will rotate by exploiting the principle of flowing water in general and tidal streams in particular. The general principle is that an 11 m diameter water turbine is lowered into the sea down a steel column drilled in the sea bed. The tidal movement of the water then rotates the turbine and generates electricity.

The prototype machines were submerged in the sea off Lynmouth in Devon. In May 2008 they installed the world's first tidal turbine in the Strangford Narrows in Northern Ireland where it is now grid-connected and generating 1.2 MW.

All the above technologies are geared to providing hydroelectric power connected to the national grid, but other micro-hydro schemes are at the planning and development stage.

3. Hydro microgenerators

The use of small hydropower (SHP) or micro-hydropower has grown over recent decades led by continuous technical developments, brought about partly in the UK by the 2010 coalition government's 'feed-in tariff' where green electricity producers are paid a premium to produce electricity from renewable sources.

The normal perception of hydropower is of huge dams, but there is a much bigger use of hydropower in smaller installations. Asia, and especially China, is set to become a leader in hydroelectric generation. Australia and New Zealand are focusing on small hydro plants. Canada, a country with a long tradition of hydropower, is developing small hydropower as a replacement for expensive diesel generation in remote off-grid communities.

Small hydropower schemes generate electricity by converting the power available in rivers, canals and streams. The object of a hydropower scheme is to convert the potential energy of a mass of water flowing in a stream with a certain fall, called the head, into electrical energy at the lower end of the stream where the powerhouse is located. The power generated is proportional to the flow, called the discharge, and to the head of water available. The fundamental asset of hydropower is that it is a clean and a renewable energy source and the fuel used, water, is not consumed in the electricity-generating process.

In the Derbyshire Peak District along the fast-flowing River Goyt there were once 16 textile mills driven by waterwheels. The last textile mill closed in the year 2000 but the Old Torr Mill has been saved. Where once the waterwheel stood is now a gigantic 12-tonne steel screw, 2.4 metres in diameter. The water now drives the Reverse Archimedian Screw, affectionately called 'Archie', to produce 130,000 kWh per year, enough electricity for 40 homes. The electricity-generating project is owned by the residents of New Mills in a sharing co-operative in which surplus electricity is sold back to the grid. See the torrshydro new mills website and mann power for interestng video of fish swimming through the turbine. The installation cost was £300,000 in 2008. See Fig. 2.8.

The type of turbine chosen for any hydro scheme will depend upon the discharge rate of the water and the head of water available. A Pelton Wheel is a water turbine in which specially shaped buckets attached to the periphery of the wheel are struck by a jet of water. The kinetic energy of the water turns the wheel which is coupled to the generator.

Axial turbines comprise a large cylinder in which a propeller-type water turbine is fixed at its centre. The water moving through the cylinder causes the turbine blade to rotate and generate electricity.

A Francis Turbine or Kaplan Turbine is also an axial turbine but the pitch of the blades can be varied according to the load and discharge rate of the water.

Small water turbines will reach a mechanical efficiency at the coupling of 90%.

Up and down the country, riverside communities must be looking at the relics of our industrial past and wondering if they might provide a modern solution for clean, green, electrical energy. However, despite the many successes and obvious potential, **there are many barriers to using waterways for electricity generation in the European countries.** It is very difficult in this country to obtain permission from the Waterways Commission to extract water from rivers,

Definition

Small hydropower schemes generate electricity by converting the power available in rivers, canals and streams.

Figure 2.8 An example of an Archimedian Screw at the River Dart country park, Devon. Credit: WRE Limited.

Key fact

The type of turbine chosen for any hydro scheme will depend upon the discharge rate of the water and the head of water available.

even though, once the water has passed through the turbine, it is put back into the river. Environmental pressure groups are opposed to micro-hydro generation because of its perceived local environmental impact on the river ecosystem and the disturbance to fishing. Therefore, once again, the value of the power produced would have to be balanced against the possible consequences.

4. Solar thermal (hot water) microgenerators

Solar thermal hot water heating systems are recognized as a reliable way to use the energy of the sun to heat water. The technology is straightforward and solar thermal panels for a three-bedroomed house cost at the time of going to press between £3,000 and £6,000 for a 3 to $6\,m^2$ panel and they will save about 260 kg of CO_2 annually.

The solar panel comprises a series of tubes containing water that is pumped around the panel and a heat exchanger in the domestic water cylinder as shown in Fig. 2.9. Solar energy heats up the domestic hot water. A solar panel of about $4\,m^2$ will deliver about 1,000 kWh per year which is about half the annual water demand of a domestic dwelling. However, most of the heat energy is generated during the summer and so it is necessary to supplement the solar system with a boiler in the winter months. Figure 2.10 shows a photo of an installed solar hot water panel.

If you travel to Germany, you will see a lot of PV and solar thermal panels on the roofs there. In the UK, planning requirements for solar thermal and PV installations have already been made much easier by 'permitted development'. A website detailing planning requirements for solar and wind may be found at www.planningportal.gov.uk/uploads/hhghouseguide.html.

Heat pumps

In applications where heat must be upgraded to a higher temperature so that it can be usefully employed, a **heat pump** must be used. Energy from a low-temperature source such as the earth, the air, a lake or river is absorbed by a gas or liquid contained in pipes, which is then mechanically compressed by an

Definition

Solar thermal hot water heating systems are recognized as a reliable way to use the energy of the sun to heat water.

Definition

In applications where heat must be upgraded to a higher temperature so that it can be usefully employed, a *heat pump* must be used.

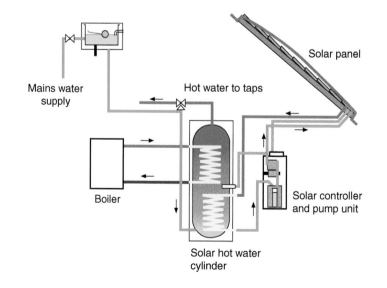

Mains water supply

Hot water to taps

Solar panel

Boiler

Solar controller and pump unit

Solar hot water cylinder

Figure 2.9 Solar-powered hot water system.

Figure 2.10 Solar hot water panel.

electric pump to produce a temperature increase. The high-temperature energy is then transferred to a heat exchanger so that it might do useful work, such as providing heat to a building. For every 1 kWh of electricity used to power the heat pump compressor, approximately 3 to 4 kWh of heating are produced.

How a heat pump works

1 A large quantity of low-grade energy is absorbed from the environment and transferred to the refrigerant inside the heat pump (called the evaporator). This causes the refrigerant temperature to rise, causing it to change from liquid to a gas.
2 The refrigerant is then compressed, using an electrically driven compressor, reducing its volume but causing its temperature to rise significantly.
3 A heat exchanger (condenser) extracts the heat from the refrigerant to heat the domestic hot water or heating system.
4 After giving up its heat energy, the refrigerant turns back into a liquid, and, after passing through an expansion valve, is once more ready to absorb energy from the environment and the cycle is repeated as shown in Fig. 2.11.

A refrigerator works on this principle. Heat is taken out of the food cabinet, compressed and passed onto the heat exchanger or radiator at the back of the fridge. This warm air then radiates by air convection currents into the room. Thus the heat from inside the cabinet is moved into the room, leaving the sealed refrigerator cabinet cold.

5. Ground source heat pump microgenerators

Ground source heat pumps extract heat from the ground by circulating a fluid through polythene pipes buried in the ground in trenches or in vertical boreholes as shown in Fig. 2.12. The fluid in the pipes extracts heat from the ground and a heat exchanger within the pump extracts heat from the fluid. These systems are most effectively used to provide underfloor radiant heating or water heating.

Calculations show that the length of pipe buried at a depth of 1.5 m required to produce 1.2 kW of heat will vary between 150 m in dry soil and 50 m in wet soil.

 **Definition**

Ground source heat pumps extract heat from the ground by circulating a fluid through polythene pipes buried in the ground in trenches or in vertical boreholes.

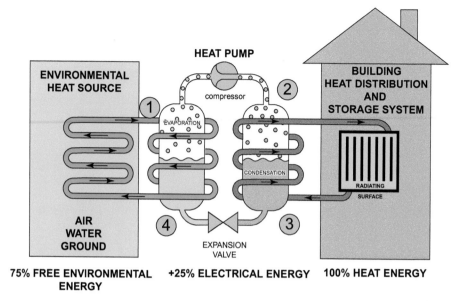

Figure 2.11 Heat pump working principle.

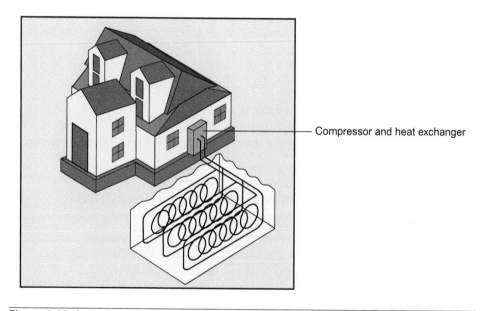

Figure 2.12 Ground source heating system.

The average heat output can be taken as 28 watts per metre of pipe. A rule of thumb guideline is that the surface area required for the ground heat exchanger should be about 2.5 times the area of the building to be heated.

This type of installation is only suitable for a new-build project and the ground heat exchanger will require considerable excavation and installation. The installer must seek Local Authority Building Control permissions before work commences.

6. Air source heat pump microgenerators

The performance and economics of heat pumps are largely determined by the temperature of the heat source and so we seek to use a high-temperature source.

The heat sources used by heat pumps may be soil, the air, ground or surface water. Unfortunately all these sources follow the external temperature, being lower in winter when demand is highest. Normal atmosphere is an ideal heat source in that it can supply an almost unlimited amount of heat although unfortunately at varying temperatures, but relatively mild winter temperatures in the UK mean excellent levels of efficiency and performance throughout the year. For every 1 kWh of electricity used to power the heat pump compressor, between 3 and 4 kWh of heating energy is produced. They also have the advantage over ground source heat pumps of lower installation costs because they do not require any groundwork. Figure 2.13 shows a commercial air source heat pump.

Figure 2.13 Air source heat pump unit.

If the air heat pump is designed to provide full heating with an outside temperature of 2 to 4 degrees centigrade, then the heat pump will provide approximately 80% of the total heating requirement with high performance and efficiency.

The point at which the output of a given heat pump meets the building heat demand is known as the 'balance point'. In the example described above, the 20% shortfall of heating capacity below the balance point must be provided by some supplementary heat. However, an air-to-air heat pump can also be operated in the reverse cycle which then acts as a cooling device, discharging cold air into the building during the summer months. So here we have a system which could be used for air conditioning in a commercial building.

7. Water source heat pump microgenerators

When we looked at the ground source heat pump at number 5 we said that heat was extracted from the ground by circulating a glycol fluid through polythene pipes buried in the ground. The wetter the soil the more conductive it is to the glycol. Water is an ideal conductor and therefore most suitable as a heat source.

Connecting the polythene pipe work together to form a raft-type system will hold the collector together. It is then floated out onto the lake using a small dinghy. When in position the collector is filled with glycol which makes it sink to the bottom of the lake. The feeder pipes are connected to the heat pump at the adjacent building and over time the collector becomes covered with silt.

The lake must be close to the building and large enough to provide a stable temperature throughout the seasons so that the pipes do not freeze at the bottom of the lake.

8. Biomass microgenerators

Biomass is derived from plant materials and animal waste. It can be used to generate heat and to produce fuel for transportation. The biomass material may be straw and crop residues, crops grown specially for energy production such as trees, or rape seed oil and waste from a range of sources including food production. The nature of the fuel will determine the way that energy can best be recovered from it.

There is a great deal of scientific research being carried out at the moment into 'biomass renewables'; that is, energy from crops. This area of research is at an early stage, but is expected to flourish in the next decade. The first renewable energy plant, which is to be located at Teesport on the River Tees in the north-east of England, has received approval from the Department for Energy and Climate Change for building to commence.

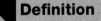

Definition

Biomass is derived from plant materials and animal waste. It can be used to generate heat and to produce fuel for transportation.

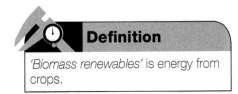

Definition

'Biomass renewables' is energy from crops.

Figure 2.14 Wood pellets being burned in a domestic biomass stove.

The facility will be one of the largest biomass plants to be built in the world and is scheduled to enter commercial operation in 2020. Young trees will be grown as a crop to produce wood chips. The plant will use 2.5 million tonnes of wood chips each year to produce 300 MW of electrical energy. The plant will operate 24 hours a day, all year round to meet some of the national grid base load. The USA have over 200 biomass plants operating which meets approximately 4% of the US energy consumption.

Domestic biomass heating can today be found in the increasing popularity of wood stoves. A wide variety of wood-burning appliances are used by households, particularly in rural areas of countries with traditional forest cultures such as Canada, Austria, Finland and Sweden. Since wood fuel was classified as a carbon-neutral fuel under the Kyoto Protocol, most countries that require winter space heating have begun to look again at wood.

Domestic and commercial boilers fuelled by wood logs, wood chips or wood pellets are available. Boiler designs and controls mean that modern boilers can be highly automated. With a water-based heat distribution system already in place, it is relatively simple for any household or commercial operation to remove an existing boiler and replace it with a wood-fuelled one that will do much the same job.

As long as they are locally and sustainably sourced, dry and well-seasoned logs are arguably the greenest of all wood fuels. Apart from drying, they require less processing, and therefore less energy than wood chips or wood pellets.

9. Combined heat and power (CHP) microgenerators

Definition

CHP is the simultaneous generation of usable heat and power in a single process. That is, heat is produced as a by-product of the power-generation process.

CHP is the simultaneous generation of usable heat and power in a single process. That is, heat is produced as a by-product of the power-generation process. A chemical manufacturing company close to where I live has a small power station which meets some of their electricity requirements using the smart meter principle. They use high-pressure steam in some of their processes and so their 100 MW turbine is driven by high-pressure steam. When the steam condenses after giving up its energy to the turbine, there remains a lot of very hot water which is then piped around the offices and some production plant buildings for space heating. Combining heat and power in this way can increase the overall efficiency of the fuel used because it is performing two operations.

CHP can also use the heat from incinerating refuse to heat a nearby school or block of flats. This is called 'district heating', or 'community heating'.

Water conservation

Conservation is the preservation of something important, especially of the natural environment. Available stored water is a scarce resource in England and Wales where there are only 1,400 cubic metres per person per year; very little compared with France, which has 3,100 cubic metres per person per year, Italy which has 2,900 and Spain 2,800. About half of the water used by an average home is used to shower, bathe and wash the laundry, and another third is used to flush the toilet.

At a time when most domestic and commercial properties have water meters installed, it saves money to harvest and reuse water.

The City and Guilds examinations board has asked us to look at two methods of water conservation: rainwater harvesting and grey water recycling.

10. Rainwater harvesting

Rainwater harvesting is the collection and storage of rainwater for future use. Rainwater has in the past been used for drinking, water for livestock and water for irrigation. It is now also being used to provide water for car cleaning and garden irrigation in domestic and commercial buildings.

Many gardeners already harvest rainwater for garden use by collecting run-off from a roof or greenhouse and storing it in a water butt or water barrel. However, a 200-litre water butt does not give much drought protection although

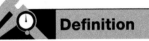

Definition

Conservation is the preservation of something important, especially of the natural environment.

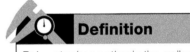

Definition

Rainwater harvesting is the collection and storage of rainwater for future use.

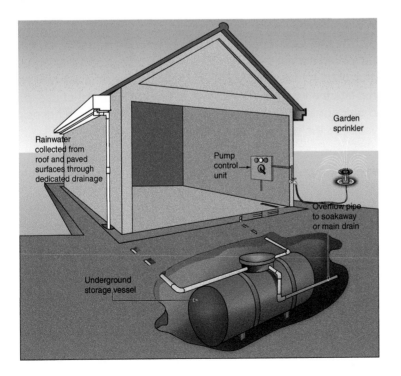

Garden sprinkler

Rainwater collected from roof and paved surfaces through dedicated drainage

Pump control unit

Overflow pipe to soakaway or main drain

Underground storage vessel

Figure 2.15 Underground rainwater storage vessel.

Figure 2.16 Rainwater harvesting can save a significant amount of water in domestic situations.

garden plants much prefer rainwater to fluoridated tap water. To make a useful contribution the rainwater storage tank should be between 2,000 and 7,000 litre capacity. The rainwater-collecting surfaces will be the roof of the building and any hard-paved surfaces such as patios. Downpipes and drainage pipes then route the water to the storage tank situated, perhaps, under the lawn. An electric pump lifts the water from the storage tank to the point of use, possibly a dedicated outdoor tap. The water is then distributed through a hosepipe or sprinkler system to the garden in the normal way.

With a little extra investment, rainwater can be filtered and used inside the house to supply washing machines and WCs. Domestic pipes and interior plumbing can be added to existing homes although it is more straightforward in a new-build home.

With the move towards more sustainable homes UK architects are becoming more likely to specify rainwater harvesting in their design to support alternatives to a mains water supply. In Germany, rainwater harvesting systems are now installed as standard in all new commercial buildings.

11. Grey water recycling

Grey water is tap water which has already performed one operation and is then made available to be used again instead of flushing it down the drain. Grey water recycling offers a way of getting double the use out of the world's most precious resource.

Definition

Grey water is tap water which has already performed one operation and is then made available to be used again instead of flushing it down the drain.

There are many products on the market such as the BRAC system which takes in water used in the shower, bath and laundry, cleans it by filtering and then reuses it for toilet flushing. It is only a matter of routing the grey waste water drainpipe from the bath, shower and laundry to the filter unit and then plumbing the sanitized grey water to the toilet tank.

These systems are easy to install, particularly in a new-build property. It is only a matter of re-routing the drainpipes. Another option for your grey water is to route it into the rainwater storage tank for further use in the garden.

Code for sustainable homes

The use of energy to provide heat for central heating and hot water in our homes is responsible for 60% of a typical family's energy bill. Heating accounts for over half of Britain's entire use of energy and carbon emissions. If Britain is to reduce its carbon footprint and achieve energy security, we must revolutionize the way we keep warm in the home.

At present 64% of our home heating comes from burning gas, 11% from oil, 3% from solid fuels such as coal and 14% from electricity which is mainly generated from these same three fossil fuels. Only 8% is currently provided by renewable sources. If Britain is to meet its clean energy targets, renewable sources will have to increase, and the revolution will have to start in the home because the country's dwellings currently provide more than half of the total demand, almost entirely for hot water and central heating.

There are about 20 million homes in the UK and a review of present buildings has found that about six million homes have inadequately lagged lofts, eight million have uninsulated cavity walls and a further seven million homes with solid walls would benefit from better insulation. If the country is to achieve its reduced carbon emissions targets, these existing homes must be heavily insulated to reduce energy demand and then supplied with renewable heat.

We cannot sustain the present level of carbon emissions without disastrous ecological consequences in the future. Low carbon homes are sustainable homes.

HRH The Prince of Wales has entered the debate, saying,

> *Becoming more sustainable is possibly the greatest challenge humanity has faced, and I am convinced that it is therefore, the most remarkable chance to secure a prosperous future for everyone. We must strive harder than ever before to convince people that by living sustainably we will improve our quality of life and our health; that by living in harmony with nature we will protect the intricate, delicate balance of the natural systems that ultimately sustain us.*

(*Daily Telegraph*, 31 July 2010).

The Code for Sustainable Homes (see Fig. 2.17) measures the sustainability of a home against categories of sustainable design, rating the whole home as a complete package, including building materials and services within the building. The Code uses a one-to-six star rating to communicate the overall sustainability performance of a new home and sets minimum standards for energy and water use at each level.

Since May 2008 all new homes are required to have a Code Rating and a Code Certificate. By 2016 all new homes must be built to zero-carbon standards, which will be achieved through step-by-step tightening of the Building Regulations.

Key fact

Heating accounts for over half of Britain's entire use of energy and carbon emissions.

Definition

The Code for Sustainable Homes (see Fig. 2.17) measures the sustainability of a home against categories of sustainable design, rating the whole home as a complete package, including building materials and services within the building.

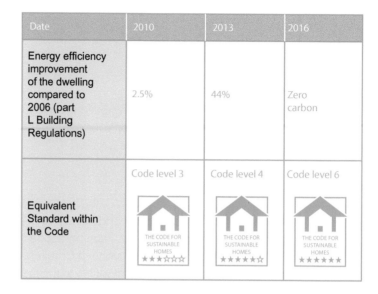

Date	2010	2013	2016
Energy efficiency improvement of the dwelling compared to 2006 (part L Building Regulations)	2.5%	44%	Zero carbon
Equivalent Standard within the Code	Code level 3	Code level 4	Code level 6

Figure 2.17 Code for Sustainable Homes.

If we look at sustainability from a manufacturing point of view, sustainable manufacture is based on the principle of meeting the needs of the current generation without compromising the ability of future generations to meet their needs.

The 11 topics which we have just considered, together with improved levels of insulation would all help our homes to be more sustainable.

Planning permission and permitted development

The Town and Country Planning Act 1990 effects control over all building development, its appearance and the layout of the building. The Public Health Acts limit the development of buildings with regard to noise, pollution and public nuisance. The Highways Act 1980 determines the layout and construction of roads and pavements. The Building Act 1984 affects the Building Regulations 2000 which enforce minimum material and design standards.

If you are considering building a new home, office or factory you will definitely require Local Authority planning permission before any work begins. However, under permitted development, small extensions to an existing building may be exempt from a formal planning application. The General Permitted Developments Amendment 2, 2008 said, for example, that porches are exempt from planning permission providing that the floor area was no more than three square metres, the height was no more than three metres high and the porch was more than two metres from the boundary wall of the property.

This same permitted developments amendment also lifted the requirements for planning permission for most domestic microgeneration technologies. However, as with the porch example given above, size limitations have been set to reduce the impact upon neighbours. For example, solar panels attached to a building must not protrude more than 200mm from the roof slope and stand-alone panels must be no higher than four metres above ground level and no nearer than five metres from the property boundary. A website detailing planning requirements may be found at www.planning portal.gov.uk/uploads/hhghouseguide.html.

Installing microgenerators – the regulations

Let us look at the regulation requirements for each of the technologies discussed earlier in this chapter.

1. Solar PV. This is probably one of the most popular forms of microgeneration in the UK. Planning permission is not required because the fitting of solar panels comes within permitted development unless the building is listed or in a conservation area.

Building regulations approval may be required because the ability of the roof to carry the weight of the panels and for the panels to remain secure in high winds must be assured. However, this will probably be waived if the installation contractor is MCS-approved because MCS-approved installers must purchase and fit only accredited fixtures, fittings and panels.

2. Small wind generators rated between 1.5 kW and 15 kW and micro wind generators rated up to 1.5 kW. Planning permission is still a genuine issue in the UK for micro and small wind generators despite the introduction in 2008 and 2009 of permitted development rights for renewable technologies. Somehow wind was not included in the permitted development right. This may soon be resolved because further legislation is expected, but it is vital to consult your local authority planning office before any work is undertaken.

3. Hydro microgeneration. It is very difficult in this country to obtain permission from the Waterways Commission to extract water from rivers, even though

Figure 2.18 A small-scale wind generator when combined with solar PV panels can produce a considerable amount of energy.

once the water has passed through the turbine it is put back into the river. Environmental pressure groups are opposed to any disturbance of any waterway because of the environmental impact on the river's ecosystem.

Planning permission will be required. The Environmental Agency must be consulted and Building Regulations approval must be obtained, in particular with regard to the electrical installation (more details at www.energysavingtrust.org. uk/generate-your-own-electricity).

4. Solar thermal. This is another popular form of microgeneration in the UK. Planning permission is not required because the installation is within permitted development unless the building is listed or in a conservation area. Building Regulations may require assurance on the roof's ability to carry the weight and withstand high winds but if an MCS installer carries out the work he must install to the standards required by the permitted development. The electrical installation must of course be installed to the satisfaction of the Building Regulations Part P.

5. Ground source heat pumps. Planning permission is not required unless the building is listed or in a conservation area. The installation of the pipe work which forms the collector will be extensive and necessitate a lot of ground work. The installation will therefore probably require Building Regulations approval.

6. Air source heat pump. Strangely, air source heat pumps currently require planning permission. However, the Government is currently consulting on allowing air source heat pumps as permitted development. So, check with the planning office of your local area before work commences.

7. Water source heat pumps. If the water is a moving watercourse or canal you will require permissions from the Waterways Commission or the Environment Agency. If the water is a big pond or lake on your own land then it probably comes under permitted development but may require Building Regulations approval. Your local planning office must be consulted before any work begins.

8. Biomass microgenerators. If the biomass microgenerator is a giant factory processing crops into fuel or the energy plant located at Teesport which I discussed earlier, then clearly it will require full planning permission and Building Regulations approval. However, if we are looking at a biomass-fuelled appliance such as a log burner in a domestic situation, then planning permission and Building Regulations approval will not be required. Of course the stove must be installed and the chimney checked by an approved installer.

9. Combined heat and power (CHP). If it is proposed to install a CHP micro-generator to provide community heating to an adjacent school or block of flats, then it will require full Planning and Building Regulations permission. However, if it is proposed to install a CHP microgenerator in a domestic situation, then planning will not be required. Building Regulations approval may be required, particularly with regard to the electrical installation and plumbing work in relation to approved documents L1A, L1B, L2A and L2B, which require that reasonable provision has been made to conserve fuel and power and, of course, this new technology must be installed by an MCS-approved installer.

10. Rainwater harvesting. If this water conservation technology is part of a new-build then it will form a part of the Planning and Building Regulations permissions for the whole building. However, if it is to be installed in an existing building, then a very large hole must be excavated to house the storage tank. A 7,000-litre tank is as big as a large skip and will require an extensive hole to bury it underground. Building Regulations may be required for such a large excavation. Check with your local office.

11. Grey water recycling. If this water conservation technology is part of a new build then it will form a part of the Planning and Building Regulations permissions for the whole building. Where it is to be installed in an existing building and incorporating a water filtration unit, then it is little more work than is required to install a washer or dishwasher and permissions will not be required.

The potential to install microgeneration technology – the building location and features

Let us look at each of the technologies discussed earlier.

1. Solar PV. The sun rises in the east and sinks in the west in the northern hemisphere and so solar panels facing south will be the best installation position. Adjacent trees or other buildings which cast a shadow on the panels will reduce the energy generated.

2. Small wind generators 1.5kW to 15kW but not small boat battery chargers. Small wind generators work best in exposed locations where there is a constant high wind speed. For this reason the site of the turbine is most important. Average wind speed for a specific location can be obtained from the internet by entering the post code or ordnance survey map reference. Alternatively the average wind speed can be measured on-site during the planning stage.

A small wind generator should not be fixed to a building because the wind blowing onto the building causes turbulence which reduces the efficiency of generation. Similarly, nearby buildings or trees will reduce efficiency. A turbine fitted to a mast or tower is by far the best method of installation and the higher the turbine the better it will perform.

All turbine towers must be robustly installed to withstand very high winds without becoming a hazard. Every mast should be designed to suit the location in which it is to be installed and for the type of turbine which it will support.

3. Hydro microgenerators. The available watercourse must have sufficient flow and head to maintain the flow rate throughout the seasons. However, hydropower is available 24 hours a day, seven days a week, month after month, year after year, unlike solar and wind energy. Hydropower has the most continuous and predictable output of any renewable energy. On-grid it continuously offsets electricity needs while selling any surplus energy back to the grid.

The old reputation of small hydro schemes being unreliable and requiring constant maintenance has been cast aside in the past 20 years, initially with the introduction of automatic electronic load controllers and more recently with the introduction of easy maintenance intake screens, similar to those used in domestic Koi pool installations.

Figure 2.19 Solar powered garden lighting.

If you have legitimate access to an appropriate water resource and can obtain the necessary permissions, then hydro could well be the best of all renewable energy solutions, despite being the most expensive.

4. Solar thermal. Solar thermal systems use the energy of the sun to heat water which is then stored in the domestic hot water cylinder to be used as required. The solar collector is usually placed on a south-facing roof with a slope of between 20 and 50 degrees for best results. Adjacent buildings or trees which cast a shadow on the panel will reduce the heat energy generated. A four-metre square panel will usually meet the hot water demands of the average family during the summer months.

5. Ground source heat pumps. The ground heat exchanger must be about 2.5 times the area of the building to be heated. The heat exchanger is harnessing the solar energy absorbed by the soil above it and so this area must remain uncovered and in direct sunlight for best results. Shading from other buildings or trees will reduce the energy efficiency of the system. To disguise the area dedicated to the heat exchanger will require some architectural feature such as a large lawn, a five-a-side football pitch, a tennis court or two or a 'paddock to exercise the ponies'. It is not a system to be considered for urban development.

6. Air source heat pump. You can see from Fig. 2.13 that this system can fit into an urban development or a city centre home. A suitable external wall is all that is required and almost all properties have this except high-rise buildings. The essential resource for this system is air within the temperature range 8 °C to 17 °C or above. Air source heat pumps can be used for heating or cooling, making them ideal for domestic or commercial buildings provided that the temperature range is available.

7. Water source heat pump. The heat exchanger will be large. You will need a large pond or watercourse close to the property in which to submerge the heat exchanger coils. The water must be deep enough to avoid freezing at the bottom in winter and of course you must have permission to use the water source.

8. Biomass heating. If the biomass heating system is based upon a wood-burning stove, then the property must have a chimney or flue system or the facility to install a chimney. I see lots of stainless steel chimneys being installed to provide a flue for a wood burner. A wood burner is a desirable feature in many homes these days, even very modern homes. The stove provides a focal point in the living area while the remaining rooms are heated more conventionally with underfloor heating or radiators. Wood-burning stoves may also incorporate a back boiler which was the open coal fire method of heating hot water before central heating became so popular.

The principal fuel is wood which is classified by the Kyoto Protocol as carbon neutral. Dry wood burns best and so you must also build an outside wood store to accommodate, ideally, about two to three tonnes of logs. In this way they will be air-dried before burning.

9. Combined heat and power (CHP). A micro combined heat and power unit can be a direct replacement for a conventional domestic boiler. The combined heat and power unit can use most types of fuel and are therefore described as fuel neutral. This new technology is developing quite quickly.

10. Rainwater harvesting. At a time when many homes are connected to water meters, it makes good economical sense to try to preserve this scarce natural resource. All that is required is a large storage vessel to collect all the rainwater which falls on the roof and flat paved areas around the property.

The ideal place for a large storage vessel is underground and when connected to a submersible pump and outside tap it will provide just the same amenity for garden watering and car washing as a tap connected to the water mains. However, you will not be using fluorinated water which has travelled through the water meter. Water from a storage vessel is also exempt from a summer hosepipe ban.

11. Grey water recycling. At its simplest it is only a matter of connecting the waste pipes from the bath, shower and laundry into the rainwater storage vessel. However, all commercial systems filter the water before reusing it again to, for

Figure 2.20 Small-scale wood burning stove.

Figure 2.21 Outside wood store.

example, flush the toilet. Commercial car-washing machines also recycle and filter the water for reuse.

Advantages and disadvantages of microgeneration technology

Our energy bills are set to rise above inflation for the foreseeable future. To reduce the effect of these rises upon household bills we should consider:

- using energy more efficiently;
- insulating our homes more effectively;
- the advantages and disadvantages of using microgeneration technology.

1. Solar PV. Government grants and the feed-in tariffs are available for solar PV even though the price per unit is reducing in value. The installation is simple and straightforward. PV systems up to 5 kW are exempt from planning permission under the permitted development laws. However, the panels must be fixed in a south-facing situation for best results. Home owners should not consider the 'rent a roof scheme' because this may reduce the value of their home. Financial payback is about 30 years if no grants or feedback tariff are available.

2. Small wind generators. People living in isolated hill farms may find benefit from a wind turbine, but not everyone will live or work in a location which has sufficient wind resources. There are issues of local wind turbulence from buildings and trees which will impede the efficient operation of a wind turbine.

Lots of people will object to the erection of a tall tower with a wind turbine at the top. Planning permission will be required. Financial payback for a 100 kW installation will be in the region of 15 to 20 years.

3. Hydro microgeneration. Hydropower is considered the renewable energy of choice if the resources are available. The civil works and capital investment required for hydropower look overwhelming when compared with solar or wind power but the benefits of hydro outweigh solar and wind because it will generate power all day every day. However, permissions will be time consuming and installation costs high. Financial payback for a 100 kW turbine installation will be in the region of 15 to 20 years.

4. Solar thermal. Solar thermal can be installed in two days by an MCS installer. The installation is simple and straightforward and requires no planning permission. Panels must be south facing for best results. Financial payback is 8 to 20 years depending upon the site chosen and the utilization of the equipment.

5. Ground source heat pump. This microgenerator is invisible, unlike solar PV, solar thermal, wind generation and hydro. This may therefore be one option for a building in a conservation area. However, a large piece of land must be given over to the heat exchanger. Financial payback is 8 to 15 years.

6. Air source heat pump. This system is not suitable for high-rise buildings. Air temperatures within the range 8 to 17 degrees or above are essential for the efficient operation of the system. Financial payback is 8 to 15 years.

7. Water source heat pump. This is another invisible microgenerator but in this case the heat exchanger must be submerged into a large mass of water. Financial payback time is 8 to 15 years.

8. Biomass heating. A wood burner gives a very cosy focal point to a living space, especially on a cold winter's day. However, unlike gas and electricity heating you must be able to store the fuel on the premises. You must also have a chimney or flue connecting the wood burner to the outside. Financial payback is not applicable here because fuel costs are about the same as coal, oil or gas and the installation of a 7 kW stove and chimney liner if necessary is less than £1,000.

9. Combined heat and power (CHP). A micro CHP domestic unit can be a straight swap for the existing boiler which may save up to 30% on the fuel bill. However, this is very new technology which is improving all the time. Financial payback is three to five years.

10. Rainwater harvesting. At a time when many people are on water meters these water-saving techniques make a lot of sense. The disadvantage is that you must find a place to put the large water storage vessel.

11. Grey water recycling. By simply connecting a few pipes in a different way you can save a precious resource and money. So why wouldn't you? It is very simple to install on a new-build property but less so on an existing property.

Financial payback

The payback period from fuel savings or income is a useful way of assessing if an investment is worthwhile and it also allows you to compare the financial merits of different technology options.

To carry out a simple estimate of the financial payback of a proposed installation you must divide the total installation costs by the annual savings. For example, let us suppose that you intend to replace your existing coal, calor gas or oil-fired

Figure 2.22 Solar powered speed warning sign.

Check your understanding

When you have completed the questions check out the answers at the back of the book.

Note: more than one multiple-choice answer may be correct.

1 Our surroundings and the world in which we live is one definition of:
   a. the Health and Safety at Work Act
   b. the Building Regulations
   c. the environment
   d. the water table.

2 Environmental technology systems:
   a. are eco friendly
   b. use renewable energy
   c. use fossil fuel
   d. use nuclear energy.

3 Identify the hazardous materials below:
   a. old glass bottles
   b. old fluorescent tubes
   c. used batteries
   d. offcuts of trunking and conduit.

4 Identify the recyclable materials below:
   a. old glass bottles
   b. old fluorescent tubes
   c. used batteries
   d. offcuts of trunking and conduit.

5 The Packaging Regulations tell us that all packaging must be designed and manufactured so that the:
   a. goods can never be broken
   b. volume and weight are at a minimum
   c. packaged goods can be moved with a fork lift truck and so avoid manual handling
   d. used packaging can be recycled and reused.

6 Solar photovoltaic panels are used to generate:
   a. heat
   b. light
   c. electricity
   d. combined heat and power.

7   Heat pumps are used to generate:

a.   water

b.   air

c.   heat

d.   electricity.

8   Wind turbines are used to generate:

a.   heat

b.   wind turbulence

c.   gas

d.   electricity.

9   In 2008 the Planning Laws were relaxed to allow the installation of small microgenerators on domestic properties within certain guidelines under an Act called:

a.   The Health and Safety at Work Act

b.   The Building Regulations

c.   Permitted Development

d.   The microgeneration certification scheme.

10   The collection and storage of rainwater for future use is one definition of:

a.   grey water recycling

b.   brown water harvesting

c.   rainwater harvesting

d.   brown water recycling.

11   The Building Regulations lay down the design and build standards for construction work in the building industry. The particular section of the Building Regulations concerned with electrical safety is:

a.   Part A

b.   Part D

c.   Part F

d.   Part P.

12   The best type of container to be used for rainwater harvesting in a domestic household would be a:

a.   30-litre bucket

b.   200-litre rainwater butt

c.   5,000-litre underground storage vessel

d.   small garden pond.

13   Tap water which has already performed one operation and is then made available to be used again is one definition of:

a.   grey water recycling

b.   brown water harvesting

c.   rainwater harvesting

d.   brown water recycling.

14   Identify four things you use at work that would require to be disposed of as hazardous waste.

15   Identify six pieces of equipment that would require to be disposed of correctly under the WEEE Regulations.

16   How do the Noise at Work Regulations protect workers?

17   Use bullet points to state the basic operating principles of a solar hot water heating system.

18   Use bullet points to state the applications and limitations of a solar hot water heating system.

19   Use bullet points to state the basic operating principle of a solar photovoltaic system.

20   Use bullet points to state the applications and limitations of a solar photovoltaic system.

21   State the advantages and disadvantages of wind energy generation.

22   Very briefly, in three sentences, describe the basic principle of heat pumps.

23   Very briefly, in three sentences, describe the basic principle of CHP systems.

24   In one sentence describe biomass heating.

25   In one sentence describe hydro microgeneration.

26   What is a 'smart' electricity meter?

27   What is 'rainwater harvesting'?

28   What is 'grey water recycling'?

Principles of electrical science

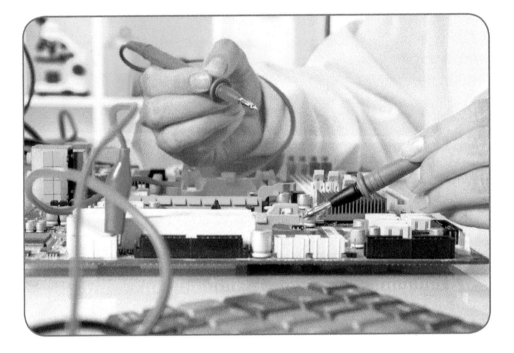

Advanced Electrical Installation Work. 978-0-367-35976-8
© 2019 Trevor Linsley. Published by Taylor & Francis. All rights reserved.

Unit 302 of the City and Guilds 2365-03 syllabus

Learning outcomes – when you have completed this chapter you should:

- understand the principles of a.c. theory;
- understand the principles of electrical machines;
- understand electrical quantities in Star and Delta configurations;
- understand the principles of electrical devices;
- understand the principles of lighting systems;
- understand the principles of heating systems;
- understand the principles of electronic components in electrical systems.

Simple machines

Our physical abilities in the field of lifting and moving heavy objects are limited. However, over the centuries we have used our superior intelligence to design tools, mechanisms and machines which have overcome this physical inadequacy. This concept is shown in Fig. 3.1.

By definition, a **machine** is an assembly of parts, some fixed, others movable, by which motion and force are transmitted. With the aid of a machine we are able to magnify the effort exerted at the input and lift or move large loads at the output.

Efficiency of any machine

In any machine the power available at the output is less than that which is put in because losses occur in the machine. The losses may result from friction in the bearings, wind resistance to moving parts, heat, noise or vibration.

The ratio of the output power to the input power is known as the **efficiency** of the machine. The symbol for efficiency is the Greek letter 'eta' (η). In general,

$$\eta = \frac{\text{Power output}}{\text{Power input}}$$

Since efficiency is usually expressed as a percentage we modify the general formula as follows:

$$\eta = \frac{\text{Power output}}{\text{Power input}} \times 100$$

> ### Definition
>
>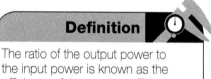
>
> By definition, a *machine* is an assembly of parts, some fixed, others movable, by which motion and force are transmitted. With the aid of a machine we are able to magnify the effort exerted at the input and lift or move large loads at the output.

> ### Definition
>
> The ratio of the output power to the input power is known as the *efficiency* of the machine. The symbol for efficiency is the Greek letter 'eta' (η). In general,
>
> $$\eta = \frac{\text{Power output}}{\text{Power input}}$$

Example

A transformer feeds the 9.81 kW motor driving a mechanical hoist. The input power to the transformer was found to be 10.9 kW. Find the efficiency of the transformer.

$$\eta = \frac{\text{Power output}}{\text{Power input}} \times 100$$

$$\eta = \frac{9.81\,\text{kW}}{10.9\,\text{kW}} \times 100 = 90\%$$

Thus the transformer is 90% efficient. Note that efficiency has no units, but is simply expressed as a percentage.

Electrical machines

Electrical machines are energy converters. If the machine input is mechanical energy and the output electrical energy then that machine is a generator, as

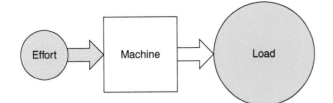

Figure 3.1 Simple machine concept.

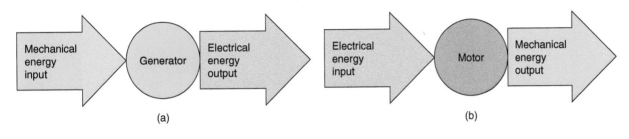

Figure 3.2 Electrical machines as energy converters.

shown in Fig. 3.2(a). Alternatively, if the machine input is electrical energy and the output mechanical energy, then the machine is a motor, as shown in Fig. 3.2(b).

An electrical machine may be used as a motor or a generator, although in practice the machine will operate more efficiently when operated in the mode for which it was designed.

Simple a.c. generator or alternator

If a simple loop of wire is rotated between the poles of a permanent magnet, as shown in Fig. 3.3, the loop of wire will cut the lines of magnetic flux between the north and south poles. This flux cutting will induce an electromotive force (e.m.f.) in the wire by **Faraday's law**, which states that *when a conductor cuts or is cut by a magnetic field, an e.m.f. is induced in that conductor*. If the generated e.m.f. is collected by carbon brushes at the slip rings and displayed on the screen of a cathode ray oscilloscope, the waveform will be seen to be approximately sinusoidal. Alternately changing, first positive and then negative, then positive again, gives an alternating output.

> **Definition**
>
> *Faraday's law* states that *when a conductor cuts or is cut by a magnetic field, an e.m.f. is induced in that conductor.*

Simple d.c. generator or dynamo

If the slip rings of Fig. 3.3 are replaced by a single split ring, called a commutator, the generated e.m.f. will be seen to be in one direction, as shown in Fig. 3.4. The action of the commutator is to reverse the generated e.m.f. every half-cycle, rather like an automatic changeover switch. However, this simple arrangement produces a very bumpy d.c. output. In a practical machine, the commutator would contain many segments and many windings to produce a smoother d.c. output similar to the unidirectional battery supply shown in Fig. 3.5.

Alternating current theory

The supply which we obtain from a car battery is a unidirectional or d.c. supply, whereas the mains electricity supply is alternating or a.c. (see Fig. 3.5).

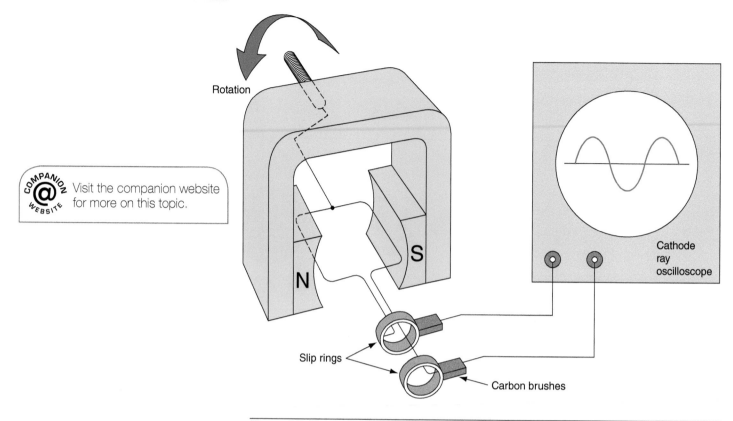

Figure 3.3 Simple a.c. generator or alternator.

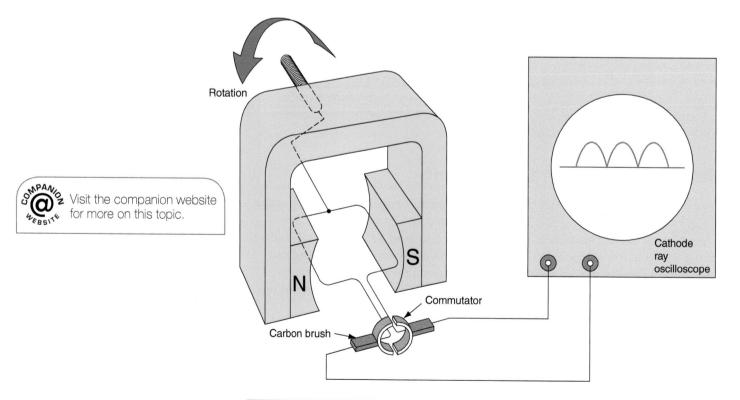

Figure 3.4 Simple d.c. generator or dynamo.

One of the reasons for using alternating supplies for the electricity mains supply is because we can very easily change the voltage levels by using a transformer which will only work on an a.c. supply.

The generated alternating supply at the power station is transformed up to 132,000 V, or more, for efficient transmission along the national grid conductors.

Most electrical equipment makes use of alternating current supplies, and for this reason knowledge of alternating waveforms is necessary for all practising electricians.

When a coil of wire is rotated inside a magnetic field as shown in Fig. 3.3, a voltage is induced in the coil. The induced voltage follows a mathematical law known as the sinusoidal law and, therefore, we can say that a sine wave has been generated. Such a waveform has the characteristics displayed in Fig. 3.6.

In the United Kingdom we generate electricity at a frequency of 50 Hz and the time taken to complete each cycle is given by

$$T = \frac{1}{f}$$

$$\therefore T = \frac{1}{50\text{Hz}} = 0.02\,\text{s}$$

An alternating waveform is constantly changing from zero to a maximum, first in one direction, then in the opposite direction, and so the instantaneous values of the generated voltage are always changing. A useful description of the electrical effects of an a.c. waveform can be given by the maximum, average and r.m.s. values of the waveform.

The maximum or peak value is the greatest instantaneous value reached by the generated waveform. Cable and equipment insulation levels must be equal to or greater than this value.

The average value is the average over one half-cycle of the instantaneous values as they change from zero to a maximum and can be found from the following formula applied to the sinusoidal waveform shown in Fig. 3.7.

$$V_{av} = \frac{V_1 + V_2 + V_3 + V_4 + V_5 + V_6}{6} = 0.637 V_{max}$$

For any sinusoidal waveform the average value is equal to 0.637 of the maximum value.

Battery supply d.c.

Mains supply a.c.

Figure 3.5 Unidirectional and alternating supply.

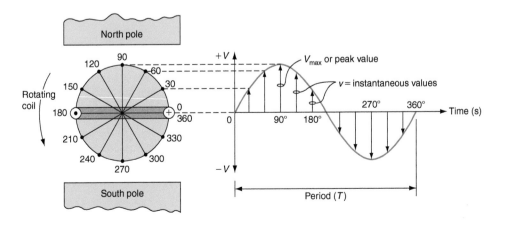

Figure 3.6 Characteristics of a sine wave.

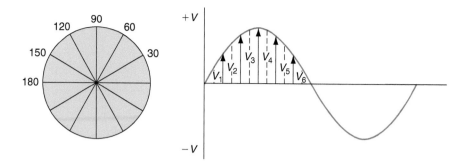

Figure 3.7 Sinusoidal waveform showing instantaneous values of voltage.

The r.m.s. value is the square root of the mean of the individual squared values and is the value of an a.c. voltage which produces the same heating effect as a d.c. voltage. The value can be found from the following formula applied to the sinusoidal waveform shown in Fig. 3.7.

$$V_{r.m.s.} = \sqrt{\frac{V_1^2 + V_2^2 + V_3^2 + V_4^2 + V_5^2 + V_6^2}{6}}$$
$$= 0.7071 V_{max}$$

For any sinusoidal waveform the r.m.s. value is equal to 0.7071 of the maximum value.

Example

The sinusoidal waveform applied to a particular circuit has a maximum value of 325.3 V. Calculate the average and r.m.s. value of the waveform.

$$\text{Average value } V_{av} = 0.637 \times V_{max}$$
$$\therefore V_{av} = 0\,637 \times 325\,3 = 207.2V$$
$$\text{r.m.s. value } V_{r.m.s.} = 0.7071 \times V_{max}$$
$$V_{r.m.s.} = 0.7071 \times 325.3 = 230V$$

When we say that the main supply to a domestic property is 230 V, we really mean 230 $V_{r.m.s.}$. Such a waveform has an average value of about 207.2 V and a maximum value of almost 325.3 V but because the r.m.s. value gives the d.c. equivalent value we almost always give the r.m.s. value without identifying it as such.

Alternating current theory

Commercial quantities of electricity for industry, commerce and domestic use are generated as a.c. in large power stations and distributed around the United Kingdom on the national grid to the end user. The d.c. electricity has many applications where portability or an emergency standby supply is important but for large quantities of power it has to be an a.c. supply because it is so easy to change the voltage levels using a transformer.

Key fact

Definitions
- Try to remember these a.c. circuit definitions
- Write them down if it helps.

Rotating a simple loop of wire or coils of wire between the poles of a magnet, such as that shown simplified in Fig. 3.3, will cut the north–south lines of magnetic flux and induce an a.c. voltage in the loop or coils of wire as shown by the display on a cathode ray oscilloscope.

This is an a.c. supply, an alternating current supply. The basic principle of the a.c. supply generated in a power station is exactly the same as Fig. 3.3 except that powerful electromagnets are used and the power for rotation comes from a steam turbine.

In this section we will first of all consider the theoretical circuits of pure resistance, inductance and capacitance acting alone in an a.c. circuit before going on to consider the practical circuits of resistance, inductance and capacitance acting together. Let us first define some of our terms of reference.

Definition

In any circuit, *resistance* is defined as opposition to current flow.

Resistance

In any circuit, **resistance** is defined as opposition to current flow. From Ohm's law:

$$R = \frac{V_R}{I_R} \, (\Omega)$$

Visit the companion website for more on this topic.

However, in an a.c. circuit, resistance is only part of the opposition to current flow. The inductance and capacitance of an a.c. circuit also cause an opposition to current flow, which we call *reactance*.

Inductive reactance (X_L) is the opposition to an a.c. current in an inductive circuit. It causes the current in the circuit to lag behind the applied voltage, as shown in Fig. 3.8. It is given by the formula:

$$X_L = 2\pi f L \, (\Omega)$$

where

$\pi = 3.142$ (a constant)
f = the frequency of the supply
L = the inductance of the circuit, or by:

$$X_L = \frac{V_L}{I_L}$$

Definition

Inductive reactance (X_L) is the opposition to an a.c. current in an inductive circuit. It causes the current in the circuit to lag behind the applied voltage.

Capacitive reactance (X_C) is the opposition to an a.c. current in a capacitive circuit. It causes the current in the circuit to lead ahead of the voltage, as shown in Fig. 3.8. It is given by the formula:

$$X_C = \frac{1}{2\pi f C} \, (\Omega)$$

where π and f are defined as before and C is the capacitance of the circuit. It can also be expressed as:

$$X_C = \frac{V_C}{I_C}$$

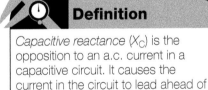

Definition

Capacitive reactance (X_C) is the opposition to an a.c. current in a capacitive circuit. It causes the current in the circuit to lead ahead of the voltage.

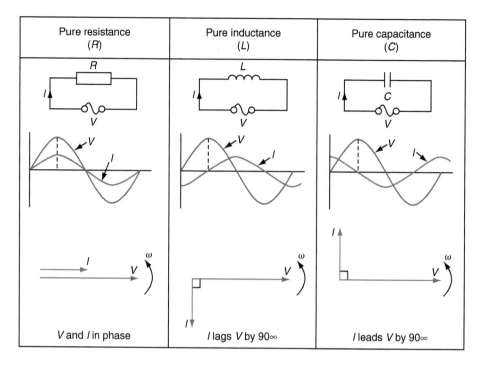

Figure 3.8 Voltage and current relationships in resistive, capacitive and inductive circuits.

Example

Calculate the reactance of a 150 μF capacitor and a 0.05 H inductor if they were separately connected to the 50 Hz mains supply.

For capacitive reactance:

$$X_C = \frac{1}{2\pi f C}$$

where $f = 50\,Hz$ and $C = 150\,\mu F = 150 \times 10^{-6}\,F$

$$\therefore X_C = \frac{1}{2 \times 3.142 \times 50\,Hz \times 150 \times 10^{-6}\,F} = 21.2\,\Omega$$

For inductive reactance:

$$X_L = 2\pi f L$$

where $f = 50\,Hz$ and $L = 0.05\,H$

$$\therefore X_l = 2 \times 3.142 \times 50\,Hz \times 0.05\,H = 15.7\,\Omega$$

Definition

The total opposition to current flow in an a.c. circuit is called *impedance* and given the symbol Z.

Impedance

The total opposition to current flow in an a.c. circuit is called **impedance** and given the symbol Z. Thus impedance is the combined opposition to current flow of the resistance, inductive reactance and capacitive reactance of the circuit and can be calculated from the formula:

$$Z = \sqrt{R^2 + X^2} \ (\Omega)$$

or

$$Z = \frac{V_T}{I_T}$$

Example 1

Calculate the impedance when a $5\,\Omega$ resistor is connected in series with a $12\,\Omega$ inductive reactance.

$$Z = \sqrt{R^2 + X_L^2} \ (\Omega)$$
$$\therefore Z = \sqrt{5^2 + 12^2}$$
$$Z = \sqrt{25 + 144}$$
$$Z = \sqrt{169}$$
$$Z = 13\,\Omega$$

Example 2

Calculate the impedance when a $48\,\Omega$ resistor is connected in series with a $55\,\Omega$ capacitive reactance.

$$Z = \sqrt{R^2 + X_C^2} \ (\Omega)$$
$$\therefore Z = \sqrt{48^2 + 55^2}$$
$$Z = \sqrt{2304 + 3025}$$
$$Z = \sqrt{5329}$$
$$Z = 73\,\Omega$$

Resistance, inductance and capacitance in an a.c. circuit

When a resistor only is connected to an a.c. circuit the current and voltage waveforms remain together, starting and finishing at the same time. We say that the waveforms are *in phase*.

When a pure inductor is connected to an a.c. circuit the current lags behind the voltage waveform by an angle of 90°. We say that the current *lags* the voltage by 90°. When a pure capacitor is connected to an a.c. circuit the current *leads* the voltage by an angle of 90°. These various effects can be observed on an oscilloscope, but the circuit diagram, waveform diagram and phasor diagram for each circuit are shown in Fig. 3.8.

Phasor diagrams

Phasor diagrams and a.c. circuits are an inseparable combination. Phasor diagrams allow us to produce a model or picture of the circuit under

Definition

A *phasor* is a straight line, having definite length and direction, which represents to scale the magnitude and direction of a quantity such as a current, voltage or impedance.

consideration which helps us to understand the circuit. A phasor is a straight line, having definite length and direction, which represents to scale the magnitude and direction of a quantity such as a current, voltage or impedance.

To find the combined effect of two quantities we combine their phasors by adding the beginning of the second phasor to the end of the first. The combined effect of the two quantities is shown by the resultant phasor, which is measured from the original zero position to the end of the last phasor.

Example

Find by phasor addition the combined effect of currents A and B acting in a circuit. Current A has a value of 4 A, and current B a value of 3 A, leading A by 90°. We usually assume phasors to rotate anticlockwise and so the complete diagram will be as shown in Fig. 3.9. Choose a scale of, for example, 1 A = 1 cm and draw the phasors to scale; that is, A = 4 cm and B = 3 cm, leading A by 90°.

The magnitude of the resultant phasor can be measured from the phasor diagram and is found to be 5 A acting at a phase angle ϕ of about 37° leading A. We therefore say that the combined effect of currents A and B is a current of 5 A at an angle of 37° leading A.

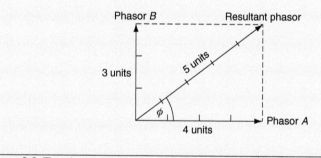

Figure 3.9 The phasor addition of currents A and B.

Phase angle ϕ

In an a.c. circuit containing resistance only, such as a heating circuit, the voltage and current are in phase, which means that they reach their peak and zero values together, as shown in Fig. 3.10(a).

In an a.c. circuit containing inductance, such as a motor or discharge lighting circuit, the current often reaches its maximum value after the voltage, which means that the current and voltage are out of phase with each other, as shown in Fig. 3.10(b). The phase difference, measured in degrees between the current and voltage, is called the phase angle of the circuit, and is denoted by the symbol ϕ, the lower-case Greek letter phi.

When circuits contain two or more separate elements, such as RL, RC or RLC, the phase angle between the total voltage and total current will be neither 0° nor 90° but will be determined by the relative values of resistance and reactance in the circuit. In Fig. 3.11 the phase angle between applied voltage and current is some angle ϕ.

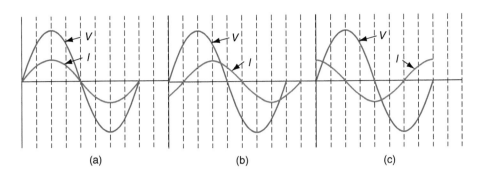

Figure 3.10 Phase relationship of a.c. waveform: (a) V and I in phase, phase angle $\phi = 0°$ and power factor $= \cos \phi = 1$; (b) V and I displaced by 45°, $\phi = 45°$ and p.f. $= 0.707$; and (c) V and I displaced by 90°, $\phi = 90°$ and p.f. $= 0$.

Alternating current series circuits

In a circuit containing a resistor and inductor connected in series as shown in Fig. 3.11, the current I will flow through the resistor and the inductor causing the voltage V_R to be dropped across the resistor and V_L to be dropped across the inductor. The sum of these voltages will be equal to the total voltage V_T but because this is an a.c. circuit the voltages must be added by phasor addition. The result is shown in Fig. 3.11, where V_R is drawn to scale and in phase with the current and V_L is drawn to scale and leading the current by 90°. The phasor addition of these two voltages gives us the magnitude and direction of V_T, which leads the current by some angle ϕ.

In a circuit containing a resistor and capacitor connected in series as shown in Fig. 3.12, the current I will flow through the resistor and capacitor causing

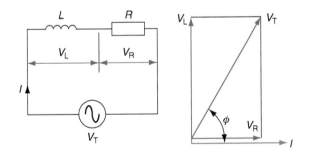

Figure 3.11 A series RL circuit and phasor diagram.

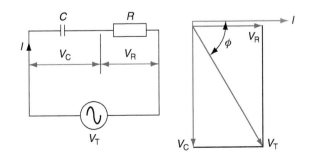

Figure 3.12 A series RC circuit and phasor diagram.

voltage drops V_R and V_C. The voltage V_R will be in phase with the current and V_C will lag the current by 90°. The phasor addition of these voltages is equal to the total voltage V_T which, as can be seen in Fig. 3.12, is lagging the current by some angle ϕ.

The impedance triangle

We have now established the general shape of the phasor diagram for a series a.c. circuit. Figures 3.11 and 3.12 show the voltage phasors, but we know that $V_R = IR$, $V_L = IX_L$, $V_C = IX_C$ and $V_T = IZ$, and therefore the phasor diagrams (a) and (b) of Fig. 3.13 must be equal. From Figure 3.13(b), by the theorem of Pythagoras, we have:

$$(IZ)^2 = (IR)^2 + (IX)^2$$
$$I^2Z^2 = I^2R^2 + I^2X^2$$

If we now divide throughout by I^2 we have:

$$Z^2 = R^2 + X^2$$
$$\text{or } Z = \sqrt{R^2 + X^2} \ \Omega$$

The phasor diagram can be simplified to the impedance triangle given in Fig. 3.13(c).

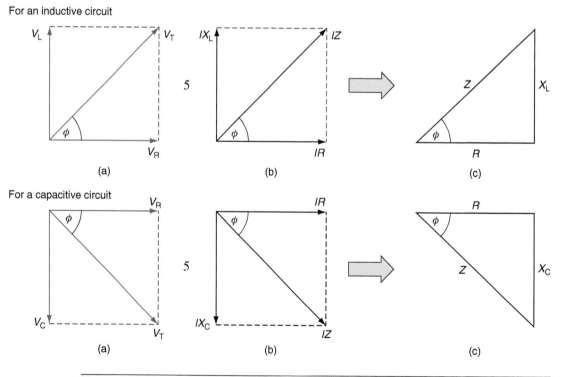

Figure 3.13 Phasor diagram and impedance triangle.

Example 1

A coil of 0.15 H is connected in series with a 50 Ω resistor across a 100 V 50 Hz supply. Calculate (a) the reactance of the coil, (b) the impedance of the circuit, and (c) the current.

For (a)

$$X_L = 2\pi f L \ (\Omega)$$
$$\therefore X_L = 2 \times 3.142 \times 50\,\text{Hz} \times 0.15\,\text{H} = 47.1\,\Omega$$

For (b)

$$Z = \sqrt{R^2 + X^2} \ (\Omega)$$
$$\therefore Z = \sqrt{(50\,\Omega)^2 + (47.1\,\Omega)^2} = 68.69\,\Omega$$

For (c)

$$I = \frac{V}{Z} \ (\text{A})$$
$$\therefore I = \frac{100\,V}{68.69\,\Omega} = 1.46\,\text{A}$$

Example 2

A 60 μF capacitor is connected in series with a 100 Ω resistor across a 230 V 50 Hz supply. Calculate (a) the reactance of the capacitor, (b) the impedance of the circuit and (c) the current.

For (a)

$$X_C = \frac{1}{2\pi f C} \ (\Omega)$$
$$\therefore X_C = \frac{1}{2\pi \times 50\,\text{Hz} \times 60 \times 10^{-6}\,\text{F}} = 53.05\,\Omega$$

For (b)

$$Z = \sqrt{R^2 + X^2} \ (\Omega)$$
$$\therefore Z = \sqrt{(100\,\Omega)^2 + (53.05\,\Omega)^2} = 113.2\,\Omega$$

For (c)

$$I = \frac{V}{Z} \ (\text{A})$$
$$\therefore I = \frac{230\,V}{113.2\,\Omega} = 2.03\,\text{A}$$

Power and power factor

Power factor (p.f.) is defined as the cosine of the phase angle between the current and voltage:

$$\text{p.f.} = \cos\phi$$

If the current lags the voltage as shown in Fig. 3.11, we say that the p.f. is lagging, and if the current leads the voltage as shown in Fig. 3.12, the p.f. is said

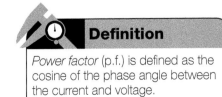

Definition

Power factor (p.f.) is defined as the cosine of the phase angle between the current and voltage.

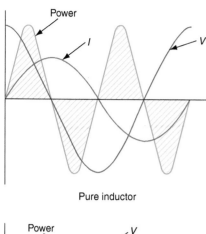

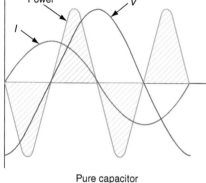

Figure 3.14 Waveform for the a.c. power in purely inductive and purely capacitive circuits.

to be leading. From the trigonometry of the impedance triangle shown in Fig. 3.13, p.f. is also equal to:

$$\text{p.f.} = \cos\phi = \frac{R}{Z} = \frac{V_R}{V_T}$$

The electrical power in a circuit is the product of the instantaneous values of the voltage and current. Figure 3.14 shows the voltage and current waveform for a pure inductor and pure capacitor. The power waveform is obtained from the product of V and I at every instant in the cycle. It can be seen that the power waveform reverses every quarter cycle, indicating that energy is alternately being fed into and taken out of the inductor and capacitor. When considered over one complete cycle, the positive and negative portions are equal, showing that the average power consumed by a pure inductor or capacitor is zero. This shows that inductors and capacitors store energy during one part of the voltage cycle and feed it back into the supply later in the cycle. Inductors store energy as a magnetic field and capacitors as an electric field.

In an electric circuit more power is taken from the supply than is fed back into it, since some power is dissipated by the resistance of the circuit, and therefore:

$$P = I^2R \text{ (W)}$$

In any d.c. circuit the power consumed is given by the product of the voltage and current, because in a d.c. circuit voltage and current are in phase. In an a.c. circuit the power consumed is given by the product of the current and that part of the voltage which is in phase with the current. The in-phase component of the voltage is given by $V \cos\phi$, and so power can also be given by the equation:

$$P = VI \cos\phi \text{ (W)}$$

Example 1

A coil has a resistance of 30 Ω and a reactance of 40 Ω when connected to a 250 V supply. Calculate (a) the impedance, (b) the current, (c) the p.f., and (d) the power.

For (a)

$$Z = \sqrt{R^2 + X^2} \ (\Omega)$$
$$\therefore Z = \sqrt{(30\,\Omega)^2 + (40\,\Omega)^2} = 50\,\Omega$$

For (b)

$$I = \frac{V}{Z} \ (\text{A})$$
$$\therefore I = \frac{250\,\text{V}}{50\,\Omega} = 5\,\text{A}$$

For (c)

$$\text{p.f.} = \cos\phi = \frac{R}{Z}$$
$$\therefore \text{p.f.} = \frac{30\,\Omega}{50\,\Omega} = 0.6 \text{ lagging}$$

For (d)

$$P = VI \cos\phi \ (\text{W})$$
$$\therefore P = 250\,\text{V} \times 5\,\text{A} \times 0.6 = 750\,\text{W}$$

Example 2

A capacitor of reactance $12\,\Omega$ is connected in series with a $9\,\Omega$ resistor across a $150\,V$ supply. Calculate (a) the impedance of the circuit, (b) the current, (c) the p.f., and (d) the power.

For (a)

$$Z = \sqrt{R^2 + X^2}\ (\Omega)$$
$$\therefore Z = \sqrt{(9\,\Omega)^2 + (12\,\Omega)^2} = 15\,\Omega$$

For (b)

$$I = \frac{V}{Z}\ (A)$$
$$\therefore I = \frac{150\,V}{15\,\Omega} = 10\,A$$

For (c)

$$\text{p.f.} = \cos\phi = \frac{R}{Z}$$
$$\therefore \text{p.f.} = \frac{9\,\Omega}{15\,\Omega} = 0.6\ \text{leading}$$

For (d)

$$P = VI\cos\phi\ (W)$$
$$\therefore P = 150\,V \times 10\,A \times 0.6 = 900\,W$$

Key fact

Trigonometry and transposition of formula are an important part of the Level 3 course in Electrical Installation Work.

I always recommend a good math book to my students called Electrical Installation Calculations: Advanced level 3. by Christopher Kitcher and A.J. Watkins.

The power factor of most industrial loads is lagging because the machines and discharge lighting used in industry are mostly inductive. This causes an additional magnetizing current to be drawn from the supply, which does not produce power, but does need to be supplied, making supply cables larger.

Example 3

A $230\,V$ supply feeds three $1.84\,kW$ loads with power factors of 1, 0.8 and 0.4. Calculate the current at each power factor.

The current is given by:

$$I = \frac{P}{V\cos\phi}$$

where $P = 1.84\,kW = 1840\,W$ and $V = 230\,V$. If the p.f. is 1, then:

$$I = \frac{1840\,W}{230\,V \times 1} = 8\,A$$

For a p.f. of 0.8:

$$I = \frac{1840\,W}{230\,V \times 0.8} = 10\,A$$

For a p.f. of 0.4:

$$I = \frac{1840\,W}{230\,V \times 0.4} = 20\,A$$

(Continued)

Example 3 (Continued)

It can be seen from these calculations that a 1.84 kW load supplied at a power factor of 0.4 would require a 20 A cable, while the same load at unity power factor could be supplied with an 8 A cable. There may also be the problem of higher voltage drops in the supply cables. As a result, the supply companies encourage installation engineers to improve their power factor to a value close to 1 and sometimes charge penalties if the power factor falls below 0.8.

Power factor correction

Most installations have a low or bad power factor because of the inductive nature of the load. A capacitor has the opposite effect of an inductor, and so it seems reasonable to add a capacitor to a load which is known to have a lower or bad power factor, for example, a motor.

Figure 3.15(a) shows an industrial load with a low power factor. If a capacitor is connected in parallel with the load, the capacitor current I_C leads the applied voltage by 90°. When this capacitor current is added to the load current as shown in Fig. 3.15(b) the resultant load current has a much improved power factor. However, using a slightly bigger capacitor, the load current can be pushed up until it is 'in phase' with the voltage, as can be seen in Fig. 3.15(c).

Capacitors may be connected across the main busbars of industrial loads in order to provide power factor improvement, but smaller capacitors may also be connected across an individual piece of equipment, as is the case for fluorescent light fittings.

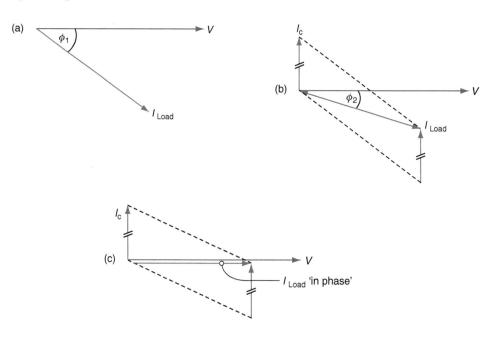

Figure 3.15 Power factor improvement using capacitors.

Operating principles of electrical machines

All electrical machines operate on the principles of magnetism. The basic rules of magnetism were laid down in Chapter 2 of *Basic Electrical Installation Work*.

Here we will look at some of the laws of magnetism as they apply to electrical machines, such as generators, motors and transformers.

A current carrying conductor maintains a magnetic field around the conductor which is proportional to the current flowing. When this magnetic field interacts with another magnetic field, forces are exerted which describe the basic principles of electric motors.

Michael Faraday demonstrated on 29 August 1831 that electricity could be produced by magnetism. He stated that 'when a conductor cuts or is cut by a magnetic field an e.m.f. is induced in that conductor. The amount of induced e.m.f. is proportional to the rate or speed at which the magnetic field cuts the conductor.' This basic principle laid down the laws of present-day electricity generation where a strong magnetic field is rotated inside a coil of wire to generate electricity.

Self- and mutual inductance

If a coil of wire is wound on to an iron core as shown in Fig. 3.16, a magnetic field will become established in the core when a current flows in the coil due to the switch being closed.

When the switch is opened the current stops flowing and, therefore, the magnetic flux collapses. The collapsing magnetic flux induces an e.m.f. into the coil and this voltage appears across the switch contacts. The effect is known as *self-inductance*, or just *inductance*, and is one property of any coil. The unit of inductance is the henry (symbol H), to commemorate the work of the American physicist Joseph Henry (1797–1878), and a circuit is said to possess an inductance of 1 henry when an e.m.f. of 1 volt is induced in the circuit by a current changing at the rate of 1 ampere per second.

Fluorescent light fittings contain a choke or inductive coil in series with the tube and starter lamp. The starter lamp switches on and off very quickly, causing rapid current changes which induce a large voltage across the tube electrodes sufficient to strike an arc in the tube.

When two separate coils are placed close together, as they are in a transformer, a current in one coil produces a magnetic flux which links with the second coil. This induces a voltage in the second coil and is the basic principle of the transformer action which is described later in this chapter. The two coils in this case are said to possess **mutual inductance**, as shown in Fig. 3.17. A mutual inductance of 1 henry exists between two coils when a uniformly varying current of 1 ampere per second in one coil produces an e.m.f. of 1 volt in the other coil.

Definition

A *mutual inductance* of 1 henry exists between two coils when a uniformly varying current of 1 ampere per second in one coil produces an e.m.f. of 1 volt in the other coil.

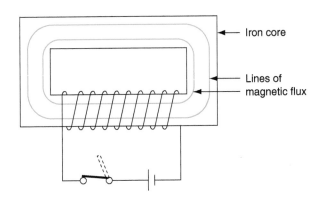

Iron core

Lines of magnetic flux

Figure 3.16 An inductive coil or choke.

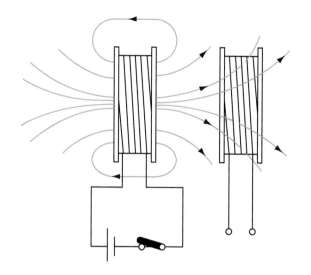

Figure 3.17 Mutual inductance between two coils.

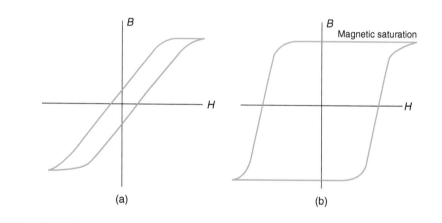

Figure 3.18 Magnetic hysteresis loops: (a) electromagnetic material, and (b) permanent magnetic material.

The e.m.f. induced in a coil such as that shown on the right-hand side in Fig. 3.17 is dependent upon the rate of change of magnetic flux and the number of turns on the coil.

Energy stored in a magnetic field

When we open the switch of an inductive circuit such as an electric motor or fluorescent light circuit the magnetic flux collapses and produces an arc across the switch contacts. The arc is produced by the stored magnetic energy being discharged across the switch contacts.

Magnetic hysteresis

There are many different types of magnetic material and they all respond differently to being magnetized. Some materials magnetize easily, and some are difficult to magnetize. Some materials retain their magnetism, while others lose it. The result will look like the graphs shown in Fig. 3.18 and are called hysteresis loops.

Magnetic hysteresis loops describe the way in which different materials respond to being magnetized.

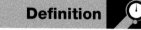

Definition

Magnetic hysteresis loops describe the way in which different materials respond to being magnetized.

Materials from which permanent magnets are made should display a wide hysteresis loop, as shown by loop (b) in Fig. 3.18.

The core of an electromagnet is required to magnetize easily, and to lose its magnetism equally easily when switched off. Suitable materials will, therefore, display a narrow hysteresis loop, as shown by loop (a) in Fig. 3.18.

When an iron core is subjected to alternating magnetization, as in a transformer, the energy loss occurs at every cycle and so constitutes a continuous power loss, and, therefore, for applications such as transformers, a material with a narrow hysteresis loop is required.

Direct current motors

All electric motors work on the principle that when a current-carrying conductor is placed in a magnetic field it will experience a force. An electric motor uses this magnetic force to turn the shaft of the electric motor. Let us try to understand this action. If a current-carrying conductor is placed into the field of a permanent magnet, as shown in Fig. 3.19(c), a force F will be exerted on the conductor to push it out of the magnetic field.

Visit the companion website for more on this topic.

To understand the force, let us consider each magnetic field acting alone. Figure 3.19(a) shows the magnetic field due to the current-carrying conductor only. Figure 3.19(b) shows the magnetic field due to the permanent magnet in which is placed the conductor carrying no current. Figure 3.19(c) shows the effect of the combined magnetic fields which are distorted and, because lines of magnetic flux never cross, but behave like stretched elastic bands always trying to find the shorter distance between a north and south pole, the force F is exerted on the conductor, pushing it out of the permanent magnetic field.

This is the basic motor principle, and the force F is dependent upon the strength of the magnetic field B, the magnitude of the current flowing in the conductor I and the length of conductor within the magnetic field l. The following equation expresses this relationship:

$$F = BIl \text{ (N)}$$

where B is in tesla, l is in metres, I is in amperes and F is in newtons.

Example

A coil which is made up of a conductor some 15 m in length lies at right angles to a magnetic field of strength 5 T. Calculate the force on the conductor when 15 A flows in the coil.

$$F = BIl \text{ (N)}$$

$$F = 5\text{T} \times 15\text{m} \times 15\text{A} = 1125\text{N}$$

Practical d.c. motors

Practical motors are constructed as shown in Fig. 3.20. All d.c. motors contain a field winding wound on pole pieces attached to a steel yoke. The armature winding rotates between the poles and is connected to the commutator.

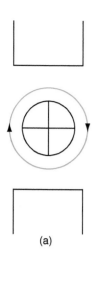

(a)

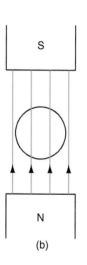

(b)

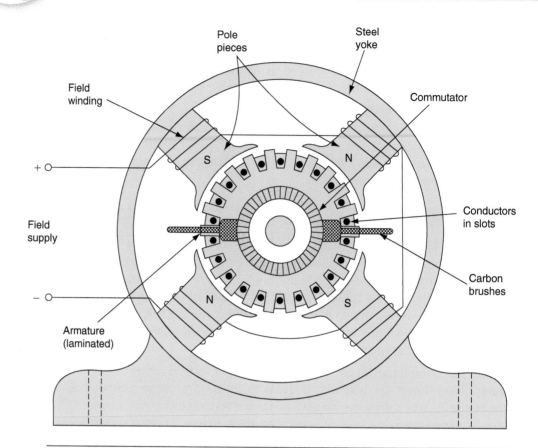

Figure 3.20 Showing d.c. machine construction.

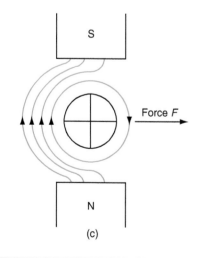

(c)

Figure 3.19 Force on a conductor in a magnetic field.

Contact with the external circuit is made through carbon brushes rubbing on the commutator segments. Direct current motors are classified by the way in which the field and armature windings are connected, which may be in series or in parallel.

Series motor

The field and armature windings are connected in series and consequently share the same current. The series motor has the characteristics of a high starting torque but a speed which varies with load. Figure 3.21 shows series motor connections and characteristics. For this reason the motor is only suitable for direct coupling to a load, except in very small motors, such as vacuum cleaners and hand drills, and is ideally suited for applications where the machine must start on load, such as electric trains, cranes and hoists.

Reversal of rotation may be achieved by reversing the connections of either the field or armature windings but not both. This characteristic means that the machine will run on both a.c. or d.c. and is, therefore, sometimes referred to as a 'universal' motor.

Three-phase a.c. motors

If a three-phase supply is connected to three separate windings equally distributed around the stationary part or stator of an electrical machine, an alternating current circulates in the coils and establishes a magnetic flux. The magnetic field established by the three-phase currents travels around the stator, establishing a rotating magnetic flux, creating magnetic forces on the rotor which turns the shaft on the motor.

Three-phase induction motor

When a three-phase supply is connected to insulated coils set into slots in the inner surface of the stator or stationary part of an induction motor, as shown in Fig. 3.22(a), a rotating magnetic flux is produced. The rotating magnetic flux cuts the conductors of the rotor and induces an e.m.f. in the rotor conductors by Faraday's law, which states that when a conductor cuts or is cut by a magnetic field, an e.m.f. is induced in that conductor, the magnitude of which is proportional to the *rate* at which the conductor cuts or is cut by the magnetic flux. This induced e.m.f. causes rotor currents to flow and establish a magnetic flux which reacts with the stator flux and causes a force to be exerted on the rotor conductors, turning the rotor, as shown in Fig. 3.22(b).

The turning force or torque experienced by the rotor is produced by inducing an e.m.f. into the rotor conductors due to the *relative* motion between the conductors and the rotating field. The torque produces rotation in the same direction as the rotating magnetic field.

Rotor construction

There are two types of induction motor rotor – the wound rotor and the cage rotor. The cage rotor consists of a laminated cylinder of silicon steel with copper or aluminium bars slotted into holes around the circumference and short-circuited at each end of the cylinder, as shown in Fig. 3.23. In small motors the rotor is cast in aluminium. Better starting and quieter running are achieved if the bars are slightly skewed. This type of rotor is extremely robust and since there are no external connections there is no need for slip rings or brushes. A machine

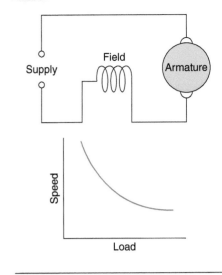

Figure 3.21 Series motor connections and characteristics.

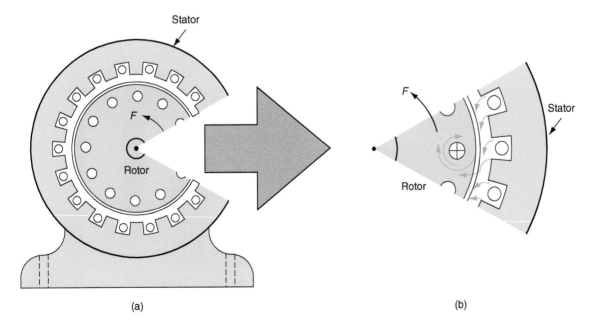

(a)   (b)

Figure 3.22 Segment taken out of an induction motor to show turning force: (a) construction of an induction motor; and (b) production of torque by magnetic fields.

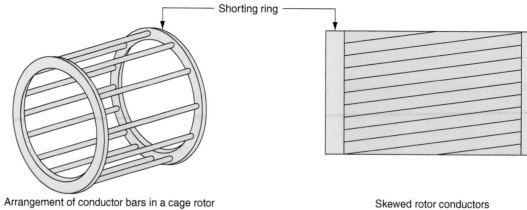

Shorting ring

Arrangement of conductor bars in a cage rotor

Skewed rotor conductors

Figure 3.23 Construction of a cage rotor.

(a)

(b)

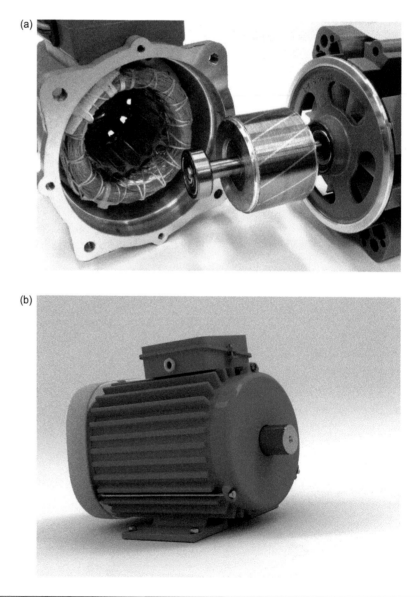

Figure 3.24 (a) The internal workings of an induction motor; (b) the external shell of an induction motor.

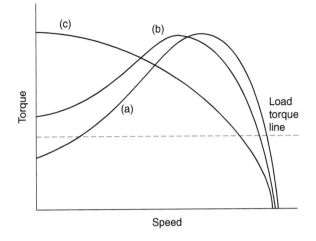

Figure 3.25 Various speed–torque characteristics for an induction motor.

fitted with a cage rotor does suffer from a low starting torque and the machine must be chosen which has a higher starting torque than the load, as shown by curve (b) in Fig. 3.25. A machine with the characteristic shown by curve (a) in Fig. 3.25 would not start since the load torque is greater than the machine starting torque. Alternatively the load may be connected after the motor has been run up to full speed.

The wound rotor consists of a laminated cylinder of silicon steel with copper coils embedded in slots around the circumference. The windings may be connected in star or delta and the end connections brought out to slip rings mounted on the shaft. Connection by carbon brushes can then be made to an external resistance to improve starting.

The cage induction motor has a small starting torque and should be used with light loads or started with the load disconnected. The speed is almost constant. Its applications are for constant speed machines such as fans and pumps. Reversal of rotation is achieved by reversing any two of the stator winding connections.

Single-phase a.c. motors

A single-phase a.c. supply produces a pulsating magnetic field, not the rotating magnetic field produced by a three-phase supply. All a.c. motors require a rotating field to start. Therefore, single-phase a.c. motors have two windings which are electrically separated by about 90°. The two windings are known as the start and run windings. The magnetic fields produced by currents flowing through these out-of-phase windings create the rotating field and turning force required to start the motor. Once rotation is established, the pulsating field in the run winding is sufficient to maintain rotation and the start winding is disconnected by a centrifugal switch which operates when the motor has reached about 80% of the full load speed.

A cage rotor is used on single-phase a.c. motors, the turning force being produced in the way described previously for three-phase induction motors and shown in Fig. 3.22. Because both windings carry currents which are out of phase with each other, the motor is known as a 'split-phase' motor. The phase

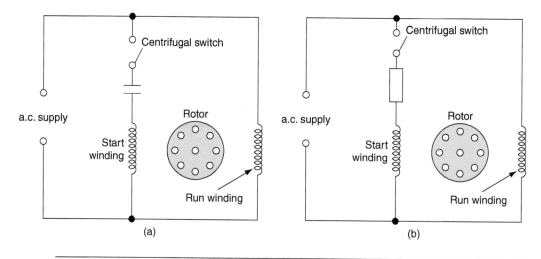

Figure 3.26 Circuit diagram of: (a) capacitor split-phase motors, and (b) resistance split-phase motors.

displacement between the currents in the windings is achieved in one of two ways:

- by connecting a capacitor in series with the start winding, as shown in Fig. 3.26(a), which gives a 90° phase difference between the currents in the start and run windings;
- by designing the start winding to have a high resistance and the run winding a high inductance, once again creating a 90° phase shift between the currents in each winding, as shown in Fig. 3.26(b).

When the motor is first switched on, the centrifugal switch is closed and the magnetic fields from the two coils produce the turning force required to run the rotor up to full speed. When the motor reaches about 80% of full speed, the centrifugal switch clicks open and the machine continues to run on the magnetic flux created by the run winding only.

Split-phase motors are constant speed machines with a low starting torque and are used on light loads such as fans, pumps, refrigerators and washing machines. Reversal of rotation may be achieved by reversing the connections to the start or run windings, but not both.

Shaded pole motors

The shaded pole motor is a simple, robust, single-phase motor, which is suitable for very small machines with a rating of less than about 50 W. Figure 3.27 shows a shaded pole motor. It has a cage rotor and the moving field is produced by enclosing one side of each stator pole in a solid copper or brass ring, called a shading ring, which displaces the magnetic field and creates an artificial phase shift.

Shaded pole motors are constant speed machines with a very low starting torque and are used on very light loads such as oven fans, record turntable motors and electric fan heaters. Reversal of rotation is theoretically possible by moving the shading rings to the opposite side of the stator pole face. However, in practice this is often not a simple process, but the motors are symmetrical and it is sometimes easier to reverse the rotor by removing the fixing bolts and reversing the whole motor.

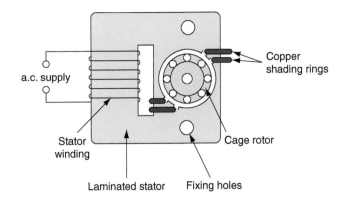

Figure 3.27 Shaded pole motor.

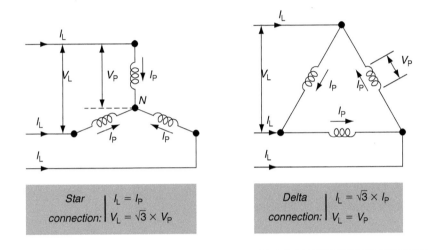

Figure 3.28 Star and delta connections.

There are more motors operating from single-phase supplies than all other types of motor added together. Most of them operate as very small motors in domestic and business machines where single-phase supplies are most common.

Star and delta connections

The three-phase windings of an a.c. generator may be star connected or delta connected, as shown in Fig. 3.28. The important relationship between phase and line currents and voltages is also shown. The square root of 3 ($\sqrt{3}$) is simply a constant for three-phase circuits, and has a value of 1.732. The delta connection is used for electrical power transmission because only three conductors are required. Delta connection is also used to connect the windings of most three-phase motors because the phase windings are perfectly balanced and, therefore, do not require a neutral connection.

Making a star connection at the local substation has the advantage that two voltages become available – a line voltage of 400 V between any two phases, and a phase voltage of 230 V between line and neutral which is connected to the star point.

In any star-connected system currents flow along the lines (I_L), through the load and return by the neutral conductor connected to the star point. In a *balanced* three-phase system all currents have the same value and, when they are added up by phasor addition, we find the resultant current is zero. Therefore, no current flows in the neutral and the star point is at zero volts. The star point of the distribution transformer is earthed because earth is also at zero potential. A star-connected system is also called a three-phase four-wire system and allows us to connect single-phase loads to a three-phase system.

Three-phase power

We know from our single-phase alternating current theory earlier in this chapter that power can be found from the following formula:

$$\text{Power} = VI \cos\phi \text{ (W)}$$

In any balanced three-phase system, the total power is equal to three times the power in any one phase.

$$\therefore \text{ Total three-phase power} = 3V_P I_P \cos\phi \text{ (W)} \qquad \text{(Equation 1)}$$

Now for a star connection,

$$V_P = \frac{V_L}{\sqrt{3}} \quad \text{and} \quad I_L = I_P \qquad \text{(Equation 2)}$$

Substituting Equation (2) into Equation (1), we have:

$$\text{Total three-phase power} = \sqrt{3} \, V_L I_L \cos\phi \text{ (W)}$$

Now consider a delta connection:

$$V_P = V_L \quad \text{and} \quad I_P = \frac{I_L}{\sqrt{3}} \qquad \text{(Equation 3)}$$

Substituting Equation (3) into Equation (1) we have, for any balanced three-phase load,

$$\text{Total three-phase power} = \sqrt{3} \, V_L I_L \cos\phi \text{ (W)}$$

So, a general equation for three-phase power is:

$$\text{Power} = \sqrt{3} \, V_L I_L \cos\phi$$

Example 1

A balanced star-connected three-phase load of $10\,\Omega$ per phase is supplied from a 400 V, 50 Hz mains supply at unity power factor. Calculate (a) the phase voltage, (b) the line current, and (c) the total power consumed.

For a star connection:

$$V_L = \sqrt{3} \, V_P \text{ and } I_L = I_P$$

For (a)

$$V_P = \frac{V_L}{\sqrt{3}} \text{ (V)}$$

$$V_P = \frac{400\,\text{V}}{1.732} = 230.9\,\text{V}$$

(Continued)

Example 1 (Continued)

For (b)

$$I_L = I_P = \frac{V_P}{R_P} \text{ (A)}$$

$$I_L = I_P = \frac{230.9\,V}{10\,\Omega} = 23.09\,A$$

For (c)

$$\text{Power} = \sqrt{3}\ V_L\ I_L\ \cos\phi \text{ (W)}$$

$$\therefore \text{Power} = 1.732 \times 400\,V \times 23.09\,A \times 1 = 16 \text{ kW}$$

Example 2

A 20 kW, 400 V balanced delta-connected load has a power factor of 0.8. Calculate (a) the line current, and (b) the phase current.

We have that:

$$\text{Three-phase power} = \sqrt{3}\ V_L\ I_L\ \cos\phi \text{ (W)}$$

For (a)

$$I_L = \frac{\text{Power}}{\sqrt{3}\ V_L\ \cos\phi} \text{ (A)}$$

$$\therefore I_L = \frac{20{,}000\,W}{1.732 \times 400\,V \times 0.8}$$

$$I_L = 36.08 \text{ (A)}$$

For delta connection

$$I_L = \sqrt{3}\ I_P \text{ (A)}$$

Thus, for (b)

$$I_P = I_L / \sqrt{3} \text{ (A)}$$

$$\therefore I_P = \frac{36.08\,A}{1.732} = 20.83\,A$$

Example 3

Three identical loads each having a resistance of 30 Ω and inductive reactance of 40 Ω are connected first in star and then in delta to a 400 V three-phase supply. Calculate the phase currents and line currents for each connection.

For each load

$$Z = \sqrt{R^2 + X_L^2} \text{ (}\Omega\text{)}$$

$$\therefore Z = \sqrt{30^2 + 40^2}$$

$$Z = \sqrt{2500} = 50\,\Omega$$

(Continued)

Example 3 (Continued)

For star connection

$$V_L = \sqrt{3}\, V_P \quad \text{and} \quad I_L = I_P$$

$$V_P = \frac{V_L}{\sqrt{3}}\,(V)$$

$$\therefore\ V_P = \frac{400\,V}{1.732} = 230.9\,V$$

$$I_P = \frac{V_P}{Z_P}\,(A)$$

$$\therefore\ I_P = \frac{230.9\,V}{50\,\Omega} = 4.62\,A$$

$$I_P = I_L$$

Therefore phase and line currents are both equal to 4.62 A.

For delta connection

$$V_L = V_P \quad \text{and} \quad I_L = \sqrt{3}\, I_P$$

$$V_L = V_P = 400\,V$$

$$I_P = V_P / Z_P \,(A)$$

$$\therefore\ I = \frac{400\,V}{50\,\Omega} = 8\,A$$

$$I_L = \sqrt{3}\, I_P \,(A)$$

$$\therefore\ I_L = 1.732 \times 8\,A = 13.86\,A$$

Motor starters

When the armature conductors of an electric motor cut the magnetic flux of the main stator field, an e.m.f. is induced in the armature winding as described by Faraday's Law: 'when a conductor cuts, or is cut by a magnetic field, an e.m.f. is induced in that conductor'. The induced e.m.f. is known as a back e.m.f., because it acts in opposition to the supply voltage. During normal running the back e.m.f. is always a little smaller than the supply voltage and acts as a limit to the motor current. However, at the instant that the motor is first switched on, the back e.m.f. does not exist because the conductors are stationary and so a very large current flows for just a few milliseconds.

The magnetic flux generated in the stator of an induction motor, such as that shown in Fig. 3.22 earlier in this chapter, rotates immediately the supply is switched on, and therefore the machine is self-starting. The purpose of the motor starter is not to start the machine as the name implies, but to reduce heavy starting currents and to provide overload and no-volt protection as required by IET Regulations 552.

Thermal overload protection is usually provided by means of a bimetal strip bending under overload conditions and breaking the starter contactor coil circuit.

This de-energizes the coil and switches off the motor under fault conditions such as overloading or single phasing. Once the motor has automatically switched off under overload conditions or because a remote stop/start has been operated, it is an important safety feature that the motor cannot restart without the operator going through the normal start-up procedure. Therefore, no-volt protection is provided by incorporating the safety devices into the motor starter control circuit which energizes the contactor coil.

Electronic thermistors (thermal transistors) provide an alternative method of sensing if a motor is overheating. These tiny heat-sensing transistors, about the size of a matchstick head, are embedded in the motor windings to sense the internal temperature; the thermistor then either trips out the contactor coil as described above or operates an alarm.

All electric motors with a rating above 0.37 kW must be supplied from a suitable motor starter and we will now consider the more common types.

Direct online (d.o.l.) starters

The d.o.l. starter switches the main supply directly onto the motor. Since motor starting currents can be seven or eight times greater than the running currents, the d.o.l. starter is only used for small motors of less than about 5 kW rating.

When the start button is pressed, current will flow from the brown phase through the control circuit and contactor coil to the grey phase which energizes the contactor coil and the contacts close, connecting the three-phase supply to the motor, as can be seen in Fig. 3.29. If the start button is released the control circuit is maintained by the hold on contact. If the stop button is pressed or the overload coils operate, the control circuit is broken and the contactor drops out, breaking the supply to the load. Once the supply is interrupted, the supply to the motor can only be reconnected by pressing the start button. Therefore, this type of arrangement also provides no-volt protection.

When large industrial motors have to be started, a way of reducing the excessive starting currents must be found. One method is to connect the motor to a star delta motor starter.

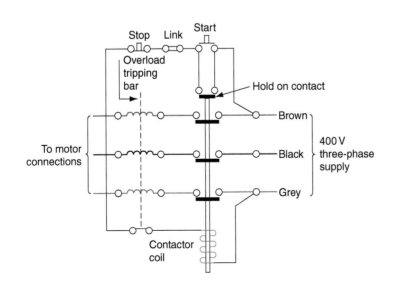

Figure 3.29 Three-phase d.o.l. starter.

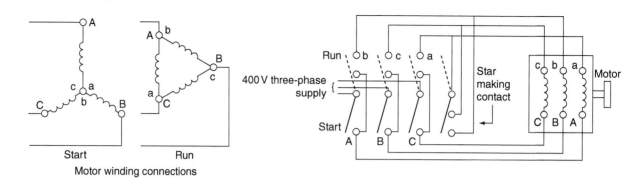

Motor winding connections

Figure 3.30 Star delta starter.

Star delta motor starters

When three loads, such as the three windings of a motor, are connected in star, the line current has only one-third of the value it has when the same load is connected in delta. A starter which can connect the windings in star during the initial starting period and then switch to a delta connection for normal running will reduce the problem of an excessive starting current. This arrangement is shown in Fig. 3.30 where the six connections to the three stator phase windings are brought out to the starter. For starting, the motor windings are star connected at the a-b-c end of the winding by the star making contacts. This reduces the phase voltage to about 58% of the running voltage, which reduces the current and the motor's torque. Once the motor is running, a double-throw switch makes the changeover from star starting to delta running, thereby achieving a minimum starting current and maximum running torque. The starter will incorporate overload and no-volt protection, but these have not been shown in Fig. 3.30 in the interests of showing more clearly the principle of operation.

Auto-transformer motor starter

An auto-transformer motor starter provides another method of reducing the starting current by reducing the voltage during the initial starting period. Since this also reduces the starting torque, the voltage is only reduced by a sufficient amount to reduce the starting current, being permanently connected to the tapping found to be most appropriate by the installing electrician. Switching the changeover switch to start position connects the auto-transformer windings in series with the delta connected motor starter winding. When sufficient speed has been achieved by the motor, the changeover switch is moved to the run connections which connect the three-phase supply directly onto the motor, as shown in Fig. 3.31.

This starting method has the advantage of only requiring three connecting conductors between the motor starter and the motor. The starter will incorporate overload and no-volt protection in addition to some method of preventing the motor from being switched to the run position while the motor is stationary. These protective devices are not shown in Fig. 3.31 in order to show more clearly the principle of operation.

Rotor resistance motor starter

Any motor containing a cage rotor such as that shown in Fig. 3.23 does suffer from a low starting torque. Therefore an induction motor must be matched to the

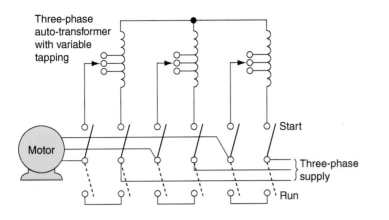

Figure 3.31 Auto-transformer starting.

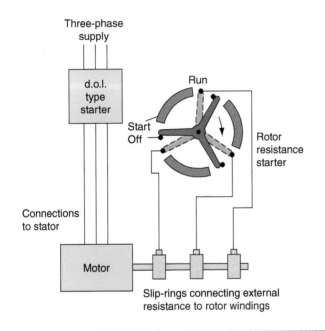

Figure 3.32 Rotor resistance starter for a wound rotor machine.

load in a way that the starting torque of the motor is greater than the load torque, as shown by curve (b) in Fig. 3.25. An induction motor with the characteristic shown by curve (a) in Fig. 3.25 would not self-start because the load torque is greater than the motor starting torque.

There are three ways to overcome this problem: first, by matching the motor to the load torque as shown by curve (b) in Fig. 3.25; second, by connecting the load after the load has run up to full speed; or third, extra resistance can be added to a wound rotor induction motor through slip rings and brushes. The extra resistance of the rotor increases the starting torque, as shown by curve (c) in Fig. 3.25.

When the motor is first switched on the external rotor resistance is at a maximum. As the motor speed increases the resistance is reduced until at full speed the external resistance is cut out and the machine continues to run as a cage induction motor. The starter is provided with overload and no-volt protection and an interlock to prevent the machine from being switched on with no rotor resistance connected. These are not shown in Fig. 3.32 because the purpose of the diagram is to show the principle of operation.

Operating principle of electrical devices

Electricity and magnetism have been inseparably connected since the experiments by Oersted and Faraday in the early nineteenth century. An electric current flowing in a conductor produces a magnetic field 'around' the conductor which is proportional to the current. Thus a small current produces a weak magnetic field, while a large current will produce a strong magnetic field.

The electrical solenoid

A current flowing in a *coil* of wire or solenoid establishes a magnetic field which is very similar to that of a bar magnet. Winding the coil around a soft iron core increases the flux density because the lines of magnetic flux concentrate on the magnetic material. The advantage of the electromagnet when compared with the permanent magnet is that the magnetism of the electromagnet can be switched on and off by a functional switch controlling the coil current. This effect is put to practical use in the contactor coil as used in a motor starter or alarm circuit. Figure 3.33 shows the structure and one application of the solenoid, a simple relay.

The electrical relay

A **relay** is an electromagnetic switch operated by a solenoid. The solenoid in a relay operates a number of switch contacts as it moves under the electromagnetic forces. Relays can be used to switch circuits on or off at a distance remotely. The energizing circuit, the solenoid, is completely separate from the switch contacts and, therefore, the relay can switch high-voltage, high-power circuits from a low-voltage switching circuit. This gives the relay many applications in motor control circuits, electronics and instrumentation systems. Figure 3.34 shows a simple relay.

Overcurrent protective devices

The consumer's mains equipment must provide protection against overcurrent; that is, a current exceeding the rated value (IET Regulation 430.3). Fuses provide

Definition

A *relay* is an electromagnetic switch operated by a solenoid.

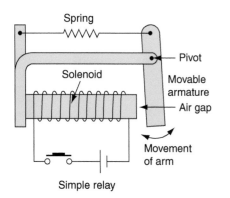

Spring

Pivot

Solenoid

Movable armature

Air gap

Movement of arm

Simple relay

Figure 3.33 The solenoid and one practical application: the relay.

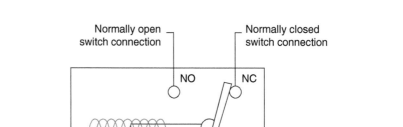

Normally open switch connection

Normally closed switch connection

NO

NC

C

Solenoid coil

Common switch connection

Figure 3.34 A simple relay.

overcurrent protection when situated in the live conductors; they must not be connected in the neutral conductor. Circuit-breakers may be used in place of fuses, in which case the circuit-breaker may also provide the means of isolation, although a further means of isolation is usually provided so that maintenance can be carried out on the circuit-breakers themselves.

An overcurrent may be an overload current, or a short-circuit current. An **overload current** can be defined as a current which exceeds the rated value in an otherwise healthy circuit. Overload currents usually occur because the circuit is abused or because it has been badly designed or modified. A **short-circuit** is an overcurrent resulting from a fault of negligible impedance connected between conductors. Short-circuits usually occur as a result of an accident which could not have been predicted before the event.

An overload may result in currents of two or three times the rated current flowing in the circuit. Short-circuit currents may be hundreds of times greater than the rated current. In both cases the basic requirements for protection are that the fault currents should be interrupted quickly and the circuit isolated safely before the fault current causes a temperature rise or mechanical effects which might damage the insulation, connections, joints and terminations of the circuit conductors or their surroundings (IET Regulations 131).

The selected protective device should have a current rating which is not less than the full load current of the circuit but which does not exceed the cable current rating. The cable is then fully protected against both overload and short-circuit faults (IET Regulation 435.1). Devices which provide overcurrent protection are:

- High breaking capacity (HBC) fuses to BS 88-2:2010. These are for industrial applications having a maximum fault capacity of 80 kA.
- Cartridge fuses to BS 88-3:2010. These are used for a.c. circuits on industrial and domestic installations having a fault capacity of about 30 kA.
- Cartridge fuses to BS 1362. These are used in 13 A plug tops and have a maximum fault capacity of about 6 kA.
- Semi-enclosed fuses to BS 3036. These were previously called rewirable fuses and are used mainly on domestic installations having a maximum fault capacity of about 4 kA.
- MCBs to BS EN 60898. These are miniature circuit-breakers (MCBs) which may be used as an alternative to fuses for some installations. The British Standard includes ratings up to 100 A and maximum fault capacities of 9 kA. They are graded according to their instantaneous tripping currents – that is, the current at which they will trip within 100 ms. This is less than the time taken to blink an eye.

The part played by fuses, MCBs, RCDs and RCBOs in protecting electrical equipment, circuits and people is discussed in Chapter 3 of *Basic Electrical Installation Work 8th Edition*. In this chapter we are looking at their operating principles.

Semi-enclosed fuses (BS 3036)

The semi-enclosed fuse consists of a fuse wire, called the fuse element, secured between two screw terminals in a fuse carrier. The fuse element is connected in series with the load, and the thickness of the element is sufficient to carry the normal rated circuit current. When a fault occurs an overcurrent flows and the fuse element becomes hot and melts or 'blows'. By definition a fuse is the weakest link in the circuit. Under fault conditions it will melt when an overcurrent flows, protecting the circuit conductors from damage.

Definition

An *overload current* can be defined as a current which exceeds the rated value in an otherwise healthy circuit.

A *short-circuit* is an overcurrent resulting from a fault of negligible impedance connected between conductors.

Definition

By definition a *fuse* is the weakest link in the circuit. Under fault conditions it will melt when an overcurrent flows, protecting the circuit conductors from damage.

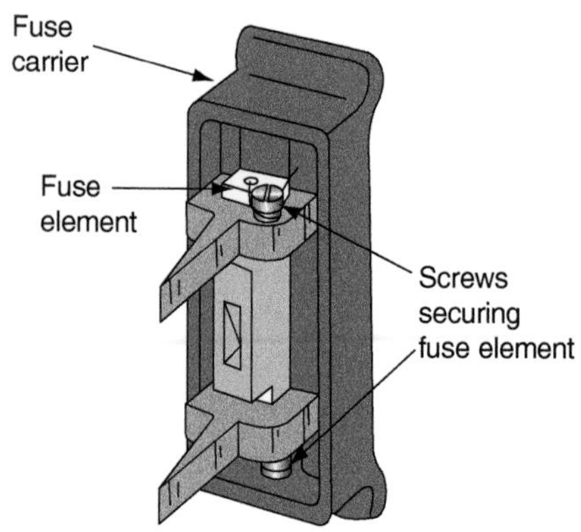

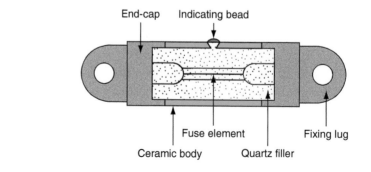

Figure 3.35 labels: Fuse carrier, Fuse element, Screws securing fuse element

Figure 3.35 A semi-enclosed fuse.

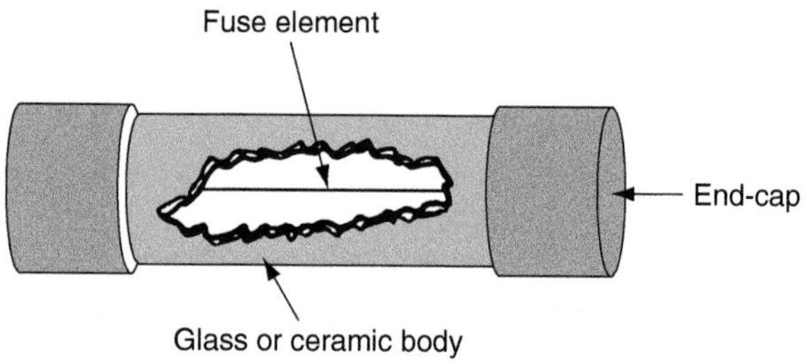

Figure 3.36 labels: Fuse element, End-cap, Glass or ceramic body

Figure 3.36 Cartridge fuse.

This type of fuse is illustrated in Fig. 3.35. The fuse element should consist of a single strand of plain or tinned copper wire having a diameter appropriate to the current rating of the fuse. *This type of fuse was very popular in domestic installations, but less so these days because of the popularity of MCBs.*

Cartridge fuses (BS 88-3:2010)

The cartridge fuse breaks a faulty circuit in the same way as a semi-enclosed fuse, but its construction eliminates some of the disadvantages experienced with an open-fuse element. The fuse element is encased in a glass or ceramic tube and secured to end-caps which are firmly attached to the body of the fuse so that they do not blow off when the fuse operates. Cartridge fuse construction is illustrated in Fig. 3.36. With larger sized cartridge fuses, lugs or tags are sometimes brazed on the end-caps to fix the fuse cartridge mechanically to the carrier. They may also be filled with quartz sand to absorb and extinguish the energy of the arc when the cartridge is brought into operation.

HBC fuses (BS 88-2:2010)

As the name might imply, these **HBC (High Breaking Capacity) cartridge fuses** are for protecting circuits where extremely high-fault currents may develop such as on industrial installations or distribution systems.

The fuse element consists of several parallel strips of pure silver encased in a substantial ceramic cylinder, the ends of which are sealed with tinned brass end-caps incorporating fixing lugs. The cartridge is filled with silica sand to ensure quick arc extraction. Incorporated on the body is an indicating device to show when the fuse has blown. HBC fuse construction is shown in Fig. 3.37.

Definition
As the name might imply, these *HBC cartridge fuses* are for protecting circuits where extremely high-fault currents may develop such as on industrial installations or distribution systems.

Figure 3.37 labels: End-cap, Indicating bead, Fuse element, Fixing lug, Ceramic body, Quartz filler

Figure 3.37 HBC fuse.

Figure 3.38 MCBs – B Breaker, fits Wylex standard consumer unit (courtesy of Wylex).

Miniature circuit-breakers (BS EN 60898)

The disadvantage of all fuses is that when they have operated they must be replaced. An MCB overcomes this problem since it is an automatic switch which opens in the event of an excessive current flowing in the circuit and can be closed when the circuit returns to normal.

An MCB of the type shown in Fig. 3.38 incorporates a thermal and magnetic tripping device. The load current flows through the thermal and the electromagnetic mechanisms. In normal operation the current is insufficient to operate either device but when an overload occurs, the thermal tripping device, that is, a bi-metal strip, heats up, bends and trips the mechanism. The time taken for this to occur provides the MCB with the ability to discriminate between an overload which persists for a very short time, for example, the starting current of an electric motor, and a persistent overload due to a fault. This device only trips when a fault occurs.

This slow operating time is ideal for overloads of short duration but when a short-circuit occurs it is essential to disconnect the fault very quickly. This is carried out by the electromagnetic tripping device.

When a large fault current (above about eight times the rated current) flows through the electromagnetic coil a strong magnetic flux is established which trips the mechanism almost instantly. The circuit can be restored when the fault is removed by activating the on-off toggle. This latches the various mechanisms within the MCB and 'makes' the switch contact. The toggle switch can also be used to disconnect the circuit for maintenance or isolation.

Two of the main reasons for the increased popularity of MCBs is that an 'ordinary person' can safely restore the circuit after a fault has occurred without the use of any tools. Second, the characteristics of the circuit protection installed by the professional electrician cannot be interfered with by an 'ordinary person', let us say the householder replacing a fuse with one of an incorrect size.

Residual current device (RCD)

When it is required to provide the very best protection from electric shock and fire risk, earth fault protection devices are incorporated into the installation. The object of the regulations concerning these devices (411.3.2 to 411.3.4) is

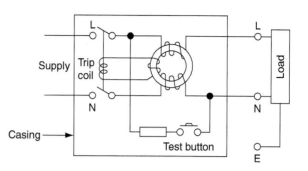

Figure 3.39 Construction of an RCD.

to remove an earth fault current very quickly, less than 0.4 seconds for all final circuits not exceeding 32 A, and limit the voltage which might appear on any exposed metal parts under fault conditions to not more than 50 V. They will continue to provide adequate protection throughout the life of the installation even if the earthing conditions deteriorate. This is in direct contrast to the protection provided by overcurrent devices, which require a low resistance earth loop impedance path.

The regulations recognize RCDs as 'additional protection' in the event of failure of the provision for basic protection, fault protection or carelessness by the users of the installation (IET Regulation 415.1.1).

The basic circuit for a single-phase RCD is shown in Fig. 3.39. The load current is fed through two equal and opposing coils wound onto a common transformer core. The phase and neutral currents in a healthy circuit produce equal and opposing fluxes in the transformer core, which induces no voltage in the tripping coil. However, if more current flows in the line conductor than in the neutral conductor as a result of a fault between live and earth, an out-of-balance flux will result in an e.m.f. being induced in the trip coil which will open the double pole switch and isolate the load. Modern RCDs have tripping sensitivities between 10 and 30 mA, and therefore a faulty circuit can be isolated before the lower lethal limit to human beings (about 50 mA) is reached.

Wherever RCDs are installed a label shall be fixed near to each RCD stating ' The device must be tested 6 monthly'. (IET Regulation 514.12.2) Note; In the 17th edition the test period was three monthly.

RCBOs (residual current circuit-breakers with overload protection)

RCBOs (residual current circuit-breakers with overload protection) combine RCD protection and MCB protection into one unit.

In a split board consumer unit, about half of the total number of final circuits are protected by the RCD. A fault on any one final circuit will trip out all of the RCD protected circuits which may cause inconvenience.

The RCBO gives the combined protection of an MCB plus RCD for each final circuit so protected and in the event of a fault occurring only the faulty circuit is interrupted.

The 18th Edition of the IET Wiring Regulations tells us at Note 5 of Table 537.4 that circuit protective devices and RCDs are not intended for frequent load switching. However, infrequent switching of MCBs is permissible for the purpose of isolation or emergency switching.

AFDD (arc fault detection devices)

The 18th edition of the IET Regulations at 532.6 and 421.1.7 has introduced a new topic, Arc Fault Detection Devices. (AFDD) These are compulsory in the EU Countries but only recommended in this country by the IET and the British standards. They are recognized as giving additional protection against fires caused by arc faults. Arc faults can be formed by cable insulation defects, damage to cables by impact or penetration by nails and screws, and loose terminal connections. AFDDs detect faults which MCBs and RCDs cannot detect. They are installed at the origin of the final circuit to be protected.

Lighting laws, lamps and luminares

In ancient times, much of the indoor work done by humans depended upon daylight being available to light the interior and this is still the case today in many developing nations where communities live beyond the reach of the national grid. Today, life in the wealthy nations carries on after dark where almost all buildings have electric lighting installed. We automatically assume that we can work indoors or out of doors at any time of the day or night, and that light will always be available.

Good lighting is important in all building interiors, helping work to be done efficiently and safely and also playing an important part in creating pleasant and comfortable surroundings.

Lighting schemes are designed using many different types of light fitting or luminaire. 'Luminaire' is the modern term given to the equipment which supports and surrounds the lamp and may control the distribution of the light. The 18th Edition of the IET Regulations has introduced a new Regulation at 411.3.4. This now requires additional protection within domestic (household) premises for **all circuits supplying luminaires** to be protected by an RCD rated at 30 mA. Modern lamps use the very latest technology to provide illumination cheaply and efficiently. To begin to understand the lamps and lighting technology used today, we must first define some of the terms we will be using.

Figure 3.40 Progress of lighting: candle, tungsten, fluoresent and LED.

Luminous intensity – symbol *I*

This is the illuminating power of the light source to radiate luminous flux in a particular direction. The earliest term used for the unit of luminous intensity was the candle power because the early standard was the wax candle. The SI unit is the candela (pronounced candeela and abbreviated as cd).

Luminous flux – symbol *F*

This is the flow of light which is radiated from a source. The SI unit is the lumen, one lumen being the light flux which is emitted within a unit solid angle (volume of a cone) from a point source of 1 candela.

Illuminance – symbol *E*

This is a measure of the light falling onto a surface, which is also called the incident radiation. The SI unit is the lux (lx) and is the illumination produced by 1 lumen over an area of $1\,m^2$.

Luminance – symbol *L*

Since this is a measure of the brightness of a surface it is also a measure of the light which is reflected from a surface. The objects we see vary in appearance according to the light which they emit or reflect towards the eye.

The SI units of luminance vary with the type of surface being considered. For a diffusing surface such as blotting paper or a matt white painted surface the unit of luminance is the lumen per square metre. With polished surfaces such as a silvered glass reflector, the brightness is specified in terms of the light intensity and the unit is the candela per square metre.

Illumination laws

Rays of light falling onto a surface from some distance *d* will illuminate that surface with an illuminance of, say, 1 lx. If the distance *d* is doubled, as shown in Fig. 3.41, the illuminance of 1 lx will fall over four square units of area. Thus the illumination of a surface follows the **inverse square law**, where

$$E = \frac{I}{d^2}\,(\text{lx})$$

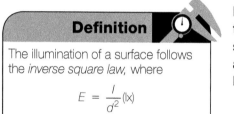

Definition

The illumination of a surface follows the *inverse square law,* where

$$E = \frac{I}{d^2}\,(\text{lx})$$

Example 1

A lamp of luminous intensity 1000 cd is suspended 2 m above a laboratory bench. Calculate the illuminance directly below the lamp:

$$E = \frac{I}{d^2}\,(\text{lx})$$

$$\therefore E = \frac{1000\ \text{cd}}{(2\ \text{m})^2} = 250\ \text{lx}$$

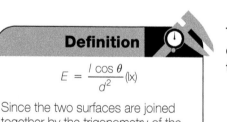

Definition

$$E = \frac{I\cos\theta}{d^2}\,(\text{lx})$$

Since the two surfaces are joined together by the trigonometry of the cosine rules this equation is known as the *cosine law.*

The illumination of surface A in Fig. 3.42 will follow the inverse square law described above. If this surface were removed, the same luminous flux would then fall onto surface B. Since the parallel rays of light falling onto the inclined

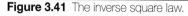

Figure 3.41 The inverse square law.

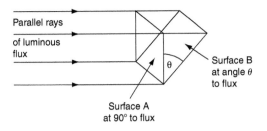

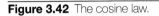

Figure 3.42 The cosine law.

surface B are spread over a larger surface area, the illuminance will be reduced by a factor θ, and therefore:

$$E = \frac{I \cos \theta}{d^2} (\text{lx})$$

Since the two surfaces are joined together by the trigonometry of the cosine rules this equation is known as the **cosine law**.

Example 2

A street lantern suspends a 2000 cd light source 4 m above the ground. Determine the illuminance directly below the lamp and 3 m to one side of the lamp base.

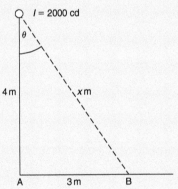

The illuminance below the lamp, E_A, is:

$$E_A = \frac{I}{d^2} (\text{lx})$$

$$\therefore E_A = \frac{2000 \text{ cd}}{(4 \text{ m})^2} = 125 \text{ lx}$$

To work out the illuminance at 3 m to one side of the lantern, E_B, we need the distance between the light source and the position on the ground at B; this can be found by Pythagoras' theorem:

$$x \text{ (m)} = \sqrt{(4 \text{ m})^2 + (3 \text{ m})^2} = \sqrt{25 \text{ m}}$$
$$x = 5 \text{ m}$$
$$\therefore E_B = \frac{I \cos \theta}{d^2} (\text{lx}) \text{ and } \cos \theta = \frac{4}{5}$$
$$\therefore E_B = \frac{2000 \text{ cd} \times 4}{(5 \text{ m})^2 \times 5} = 64 \text{ lx}$$

Example 3

A discharge lamp is suspended from a ceiling 4 m above a bench. The illuminance on the bench below the lamp was 300 lx. Find:

(a)   the luminous intensity of the lamp

(b)   the distance along the bench where the illuminance falls to 153.6 lx.

For (a)

$$E_A = \frac{I}{d^2} \text{(lx)}$$

$$\therefore I = E_A \, d^2 \text{(cd)}$$

$$I = 300 \text{ lx} \times 16 \text{ m} = 4800 \text{ cd}$$

For (b)

$$E_B = \frac{I}{d^2} \cos\theta \text{ (lx)}$$

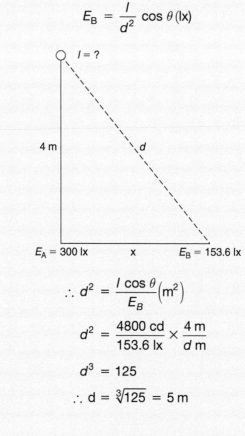

$I = ?$

4 m d

$E_A = 300$ lx x $E_B = 153.6$ lx

$$\therefore d^2 = \frac{I \cos\theta}{E_B} \text{(m}^2)$$

$$d^2 = \frac{4800 \text{ cd}}{153.6 \text{ lx}} \times \frac{4 \text{ m}}{d \text{ m}}$$

$$d^3 = 125$$

$$\therefore d = \sqrt[3]{125} = 5 \text{ m}$$

By Pythagoras

$$x = \sqrt{5^2 - 4^2} = 3 \text{ m}$$

Definition

When designing interior lighting schemes the method most frequently used depends upon a determination of the total flux required to provide a given value of illuminance at the working place. This method is generally known as the *lumen method*.

The recommended levels of illuminance for various types of installation are given by the IES (Illumination Engineers Society). Some examples are given in Table 3.1.

The activities being carried out in a room will determine the levels of illuminance required, since different levels of illumination are required for the successful operation or completion of different tasks. The assembly of electronic components in a factory will require a higher level of illumination than, say, the assembly of engine components in a garage because the electronic components are much smaller and finer detail is required for their successful assembly.

Table 3.1 Illuminance values

Task	Working situation	Illuminance (lx)
Casual vision	Storage rooms, stairs and washrooms	100
Rough assembly	Workshops and garages	300
Reading, writing and drawing	Classrooms and offices	500
Fine assembly	Electronic component assembly	1000
Minute assembly	Watchmaking	3000

The inverse square law calculations considered earlier are only suitable for designing lighting schemes where there are no reflecting surfaces producing secondary additional illumination. This method could be used to design an outdoor lighting scheme for a cathedral, bridge or public building.

Interior luminaires produce light directly on to the working surface but additionally there is a secondary source of illumination from light reflected from the walls and ceilings. When designing interior lighting schemes the method most frequently used depends upon a determination of the total flux required to provide a given value of illuminance at the working place. This method is generally known as the **lumen method**.

The lumen method

To determine the total number of luminaires required to produce a given illuminance by the lumen method we apply the following formula:

$$\text{Total number of luminaires required to provide a chosen level of illumination at a surface} = \frac{\text{Illuminance level (lx)} \times \text{Area (m}^2\text{)}}{\text{Lumen output of each luminaire (lm)} \times \text{UF} \times \text{LLF}}$$

where:

- the illuminance level is chosen after consideration of the IES code; or the architects specification.
- the area is the working area to be illuminated;
- the lumen output of each luminaire is that given in the manufacturer's specification and may be found by reference to tables such as Table 3.2;
- UF is the utilization factor;
- LLF is the light loss factor.

Utilization factor

The light flux reaching the working plane is always less than the lumen output of the lamp, since some of the light is absorbed by the various surface textures. The method of calculating the utilization factor (UF) is detailed in Chartered Institution of Building Services Engineers (CIBSE) Technical Memorandum No 5, although lighting manufacturers' catalogues give factors for standard conditions. The UF is expressed as a number which is always less than unity; a typical value might be 0.9 for a modern office building.

Light loss factor

The light output of a luminaire is reduced during its life because of an accumulation of dust and dirt on the lamp and fitting. Decorations also deteriorate with time, and this results in more light flux being absorbed by the walls and ceiling.

You can see from Table 3.2 that the output lumens of the lamp decrease with time – for example, a warm white tube gives out 4950 lumens after the first 100 hours of its life but this falls to 4600 lumens after 2000 hours.

The total light loss can be considered under four headings:

1 light loss due to luminaire dirt depreciation (LDD),
2 light loss due to room dirt depreciation (RDD),
3 light loss due to lamp failure factor (LFF),
4 light loss due to lamp lumen depreciation (LLD).

The LLF is the total loss due to these four separate factors and typically has a value between 0.8 and 0.9.

When using the LLF in lumen method calculations we always use the manufacturer's initial lamp lumens for the particular lamp because the LLF takes account of the depreciation in lumen output with time. Let us now consider a calculation using the lumen method.

Table 3.2 Characteristics of a Thorn Lighting 1500 mm 65 W bi-pin tube

Tube colour	Initial lamp lumens*	Lighting design lumens†	Colour rendering quality	Colour appearance
Artificial daylight	2600	2100	Excellent	Cool
De luxe natural	2900	2500	Very Good	Intermediate
De luxe warm white	3500	3200	Good	Warm
Natural	3700	3400	Good	Intermediate
Daylight	4800	4450	Fair	Cool
Warm white	4950	4600	Fair	Warm
White	5100	4750	Fair	Warm
Red	250*	250	Poor	Deep red

Coloured tubes are intended for decorative purposes only
*The initial lumens are the measured lumens after 100 hours of life
†The lighting design lumens are the output lumens after 2000 hours of life

Burning position — Lamp may be operated in any position
Rated life — 7500 hours
Efficacy — 30–70 lm/W depending upon the tube colour

Example

It is proposed to illuminate an electronic workshop of dimensions 9 × 8 × 3m to an illuminance of 550 lx at the bench level. The specification calls for luminaires having one 1500 mm 65 W natural tube with an initial output of 3700 lumens (see Table 3.2). Determine the number of luminaires required for this installation when the UF and LLF are 0.9 and 0.8, respectively.

(Continued)

Example 3 (Continued)

$$\text{The number of luminaires required} = \frac{E \text{ (lx)} \times \text{area } \left(\text{m}^2\right)}{\text{lumens from each luminaire} \times \text{UF} \times \text{LLF}}$$

$$\text{The number of luminaires} = \frac{550 \text{ lx} \times 9 \text{ m} \times 8 \text{ m}}{3700 \times 0.9 \times 0.8} = 14.86$$

Therefore 15 luminaires will be required to illuminate this workshop to a level of 550 lx. The electrical designer will set out the positions of the 15 luminaires as is most appropriate for the room size and layout of the work areas.

Comparison of light sources

When comparing one light source with another we are interested in the colour-reproducing qualities of the lamp and the efficiency with which the lamp converts electricity into illumination. These qualities are expressed by the lamp's efficacy and colour rendering qualities.

Lamp efficacy

The performance of a lamp is quoted as a ratio of the number of lumens of light flux which it emits to the electrical energy input which it consumes. Thus **efficacy** is measured in lumens per watt; the greater the efficacy the better is the lamp's performance in converting electrical energy into light energy.

A general lighting service (GLS) lamp, for example, has an efficacy of 14 lumens per watt, while a fluorescent tube, which is much more efficient at converting electricity into light, has an efficacy of about 50 lumens per watt. An LED lamp has an efficacy of between 90 and 110 lm/watt depending upon the colour output of the lamp.

Colour rendering

We recognize various materials and surfaces as having a particular colour because luminous flux of a frequency corresponding to that colour is reflected from the surface to our eye which is then processed by our brain. White light is made up of the combined frequencies of the colours red, orange, yellow, green, blue, indigo and violet. Colours can only be seen if the lamp supplying the illuminance is emitting light of that particular frequency. The ability to show colours faithfully as they would appear in daylight is a measure of the colour rendering property of the light source.

Building regulations for energy-efficient lighting

Part P of the Building Regulations 2006 relates to Electrical Safety in Dwellings. All new installations must comply with the Part P regulations and any other relevant parts of the Building Regulations. Approved document L1A and L1B Conservation of Fuel and Power 2006 is relevant to us as electricians because

Definition

The performance of a lamp is quoted as a ratio of the number of lumens of light flux which it emits to the electrical energy input which it consumes. Thus *efficacy* is measured in lumens per watt; the greater the efficacy the better is the lamp's performance in converting electrical energy into light energy.

Key fact

Energy efficiency
- fluorescent lamps and CFLs (compact fluorescent lamps) are energy-efficient lamps
- they give out more light for every electrical watt input.

it says that reasonable provision shall be made to provide lighting systems with energy-efficient lamps and sufficient controls so that electrical energy can be used efficiently. Part L describes methods of compliance with these regulations for both internal and external lighting. It says:

- A reasonable number of internal lighting points should be wired that will only take energy-efficient lamps such as fluorescent tubes and compact fluorescent lamps, CFLs and LEDs, which have an efficacy greater than 40 lumens/watt. They should be installed in the areas most frequently used such as hallways, landings, kitchens and sitting rooms to a number at least of:
  - one per 25 m² of dwelling floor area or
  - one per four fixed luminaires.
- External lighting fixed to the building, including lighting in porches but not lighting in garages or carports, should provide reasonable provision for energy-efficient lamps. These lamps should automatically extinguish in daylight and when not required at night, by being controlled by passive infra-red (PIR) detectors.

The traditional light bulb, called a GLS (general lighting service) lamp, is hopelessly bad in energy-efficiency terms, producing only 14 lumens of light output for every electrical watt input. Fluorescent tubes and CFLs produce more than 50 lumens of light output for every watt input and LEDs produce between 90 and 110 lumens per watt depending upon the colour output. The government calculates that if every British household was to replace three 60 or 100 W light bulbs with LEDs, the energy saving would be greater than the power used by the entire street lighting network.

Mr Hilary Benn, the then Environment Secretary, announced in 2010 that the traditional GLS light bulbs of 150, 100, 60 and 40 W would begin to be phased out. Thus, households will have to use more energy-efficient lamps in the future. September 2018 saw the final stage of an EU Directive banning the import or manufacture of GLS and halogen lamps take effect. This Directive was introduced in 2009 in order to restrict the sale of 'high energy lamps' and to create a market for 'high energy saving lamps' such as LEDs and CFLs. LED lamps are now probably the first choice for electrical designers.

Let us now look at 10 different types of lamp.

GLS lamps

GLS lamps produce light as a result of the heating effect of an electrical current. Most of the electricity goes to producing heat and a little to producing light. A fine tungsten wire is first coiled and coiled again to form the incandescent filament of the GLS lamp. The coiled coil arrangement reduces filament cooling and increases the light output by allowing the filament to operate at a higher temperature. The light output covers the visible spectrum, giving a warm white to yellow light with a colour rendering quality classified as fairly good. The efficacy of the GLS lamp is 14 lumens per watt over its intended lifespan of 1,000 h.

The filament lamp in its simplest form is a purely functional light source which is unchallenged on the domestic market despite the manufacture of more efficient lamps. One factor which may have contributed to its popularity is that lamp designers have been able to modify the glass envelope of the lamp to give a very pleasing decorative appearance, as shown in Fig. 3.43.

Key fact

Energy efficiency

Over the next few years the government will phase out GLS lamps so that we will have to use more energy-efficient lamps such as compact fluorescent lamps (CFLs).

Definition

GLS lamps produce light as a result of the heating effect of an electrical current. Most of the electricity goes to producing heat and a little to producing light. A fine tungsten wire is first coiled and coiled again to form the incandescent filament of the GLS lamp.

Tungsten halogen lamps and the halogen cycle

The high operating temperature of the tungsten filament in the GLS lamp causes some evaporation of the tungsten which is carried by convection currents onto the bulb wall. When the lamp has been in service for some time, evaporated tungsten darkens the bulb wall, the light output is reduced, and the filament becomes thinner and eventually fails.

To overcome these problems the envelope of the tungsten halogen lamp contains a trace of one of the halogen gases, iodine, chlorine, bromine or fluorine. This allows a reversible chemical action to occur between the tungsten filament and the halogen gas.

When tungsten is evaporated from the incandescent filament, some part of it spreads out towards the bulb wall but, at a point close to the wall where the temperature conditions are favourable, the tungsten combines with the halogen. This tungsten halide molecule then drifts back towards the filament where it once more separates, depositing the tungsten back onto the filament, thereby leaving the halogen available for a further reaction cycle.

Since all the evaporated tungsten is returned to the filament, the bulb blackening normally associated with tungsten lamps is completely eliminated and a high efficacy is maintained throughout the life of the lamp.

A minimum bulb wall temperature of 250°C is required to maintain the halogen cycle and consequently a small glass envelope is required. This also permits a much higher gas pressure to be used which increases the lamp life to 2000 hours and allows the filament to be operated at a higher temperature, giving more light. The lamp is very small and produces a very white intense light giving it a colour rendering classification of good and an efficacy of 20 lumens per watt.

The tungsten halogen lamp shown in Fig. 3.44 was a major development in lamp design and resulted in Thorn Lighting gaining the Queens Award for Technical Innovation in Industry in 1972.

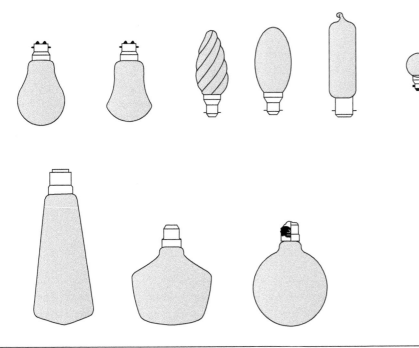

Figure 3.43 Some decorative GLS lamp shapes.

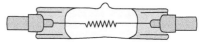

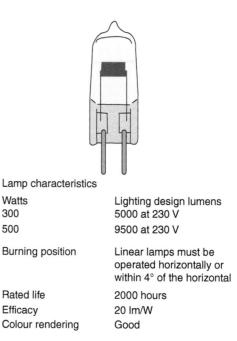

Lamp characteristics

Watts	Lighting design lumens
300	5000 at 230 V
500	9500 at 230 V
Burning position	Linear lamps must be operated horizontally or within 4° of the horizontal
Rated life	2000 hours
Efficacy	20 lm/W
Colour rendering	Good

Figure 3.44 A tungsten halogen lamp.

When installing the lamp, grease contamination of the glass envelope by touching must be avoided. Any grease present on the outer surface will cause cracking and premature failure of the lamp because of the high operating temperatures. The lamp should be installed using the paper sleeve and, if accidentally touched with bare hands, the lamp should be cleaned with metholated spirit to remove the grease.

Tungsten halogen dichroic reflector miniature spot lamps

Tungsten halogen dichroic reflector miniature spot lamps such as the one shown in Fig. 3.45 are extremely popular in the lighting schemes of the new millennium. Their small size and bright white illumination makes them very popular in both commercial and domestic installations. They are available as a 12 V bi-pin package in 20, 35 and 50 W and as a 230 V bayonet type cap (called a GU10 or GZ10 cap) in 20, 35 and 50 W. The number 10 in the GU10 description indicates that there is 10mm between the two pins. At 20 lumens of light output over its intended lifespan of 2000h they are more energy efficient than GLS lamps. However, only lamps offering more than 40 lumens of light output are considered energy efficient by the government's new criteria.

Figure 3.45 Tungsten halogen dichroic reflector lamp.

Definition

Discharge lamps do not produce light by means of an incandescent filament but by the excitation of a gas or metallic vapour contained within a glass envelope.

Discharge lamps

Discharge lamps do not produce light by means of an incandescent filament but by the excitation of a gas or metallic vapour contained within a glass envelope. A voltage applied to two terminals or electrodes sealed into the end of a glass tube containing a gas or metallic vapour will excite the contents and produce light directly. The colour of the light produced depends upon the type of gas or metallic vapour contained within the tube. Some examples are:

Gases	neon	red
	argon	green/blue
	hydrogen	pink
	helium	ivory
	mercury	blue
Metallic vapours	sodium	yellow
	magnesium	grass green

Fluorescent tubes and CFLs operate on this principle.

Fluorescent luminaires and lamps

A luminaire is equipment which supports an electric lamp and distributes or filters the light created by the lamp. It is essentially the 'light fitting'. The 18th edition of the Regulations at 411.3.4 now requires all circuits supplying luminaires to be protected by an RCD rated at 30mA.

A lamp is a device for converting electrical energy into light energy. There are many types of lamps. General lighting service (GLS) lamps and tungsten halogen lamps use a very hot wire filament to create the light and so they also become very hot in use. Fluorescent tubes operate on the 'discharge' principle; that is, the excitation of a gas within a glass tube. They are cooler in operation and very efficient in converting electricity into light. LEDs and CFLs today form the basic principle of most energy-efficient lamps.

Fluorescent lamps are linear arc tubes, internally coated with a fluorescent powder, containing a little low-pressure mercury vapour and argon gas. The lamp construction is shown in Fig. 3.46.

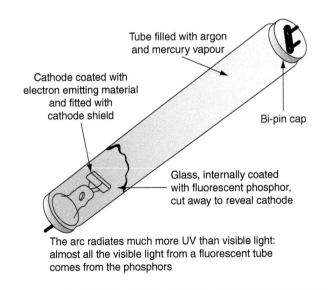

Tube filled with argon and mercury vapour

Cathode coated with electron emitting material and fitted with cathode shield

Bi-pin cap

Glass, internally coated with fluorescent phosphor, cut away to reveal cathode

The arc radiates much more UV than visible light: almost all the visible light from a fluorescent tube comes from the phosphors

Figure 3.46 Fluorescent lamp construction.

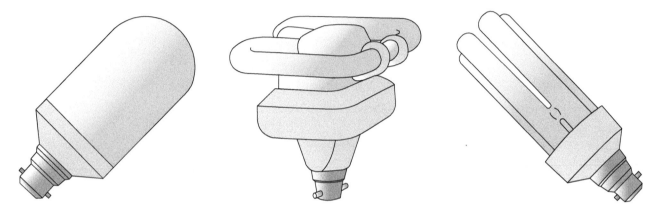

Figure 3.47 Fluorescent lamp circuit arrangement.

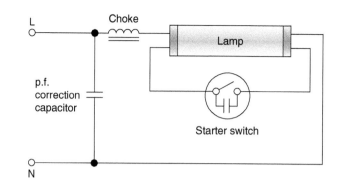

Figure 3.48 Energy-efficient lamps.

Passing a current through the electrodes of the tube produces a cloud of electrons that ionize the mercury vapour and the argon in the tube, producing invisible ultraviolet light and some blue light. The fluorescent powder on the inside of the glass tube is very sensitive to ultraviolet rays and converts this radiation into visible light.

Fluorescent luminaires require a simple electrical circuit to initiate the ionization of the gas in the tube and a device to control the current once the arc is struck and the lamp is illuminated. Such a circuit is shown in Fig. 3.47.

A typical application for a fluorescent luminaire is in suspended ceiling lighting modules used in many commercial buildings. Energy-efficient lamps use electricity much more efficiently.

Compact fluorescent lamps

CFLs are miniature fluorescent lamps designed to replace ordinary GLS lamps. They are available in a variety of shapes and sizes so that they can be fitted into existing light fittings. Figure 3.48 shows three typical shapes. The 'stick' type give most of their light output radially while the flat 'double D' type give most of their light output above and below.

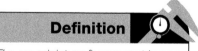

Definition

CFLs are miniature fluorescent lamps designed to replace ordinary GLS lamps.

LED lamps

Light emitting diodes (LEDs) have been around for more than fifty years as the small signal indicator devices shown in Figure 3.90, However recent scientific research and new technology, have enabled the super efficient LED lamps that

Figure 3.49 Energy-efficient lamps use less electricity.

we now know to be created since we started to phase out incandescent GLS lamps in 2010 on the grounds of improved energy efficiency. LED lamps have come such a long way in recent years and are now probably the first choice for energy efficient lamp and luminaire replacement.

A light emitting diode (LED) is a very small, semi-conductor source of illumination. It is a P-N junction diode that emits photons of light when activated, called electroeluminence.

LED lamps have many advantages over GLS and compact fluorescent lamps (CFLs) including very low energy consumption, greater robustness, smaller size and longer life and they reach full brilliance immediately at switch on. They are available in a range of shapes and sizes with Edison Screw, Bayonet Cap, Bi-Pin, GU 10 and GX 53 connections, linear tubes and modules, so that they may easily replace existing lamp types.

A recent research paper from Cambridge University titled 'Lighting for the 21st Century' tells us that lighting in the UK currently consumes 20% of all the power generated by UK power stations, and that LEDs have the potential to reduce this figure by 50%.Table 3.3 shows a comparison of the power consumed by common lamp types, at various levels of brightness. You can see how much energy is lost as heat when comparing a GLS lamp with an LED lamp for the same levels of brightness.

Table 3.3 The table shows a comparison of the power consumed by various lamp types. Note, the Lumen is a measure of the brightness or volume of light output from a light source.

Light output/ brightness	220-240 Lumens	400-500 Lumens	700-900 Lumens	900-1125 Lumens	1200-1500 Lumens
GLS Lamp	25 W	40 W	60 W	75 W	100 W
Halogen	18 W	28 W	42 W	53 W	70 W
CFL	6 W	9 W	12 W	15 W	20 W
LED	2-4 W	4-6 W	7-10 W	9-13 W	13-16 W

When installing LED Lamps and luminaires, the installer must take account of the colour of the light emitted by the LED lamp, cold white or warm white, for a particular application. Manufacturers describe a light source with an orange or yellow tint as a 'warm white' and those with a blue tint as a 'cool white'. The colour of the light emitted by an LED lamp is measured in Kelvin, a temperature measurement. The higher the number, the whiter and colder the light output appears to the eye. A light output of 3,000K and above gives a bright white, bluish light suitable for installation in contemporary kitchens and dentist surgeries. A light output of 2,700K and below gives a much warmer yellowish light, suitable for table lamps with shades in a domestic situation. A 2,700K LED lamp almost replicates the soft yellow/white light output of a GLS filament lamp but consumes much less energy. Table 3.4 shows the comparison between the temperature region of an LED lamp, a description of the colour and typical applications.

Table 3.4 An example of the temperature, colour and appearance of an LED lamp

Lamp temperature (K)	Colour	Colour Appearance	Applications
2000 to 3000	warm white	white with Yellow tint	bedrooms, sitting rooms, lounge areas
3100 to 4500	cool white	neutral white	kitchens, bathroom
4600 to 6500	daylight	bright white/cold blue tint	Dentist surgeries, security lighting, medical/clinical applications

Although manufacturers and suppliers are claiming a very long life for LED products, electrical contractors are reporting very early failures and shorter product life. Manufacturers suggest this may be because the products are not being installed in accordance with the manufacturer's instructions. Designers and installers should note Regulation 134.1.1 which advises 'the installation of electrical equipment shall take account of the manufacturer's instructions'.

Finally, electrical designers should note that LED lamps, strips and tapes can exhibit the flickering and stroboscopic effects similar to discharge lamps which are discussed later in this chapter. So, when installing LED lighting systems close to rotating machinery, the designer should seek the advice of the lamp manufacturer to avoid the stroboscopic effect.

High-pressure mercury vapour lamp

The high-pressure mercury discharge takes place in a quartz glass arc tube contained within an outer bulb which, in the case of the lamp classified as MBF, is internally coated with fluorescent powder. The lamp's construction and characteristics are shown in Fig. 3.50.

The inner discharge tube contains the mercury vapour and a small amount of argon gas to assist starting. The main electrodes are positioned at either end of the tube and a starting electrode is positioned close to one main electrode.

When the supply is switched on the current is insufficient to initiate a discharge between the main electrodes, but ionization does occur between the starting electrode and one main electrode in the argon gas.

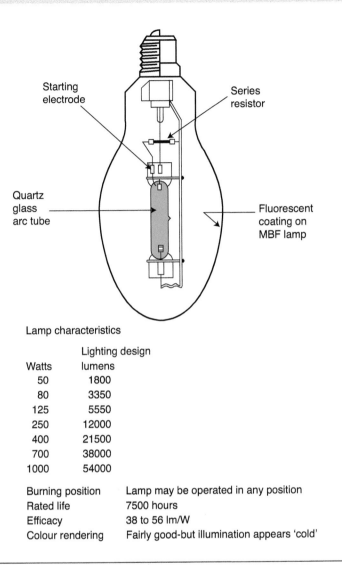

Starting electrode

Series resistor

Quartz glass arc tube

Fluorescent coating on MBF lamp

Lamp characteristics

Watts	Lighting design lumens
50	1800
80	3350
125	5550
250	12000
400	21500
700	38000
1000	54000

Burning position	Lamp may be operated in any position
Rated life	7500 hours
Efficacy	38 to 56 lm/W
Colour rendering	Fairly good-but illumination appears 'cold'

Figure 3.50 High-pressure mercury vapour lamp.

This spreads through the arc tube to the other main electrode. As the lamp warms the mercury is vaporized, the pressure builds up and the lamp achieves full brilliance after about five to seven minutes.

If the supply is switched off the lamp cannot be relit until the pressure in the arc tube has reduced. It may take a further five minutes to restrike the lamp.

The lamp is used for commercial and industrial installations, street lighting, shopping centre illumination and area floodlighting.

Metal halide lamps

Metal halide lamps are high-pressure mercury vapour lamps in which metal halide chemical compounds have been added to the arc tube. This improves the colour-rendering properties of the lamp making it a better artificial light source for photography.

Low-pressure sodium lamps

The low-pressure sodium discharge takes place in a U-shaped arc tube made of special glass which is resistant to sodium attack. This U-tube is encased in a tubular outer bulb of clear glass as shown in Fig. 3.51. Lamps classified as type

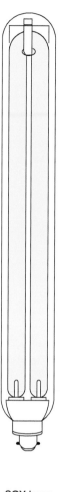

SOX lamp   SLI/H lamp

Lamp characteristics

Watts	Lighting design lumens
Type SOX	
35	4300
55	7500
90	12500
135	21500
Type SLI/H	
140	20000
200	25000
200 HO	27500
Burning position	Horizontal or within 20° of the horizontal
Rated life	6000 hours
Guaranteed life	4000 hours
Efficacy	61 to 160 lm/W
Colour rendering	Very poor – illumination very yellow

Figure 3.51 Low-pressure sodium lamp.

SOX have a BC lampholder while the SL1/H lamp has a bi-pin lampholder at each end.

Since at room temperature the pressure of sodium is very low, a discharge cannot be initiated in sodium vapour alone. Therefore, the arc tube also contains neon gas to start the lamp. The arc path of the low-pressure sodium lamp is much longer than that of mercury lamps and starting is achieved by imposing a high voltage equal to about twice the main voltage across the electrodes by means of a leakage transformer. This voltage initiates a red discharge in the neon gas which heats up the sodium. The sodium vaporizes and over a period of six to 11 minutes the lamp reaches full brilliance, changing colour from red to bright yellow.

The lamp must be operated horizontally so that when the lamp is switched off the condensing sodium is evenly distributed around the U-tube.

The light output is yellow and has poor colour-rendering properties but this is compensated by the fact that the wavelength of the light is close to that at which the human eye has its maximum sensitivity, giving the lamp a high efficacy. The main application for this lamp is street lighting where the light output meets the requirements of the Ministry of Transport.

High-pressure sodium lamp

The high-pressure sodium discharge takes place in a sintered aluminium oxide arc tube contained within a hard glass outer bulb. Until recently no suitable material was available which would withstand the extreme chemical activity of sodium at high pressure. The construction and characteristics of the high-pressure sodium lamp classified as type SON are given in Fig. 3.52.

The arc tube contains sodium and a small amount of argon or xenon to assist starting. When the lamp is switched on an electronic pulse igniter of 2 kV or more initiates a discharge in the starter gas. This heats up the sodium and in about five to seven minutes the sodium vaporizes and the lamp achieves full brilliance. Both colour and efficacy improve as the pressure of the sodium rises, giving a pleasant golden white colour to the light which is classified as having a fair colour-rendering quality.

The SON lamp is suitable for many applications. Because of the warming glow of the illuminance it is used in food halls and hotel reception areas. Also, because of the high efficacy and long lamp life it is used for high bay lighting in factories and warehouses and for area floodlighting at airports, car parks and dockyards.

Control gear for lamps

Luminaires are wired using the standard lighting circuits described in Chapter 3 of *Basic Electrical Installation Work 8th edition,* but discharge lamps require additional control gear and circuitry for their efficient and safe operation. The circuit diagrams for high-pressure mercury vapour and high- and low-pressure sodium lamps are given in Fig. 3.53. Each of these circuits requires the inclusion of a choke or transformer, creating a lagging power factor which must be corrected. This is usually achieved by connecting a capacitor across the supply to the luminaire as shown.

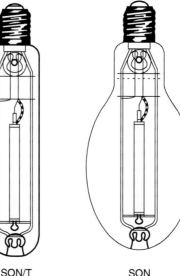

SON/T   SON

Lamp characteristics

	Lighting design lumens
Watts	
Tubular clear (SON/T)	
250	21000
400	38000
Elliptical coated (SON)	
250	19500
400	36000
Burning position	Universal
Rated life	6000 hours
Guaranteed life	4000 hours
Efficacy	100 to 120 lm/W
Colour rendering	Fair – illumination appears 'warm' and golden

Figure 3.52 High-pressure sodium lamp.

Fluorescent lamp control circuits

A fluorescent lamp requires some means of initiating the discharge in the tube, and a device to control the current once the arc is struck. Since the lamps are usually operated on a.c. supplies, these functions are usually achieved by means of a choke ballast. Three basic circuits are commonly used to achieve starting: switch-start, quick-start and semi-resonant.

Switch-start fluorescent lamp circuit

Figure 3.54 shows a switch-start fluorescent lamp circuit in which a glow-type starter switch is now standard. A glow-type starter switch consists of two bimetallic strip electrodes encased in a glass bulb containing an inert gas. The starter switch is initially open-circuit. When the supply is switched on the full mains voltage is developed across these contacts and a glow discharge takes

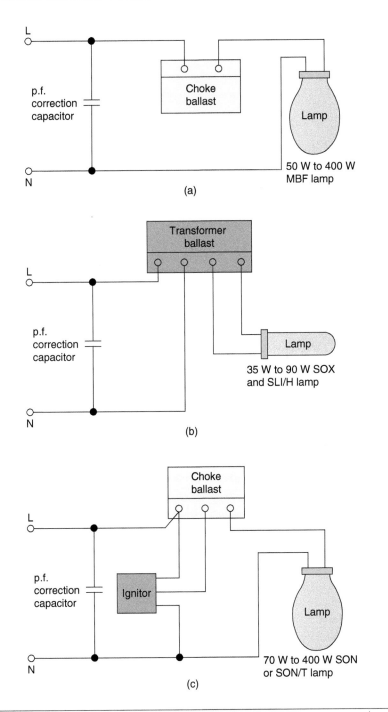

Figure 3.53 Discharge lamp control gear circuits: (a) high-pressure mercury vapour lamp; (b) low-pressure sodium lamp; (c) high-pressure sodium lamp.

place between them. This warms the switch electrodes and they bend towards each other until the switch makes contact. This allows current to flow through the lamp electrodes, which become heated so that a cloud of electrons is formed at each end of the tube, which in turn glows.

When the contacts in the starter switch are made the glow discharge between the contacts is extinguished since no voltage is developed across the switch. The starter switch contacts cool and after a few seconds spring apart. Since there is a choke in series with the lamp, the breaking of this inductive circuit causes a voltage surge across the lamp electrodes which is sufficient to strike the main arc in the tube. If the lamp does not strike first time the process is repeated.

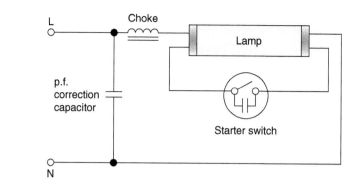

Figure 3.54 Switch-start fluorescent lamp circuit.

When the main arc has been struck in the low-pressure mercury vapour, the current is limited by the choke. The capacitor across the mains supply provides power-factor correction and the capacitor across the starter switch contact is for radio interference suppression.

Quick-start fluorescent lamp circuit

When the circuit is switched on the tube cathodes are heated by a small auto-transformer. After a short pre-heating period the mercury vapour is ionized and spreads rapidly through the tube to strike the arc. The luminaire or some earthed metal must be in close proximity to the lamp to assist in the striking of the main arc.

When the main arc has been struck the current flowing in the circuit is limited by the choke. A capacitor connected across the supply provides power-factor correction. The circuit is shown in Fig. 3.55.

Semi-resonant start fluorescent lamp circuit

In this circuit a specially wound transformer takes the place of the choke. When the circuit is switched on, a current flows through the primary winding to one cathode of the lamp, through the secondary winding and a large capacitor to the other cathode. The secondary winding is wound in opposition to the primary winding. Therefore, the voltage developed across the transformer windings is 180° out of phase.The current flowing through the electrodes causes an electron cloud to form around each cathode. This cloud spreads rapidly through the tube

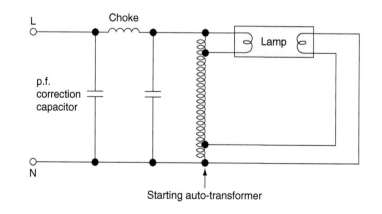

Figure 3.55 Quick-start fluorescent lamp circuit.

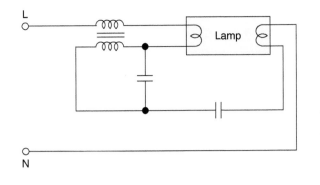

Figure 3.56 Semi-resonant start fluorescent lamp circuit.

due to the voltage across the tube being increased by winding the transformer windings in opposition. When the main arc has been struck the current is limited by the primary winding of the transformer which behaves as a choke. A power-factor correction capacitor is not necessary since the circuit is predominantly capacitive and has a high power factor. With the luminaire earthed to assist starting this circuit will start very easily at temperatures as low as –5°C. The circuit is shown in Fig. 3.56.

Installation of discharge luminaires

Operating position

Some lamps, particularly discharge lamps, have limitations placed upon their operating position. Since the luminaire is designed to support the lamp, any restrictions upon the operating position of the lamp will affect the position of the luminaire. Some indications of the operating position of lamps were given earlier under the individual lamp characteristics. The luminaire must be suitable for the environment in which it is to operate. This may be a corrosive atmosphere, an outdoor situation or a low-temperature zone. On the other hand, the luminaire may be required to look attractive in a commercial environment. It must satisfy all these requirements while at the same time providing adequate illumination without glare.

Many lamps contain a wire filament and delicate supports enclosed in a glass envelope. The luminaire is designed to give adequate support to the lamp under normal conditions, but a luminaire subjected to excessive vibration will encourage the lamp to fail prematurely by either breaking the filament, cracking the glass envelope or breaking the lamp holder seal.

Control gear for discharge luminaires

Chokes and ballasts for discharge lamps have laminated sheet steel cores in which a constant reversal of the magnetic field due to the a.c. supply sets up vibrations. In most standard chokes the noise level is extremely low, and those manufactured to BS EN 60598 have a maximum permitted noise level of 30 dB. This noise level is about equal to the sound produced by a Swatch watch at a distance of one metre in a very quiet room.

Chokes must be rigidly fixed, otherwise metal fittings can amplify choke noise. Plasterboard, hardboard or wooden panels can also act as a sounding board for control gear or luminaires mounted upon them, thereby amplifying choke noise. The background noise will obviously affect people's ability to detect choke noise, and so control gear and luminaires which would be considered noisy in a library or church may be unnoticeable in a busy shop or office.

The cable, accessories and fixing box must be suitable for the mass suspended and the ambient temperature in accordance with IET Regulations Section 559. Self-contained luminaires must have an adjacent means of isolation provided in addition to the functional switch to facilitate safe maintenance and repair (Regulation 537.2) All circuits supplying luminairs must be protected by a 30mA RCD.

Control gear should be mounted as close as possible to the lamp. Where it is liable to cause overheating it must be either:

- enclosed in a suitably designed non-combustible enclosure; or
- mounted so as to allow heat to dissipate; or
- placed at a sufficient distance from adjacent materials to prevent the risk of fire (Regulations 559.4.1).

Discharge lighting may also cause a stroboscopic effect where rotating or reciprocating machinery is being used. This effect causes rotating machinery to appear stationary, and we will consider the elimination of this dangerous effect next.

Stroboscopic effect (IET Regulation 559.9)

The flicker effect of any light source can lead to the risk of a stroboscopic effect. This causes rotating or reciprocating machinery to appear to be running at speeds other than their actual speed, and in extreme cases a circular saw or lathe chuck may appear stationary when rotating. A stroboscopic light is used to good effect when electronically 'timing' a car, by making the crank shaft appear stationary when the engine is running so that the top dead centre position (TDC) may be found.

All discharge lamps used on a.c. circuits flicker, often unobtrusively, due to the arc being extinguished every half-cycle as the lamp current passes through zero. This variation in current and light output is illustrated in Fig. 3.57.

The elimination of this flicker is desirable in all commercial installations and particularly those which use rotating machinery. Fluorescent phosphors with a long afterglow help to eliminate the brightness peaks. Where a three-phase supply is available, discharge lamps may be connected to a different phase, so that they reach their brightness peaks at different times.The combined effect is to produce a much more uniform overall level of illumination, which eliminates the flicker effect.

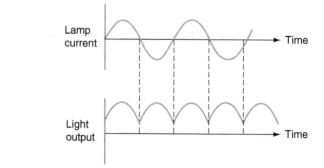

Figure 3.57 Variation of current and light output for a discharge lamp.

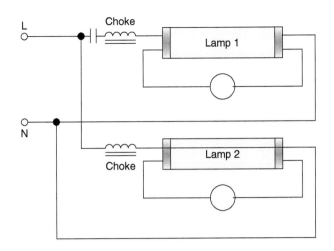

Figure 3.58 Circuit diagrams for lead-lag fluorescent lamp circuits.

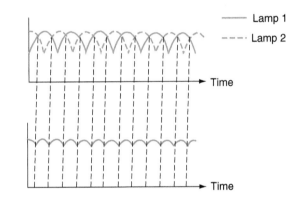

Figure 3.59 Elimination of stroboscopic effect.

When only a single-phase supply is available, adjacent fluorescent tubes in twin fittings may be connected to a lead-lag circuit, as illustrated in Fig. 3.58. This is one lamp connected in series with a choke and the other in series with a choke and capacitor, which causes the currents in the two lamps to differ by between 120 and 180 degrees. The lamps flicker out of phase, producing a more uniform level of illumination which eliminates the stroboscopic effect, as shown in Fig. 3.59.

GLS lamps do not flicker because the incandescent filament produces a carrying over of the light output as the current reverses. Thus the light remains almost constant and, therefore, the stroboscopic flicker effect is not apparent.

Loading and switching of discharge circuits

Discharge circuits must be capable of carrying the total steady current (the current required by the lamp plus the current required by any control gear). Appendix 1 of the *On Site Guide* states that where more exact information is not available, the rating of the final circuits for discharge lamps may be taken as the rated lamp wattage multiplied by 1.8. Therefore, an 80 W fluorescent lamp luminaire will have an assumed demand of $80 \times 1.8 = 144$ W.

All discharge lighting circuits are inductive and will cause excessive wear of the functional switch contacts. Where discharge lighting circuits are to be switched, the rating of the functional switch must be suitable for breaking an inductive load. To comply with this requirement we usually assume that the rating of the functional switch should be *twice* the total steady current of the inductive circuit from information previously given in the 15th edition of the IET Regulations. The 18th edition of the Regulations at 537.3.1.2 says it must be suitable for the most onerous duty it is intended to perform.

Maintenance of discharge lighting installations

The extent of the maintenance required will depend upon the size and type of lighting installation and the installed conditions – for example, whether the environment is dusty or clean. However, lamps and luminaires will need cleaning, and lamps will need to be replaced. They can be replaced either by 'spot replacement' or 'group replacement'.

Spot replacement is the replacement of individual lamps as and when they fail. This is probably the most suitable method of maintaining small lighting installations in shops, offices and nursing homes, and is certainly the preferred method in domestic property. But each time a lamp is replaced or cleaned, a small disturbance occurs to the normal environment. Access equipment must be set up, furniture must be moved and the electrician chats to the people around him. In some environments this small disturbance and potential hazard from someone working on steps or a mobile tower is not acceptable and, therefore, group replacement of lamps must be considered.

Group replacement is the replacement of all the lamps at the same time. The time interval for lamp replacement is determined by taking into account the manufacturer's rated life for the particular lamp and the number of hours each week that the lamp is illuminated.

Group replacement and cleaning can reduce labour costs and inconvenience to customers and is the preferred method for big stores and major retail outlets. Group replacement in these commercial installations is often carried out at night or at the weekend to avoid a disturbance of the normal working environment.

Water heating circuits

A small, single-point over-sink-type water heater may be considered as a permanently connected appliance and so may be connected to a ring circuit through a fused connection unit. A water heater of the immersion type is usually rated at a maximum of 3 kW, and could be considered as a permanently connected appliance, fed from a fused connection unit. However, many immersion heating systems are connected into storage vessels of about 150 litres in domestic installations, and the *On Site Guide* states that immersion heaters fitted to vessels in excess of 15 litres should be supplied by their own circuit.

Therefore, immersion heaters must be wired on a separate radial circuit when they are connected to water vessels which hold more than 15 litres. Figure 3.60 shows the wiring arrangements for an immersion heater. Every switch must be a double-pole (DP) switch and out of reach of anyone using a fixed bath or shower when the immersion heater is fitted to a vessel in a bathroom.

Supplementary equipotential bonding to pipework *will only be required* as an addition to fault protection (IET Regulation 415.2) if the immersion heater vessel is in a bathroom that *does not have*:

- all circuits protected by a 30 mA RCD; **and**
- protective equipotential bonding (IET Regulation 701.415.2), as shown in Fig. 5.15.

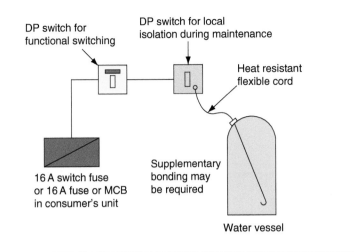

DP switch for
functional switching

DP switch for local
isolation during maintenance

Heat resistant
flexible cord

16 A switch fuse
or 16 A fuse or MCB
in consumer's unit

Supplementary
bonding may
be required

Water vessel

Figure 3.60 Immersion heater wiring.

Electric space heating circuits

Electrical heating systems can be broadly divided into two categories: unrestricted local heating and off-peak heating.

Unrestricted local heating may be provided by portable electric radiators which plug into the socket outlets of the installation. Fixed heaters that are wall mounted or inset must be connected through a fused connection and incorporate a local switch, either on the heater itself or as a part of the fuse connecting unit. Heating appliances where the heating element can be touched must have a DP switch which disconnects all conductors. This requirement includes radiators which have an element inside a silica-glass sheath.

Off-peak heating systems may provide central heating from storage radiators, ducted warm air or underfloor heating elements. All three systems use the thermal storage principle, whereby a large mass of heat-retaining material is heated during the off-peak period and allowed to emit the stored heat throughout the day. The final circuits of all off-peak heating installations must be fed from a separate supply controlled by the electricity supplier's time clock.

When calculating the size of cable required to supply a single-storage radiator, it is good practice to assume a current demand equal to 3.4 kW at each point. This will allow the radiator to be changed at a future time with the minimum disturbance to the installation. Each radiator must have a 20 A DP means of isolation adjacent to the heater and the final connection should be via a flex outlet. See Fig. 3.61 for wiring arrangements.

Ducted warm air systems have a centrally sited thermal storage heater with a high storage capacity. The unit is charged during the off-peak period, and a fan drives the stored heat in the form of warm air through large air ducts to outlet grilles in the various rooms. The wiring arrangements for this type of heating are shown in Fig. 3.62.

The single-storage heater is heated by an electric element embedded in bricks and rated between 6 and 15 kW depending upon its thermal capacity. A radiator of this capacity must be supplied on its own circuit, in cable capable of carrying the maximum current demand and protected by a fuse or miniature

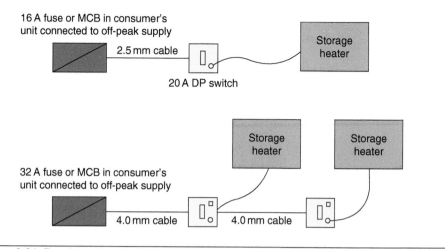

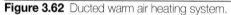

Figure 3.61 Possible wiring arrangements for storage heaters.

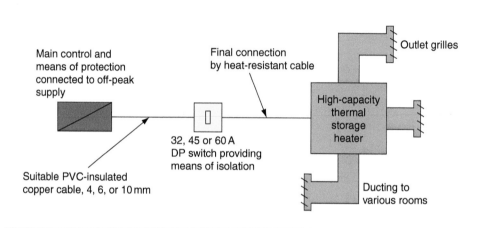

Figure 3.62 Ducted warm air heating system.

circuit-breaker (MCB) of 30, 45 or 60 A as appropriate. At the heater position, a DP switch must be installed to terminate the fixed heater wiring. The flexible cables used for the final connection to the heaters must be of the heat-resistant type.

Floor-warming installations use the thermal storage properties of concrete. Special cables are embedded in the concrete floor screed during construction. When current is passed through the cables they become heated, the concrete absorbs this heat and radiates it into the room. The wiring arrangements are shown in Fig. 3.63. Once heated, the concrete will give off heat for a long time after the supply is switched off and is, therefore, suitable for connection to an off-peak supply.

Underfloor heating cables installed in bathrooms or shower rooms must incorporate an earthed metallic sheath or be covered by an earthed metallic grid connected to the protective conductor of the supply circuit (IET Regulation 701.753).

Electrical heating controls

Water heating or space heating systems can be controlled by thermostats, time switches, programmers, or a combination of the three. Let us first look at the science of temperature and heat.

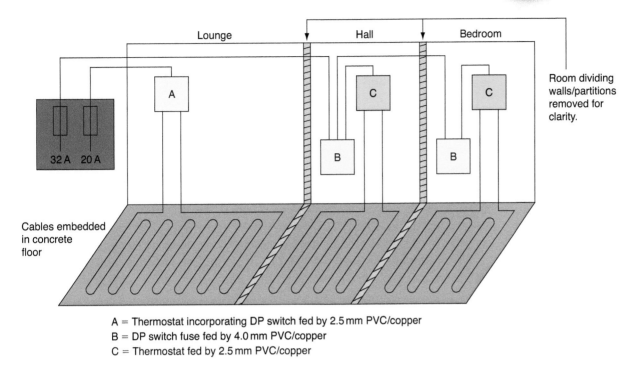

A = Thermostat incorporating DP switch fed by 2.5 mm PVC/copper
B = DP switch fuse fed by 4.0 mm PVC/copper
C = Thermostat fed by 2.5 mm PVC/copper

Figure 3.63 Floor-warming installations.

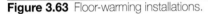

Temperature and heat transfer

Heat has the capacity to do work and is a form of energy. Temperature is not an energy unit but describes the hotness or coldness of a substance or material.

Heat energy is transferred by three separate processes which can occur individually or in combination. The processes are convection, radiation and conduction.

Convection

Air which passes over a heated surface expands, becomes lighter and warmer and rises, being replaced by descending cooler air. These circulating currents of air are called *convection currents*. In circulating, the warm air gives up some of its heat to the surfaces over which it passes and so warms a room and its contents. The same process takes place in water. The immersion heater in a water vessel such as that shown in Fig. 3.60 heats up the water in contact with the heater element which rises and is replaced by cold water, which then heats up and rises and so on. This process sets up circulating 'convection' currents in the water which heats up the contents of the whole vessel.

Radiation

Molecules on a metal surface vibrating with thermal energy generate electromagnetic waves. The waves travel away from the surface at the speed of light, taking energy with them and leaving the surface cooler. If the electromagnetic waves meet another body or material they produce a disturbance of the surface molecules; this then raises the temperature of that body or material. Radiated heat requires no intervening medium between the transmitter and receiver, obeys the same laws as light energy and is the method by which the energy from the sun reaches the earth.

Conduction

Heat transferred through a material by conduction occurs because there is direct contact between the vibrating molecules of the material. The application of a heat source to the end of a metal bar causes the atoms to vibrate rapidly within the lattice framework of the material. This violent shaking causes adjacent atoms to vibrate and liberates any loosely bound electrons which also pass on the heat energy. Thus the heat energy travels through the material by *conduction*.

Temperature scales

When planning any scale of measurement a set of reference points must first be established. Two obvious reference points on any temperature scale are the temperature at which ice melts and water boils. Between these two points the scale is divided into a convenient number of divisions. In the case of the Celsius or Centigrade scale the lower fixed point is zero degrees and the upper fixed point 100 degrees Celsius. The Kelvin takes as its lower fixed point the lowest possible temperature which has a value of 273°C, called absolute zero. A temperature change of one Kelvin is exactly the same as one degree Celsius and so we can say that

$$0°C = 273 \text{ K} \qquad \text{or} \qquad 0 \text{ K} = -273°C$$

Temperature measurement

One instrument which measures temperature is a thermometer. This uses the properties of an expanding liquid in a glass tube to indicate a temperature level. Most materials change their dimensions when heated and this property is often used to give a measure of temperature. Many materials expand with an increase in temperature and the *rate* of expansion varies with different materials.

A bimetal strip is formed from two dissimilar metals joined together. When the temperature increases the metals expand at different rates and the bimetal strip bends or bows.

Thermostats

A thermostat is a device for maintaining a constant temperature at some predetermined value. The operation of a thermostat is often based on the principle of differential expansion between dissimilar metals; that is, a bimetal strip, which causes a contact to make or break at a chosen temperature. Figure 3.64 shows the principle of a rod-type thermostat which is often used with water heaters. An invar rod, which has minimal expansion when heated, is housed within a copper tube and the two metals are brazed together at one end. The other end of the copper tube is secured to one end of the switch mechanism. As the copper expands and contracts under the influence of a varying temperature, the switch contacts make and break. In the case of a thermostat such as this, the electrical circuit is broken when the temperature setting is reached. Figure 3.65 shows a room thermostat which works on the same principle.

Programmers

A heating system may be fuelled by gas, oil, electricity or solar, but whatever the fuel system providing the heat energy, it is very likely that the system will be controlled by an electrical or electronic programmer, thermostats and a circulating pump for a water-based system. Fig 3.66 shows a block diagram for a typical domestic space heating and water heating system.

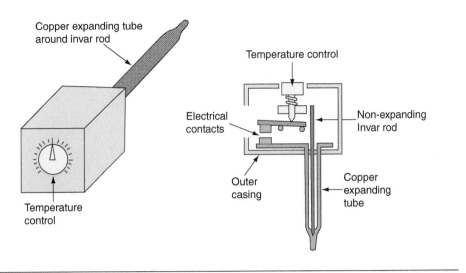

Figure 3.64 A rod-type thermostat.

Figure 3.65 Room thermostat.

The programmer will incorporate a time clock so that the customer or client can select the number of hours each day that the system will operate. For a working family in a domestic property this will probably be a couple of hours before the working day begins, and then five to seven hours in the evening. In a commercial situation it will probably switch on a couple of hours before the working day begins and remain on until the working day ends. In both situations the programmer unit often incorporates an override or boost facility for changing circumstances and weather conditions.

During the period when the space heating or water heating is in the 'on position', thermostats based on the bimetal strip principle will maintain the room and water temperature at some predetermined level. Figures 3.67 and 3.68 show a wiring diagram for a domestic space and water heating system which circulates water

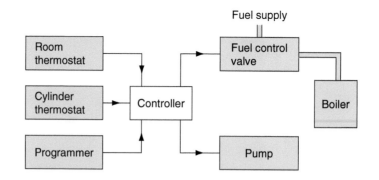

Figure 3.66 Block diagram: space heating control system (Honeywell Y plan).

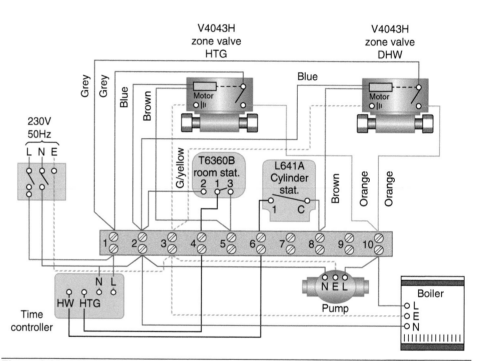

Figure 3.67 Wiring diagram – space heating control system (Honeywell S Plan).

through radiators to warm individual rooms. The Y plan uses one, three way water control valve. With this system the user may select hot water only or, hot water plus space heating. The S plan uses two separate control valves and therefore the user may select hot water only, space heating only or both together. The S plan gives the user more control and choice. In this type of system, energy conservation now recommends that each radiator be additionally fitted with a TRV (thermostatic radiator valve) to control the temperature in each room.

Basic electronics

There are numerous types of electronic component – diodes, transistors, thyristors and integrated circuits (ICs) – each with its own limitations, characteristics and designed application. When repairing electronic circuits it is important to replace a damaged component with an identical or equivalent component. Manufacturers issue comprehensive catalogues with details of working voltage, current, power dissipation, etc., and the reference numbers of equivalent components. These catalogues of information, together with a

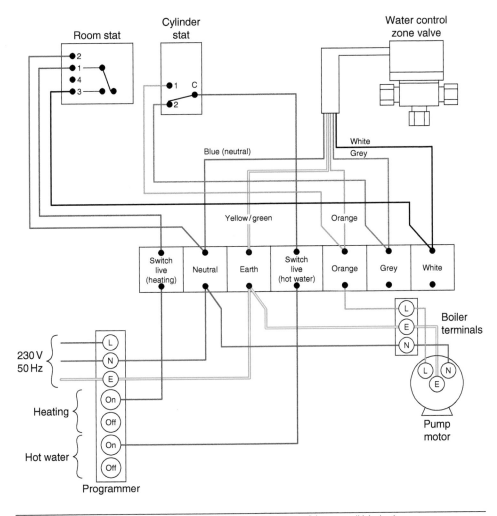

Room stat
Cylinder stat
Water control zone valve

Figure 3.68 Wiring diagram: space heating control system (Honeywell Y plan).

high-impedance multimeter, should form a part of the extended toolkit for anyone in the electrical industries proposing to repair electronic circuits.

Electronic circuit symbols

The British Standard BS EN 60617 recommends that particular graphical symbols be used to represent a range of electronic components on circuit diagrams. The same British Standard recommends a range of symbols suitable for electrical installation circuits with which electricians will already be familiar. Figure 3.70 shows a selection of electronic symbols.

Resistors

All materials have some resistance to the flow of an electric current but, in general, the term **resistor** describes a conductor specially chosen for its resistive properties.

Resistors are the most commonly used electronic component and they are made in a variety of ways to suit the particular type of application. They are usually manufactured as either carbon composition or carbon film. In both cases the base resistive material is carbon and the general appearance is of a small cylinder with leads protruding from each end, as shown in Fig. 3.71(a).

> **Definition**
>
> All materials have some resistance to the flow of an electric current but, in general, the term *resistor* describes a conductor specially chosen for its resistive properties.

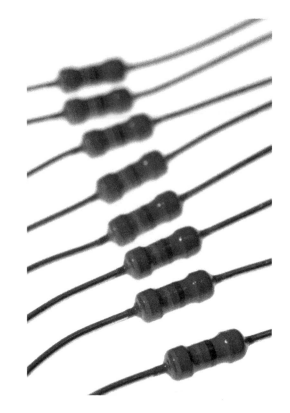

Figure 3.69 Resistors used in electronic circuits.

If subjected to overload, carbon resistors usually decrease in resistance since carbon has a negative temperature coefficient. This causes more current to flow through the resistor, so that the temperature rises and failure occurs, usually by fracturing. Carbon resistors have a power rating between 0.1 and 2 W, which should not be exceeded.

When a resistor of a larger power rating is required a wire-wound resistor should be chosen. This consists of a resistance wire of known value wound on a small ceramic cylinder which is encapsulated in a vitreous enamel coating, as shown in Fig. 3.71(b). Wire-wound resistors are designed to run hot and have a power rating up to 20 W. Care should be taken when mounting wire-wound resistors to prevent the high operating temperature from affecting any surrounding components.

A variable resistor is one which can be varied continuously from a very low value to the full rated resistance. This characteristic is required in tuning circuits to adjust the signal or voltage level for brightness, volume or tone. The most common type used in electronic work has a circular carbon track contacted by a metal wiper arm. The wiper arm can be adjusted by means of an adjusting shaft (rotary type) or by placing a screwdriver in a slot (preset type), as shown in Fig. 3.72. Variable resistors are also known as potentiometers because they can be used to adjust the potential difference (voltage) in a circuit. The variation in resistance can be either a logarithmic or a linear scale.

The value of the resistor and the tolerance may be marked on the body of the component either by direct numerical indication or by using a standard colour code. The method used will depend upon the type, physical size and manufacturer's preference, but in general the larger components have values marked directly on the body and the smaller components use the standard resistor colour code.

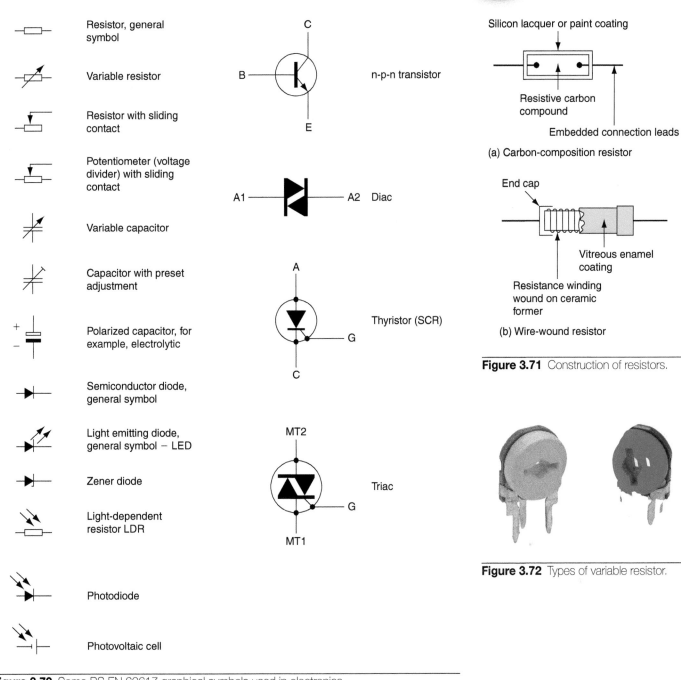

Resistor, general symbol

Variable resistor

Resistor with sliding contact

Potentiometer (voltage divider) with sliding contact

Variable capacitor

Capacitor with preset adjustment

Polarized capacitor, for example, electrolytic

Semiconductor diode, general symbol

Light emitting diode, general symbol – LED

Zener diode

Light-dependent resistor LDR

Photodiode

Photovoltaic cell

n-p-n transistor

Diac

Thyristor (SCR)

Triac

Silicon lacquer or paint coating

Resistive carbon compound

Embedded connection leads

(a) Carbon-composition resistor

End cap

Vitreous enamel coating

Resistance winding wound on ceramic former

(b) Wire-wound resistor

Figure 3.71 Construction of resistors.

Figure 3.72 Types of variable resistor.

Figure 3.70 Some BS EN 60617 graphical symbols used in electronics.

Abbreviations used in electronics

Where the numerical value of a component includes a decimal point, it is standard practice to include the prefix for the multiplication factor in place of the decimal point, to avoid accidental marks being mistaken for decimal points. Multiplication factors and prefixes are dealt with in Chapter 9.

The abbreviation   R means × 1
k means × 1000
M means × 1,000,000

Therefore, a 4.7 kΩ resistor would be abbreviated to 4k7, a 5.6 Ω resistor to 5R6 and a 6.8 MΩ resistor to 6M8.

Tolerances may be indicated by adding a letter at the end of the printed code.

The abbreviation F means ±1%, G means ±2%, J means ±5%, K means ±10% and M means ±20%. Therefore a 4.7 kΩ resistor with a tolerance of 2% would be abbreviated to 4k7G. A 5.6 Ω resistor with a tolerance of 5% would be abbreviated to 5R6J. A 6.8 MΩ resistor with a 10% tolerance would be abbreviated to 6M8K.

This is the British Standard BS 1852 code which is recommended for indicating the values of resistors on circuit diagrams and components when their physical size permits.

The standard colour code

Small resistors are marked with a series of coloured bands, as shown in Table 3.3. These are read according to the standard colour code to determine the resistance. The bands are located on the component towards one end. If the resistor is turned so that this end is towards the left, the bands are then read from left to right. Band (a) gives the first number of the component value, band (b) the second number, band (c) the number of zeros to be added after the first two numbers and band (d) the resistor tolerance. If the bands are not clearly oriented towards one end, first identify the tolerance band and turn the resistor so that this is towards the right before commencing to read the colour code as described.

The tolerance band indicates the maximum tolerance variation in the declared value of resistance. Thus a 100 Ω resistor with a 5% tolerance will have a value somewhere between 95 and 105 Ω, since 5% of 100 Ω is 5 Ω.

Table 3.5 The resistor colour code

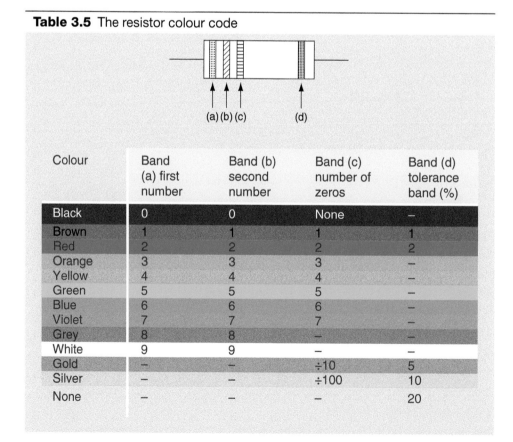

Colour	Band (a) first number	Band (b) second number	Band (c) number of zeros	Band (d) tolerance band (%)
Black	0	0	None	–
Brown	1	1	1	1
Red	2	2	2	2
Orange	3	3	3	–
Yellow	4	4	4	–
Green	5	5	5	–
Blue	6	6	6	–
Violet	7	7	7	–
Grey	8	8	–	–
White	9	9	–	–
Gold	–	–	÷10	5
Silver	–	–	÷100	10
None	–	–	–	20

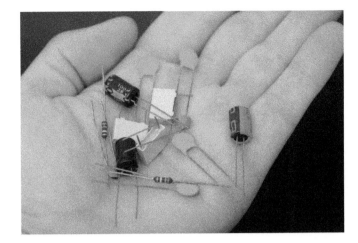

Figure 3.73 A handful of electronic resistors and capacitors.

Example 1

A resistor is colour coded yellow, violet, red, gold. Determine the value of the resistor.

Band (a) – yellow has a value of 4
Band (b) – violet has a value of 7
Band (c) – red has a value of 2
Band (d) – gold indicates a tolerance of 5%

The value is therefore 4700 ± 5%
This could be written as 4.7 kΩ ± 5% or 4 k7J.

Example 2

A resistor is colour coded green, blue, brown, silver. Determine the value of the resistor.

Band (a) – green has a value of 5
Band (b) – blue has a value of 6
Band (c) – brown has a value of 1
Band (d) – silver indicates a tolerance of 10%

The value is therefore 560 ± 10% and could be written as 560 Ω ± 10% or 560 RK.

Example 3

A resistor is colour coded blue, grey, green, gold. Determine the value of the resistor.

Band (a) – blue has a value of 6
Band (b) – grey has a value of 8
Band (c) – green has a value of 5
Band (d) – gold indicates a tolerance of 5%

The value is therefore 6,800,000 ± 5% and could be written as 6.8 M Ω ± 5% or 6 M8J.

Example 4

A resistor is colour coded orange, white, silver, silver. Determine the value of the resistor.

Band (a) – orange has a value of 3
Band (b) – white has a value of 9
Band (c) – silver indicates divide by 100 in this band
Band (d) – silver indicates a tolerance of 10%

The value is therefore 0.39 ± 10% and could be written as 0.39 Ω ± 10% or R39 K.

Try this

Electronics

Electricians are increasingly coming across electronic components and equipment. Make a list in the margin of some of the electronic components that you have come across at work.

Preferred values

It is difficult to manufacture small electronic resistors to exact values by mass-production methods. This is not a disadvantage as in most electronic circuits the value of the resistors is not critical. Manufacturers produce a limited range of *preferred* resistance values rather than an overwhelming number of individual resistance values. Therefore, in electronics, we use the preferred value closest to the actual value required.

A resistor with a preferred value of 100 Ω and a 10% tolerance could have any value between 90 and 110 Ω. The next larger preferred value which would give the maximum possible range of resistance values without too much overlap would be 120 Ω. This could have any value between 108 and 132 Ω. Therefore, these two preferred value resistors cover all possible resistance values between 90 and 132 Ω. The next preferred value would be 150 Ω, then 180, 220 Ω and so on.

There is a series of preferred values for each tolerance level, as shown in Table 3.4, so that every possible numerical value is covered. Table 3.4 indicates the values between 10 and 100, but larger values can be obtained by multiplying these preferred values by some multiplication factor. Resistance values of 47 Ω, 470 Ω, 4.7 k Ω, 470 k Ω, 4.7 M Ω, etc. are available in this way.

Testing resistors

The resistor being tested should have a value close to the preferred value and within the tolerance stated by the manufacturer. To measure the resistance of a resistor which is not connected into a circuit, the leads of a suitable ohmmeter should be connected to each resistor connection lead and a reading obtained.

If the resistor to be tested is connected into an electronic circuit it is *always necessary* to disconnect one lead from the circuit before the test leads are connected; otherwise the components in the circuit will provide parallel paths, and an incorrect reading will result.

Table 3.6 Preferred values

E6 series 20% tolerance	E12 series 10% tolerance	E24 series 5% tolerance
10	10	10
		11
	12	12
		13
15	15	15
		16
	18	18
		20
22	22	22
		24
	27	27
		30
33	33	33
		36
	39	39
		43
47	47	47
		51
	56	56
		62
68	68	68
		75
	82	82
		91

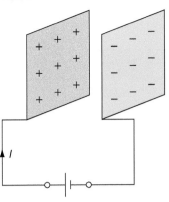

Charging condition

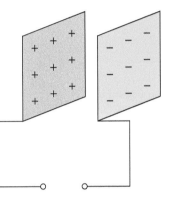

Charge trapped

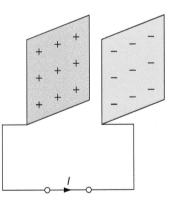

Discharge condition

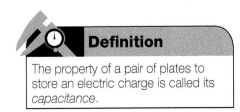

Figure 3.74 The charge on a capacitor's plates.

Electrostatics

If a battery is connected between two insulated plates, the emf of the battery forces electrons from one plate to another until the p.d. between the plates is equal to the battery emf.

The electrons flowing through the battery constitute a current, I (in amperes), which flows for a time, t (in seconds). The plates are then said to be charged.

The amount of charge transferred is given by:

$$Q = It \text{ (coulomb [Symbol C])}$$

When the voltage is removed the charge Q is trapped on the plates, but if the plates are joined together, the same quantity of electricity, $Q = It$, will flow back from one plate to the other, so discharging them. The property of a pair of plates to store an electric charge is called its **capacitance**.

Definition

The property of a pair of plates to store an electric charge is called its *capacitance*.

By definition, a **capacitor** has a capacitance (*C*) of one farad (symbol F) when a p.d. of one volt maintains a charge of one coulomb on that capacitor, or

$$C = \frac{Q}{V} \text{ (F)}$$

Collecting these important formulae together, we have:

$$Q = It = CV$$

Capacitors

In this section we shall consider the practical aspects associated with capacitors in electronic circuits.

A capacitor stores a small amount of electric charge; it can be thought of as a small rechargeable battery which can be quickly recharged. In electronics we are not only concerned with the amount of charge stored by the capacitor but in the way the value of the capacitor determines the performance of timers and oscillators by varying the time constant of a simple capacitor–resistor circuit.

Capacitors in action

If a test circuit is assembled as shown in Fig. 3.75 and the changeover switch connected to d.c., the signal lamp will only illuminate for a very short pulse as the capacitor charges. The charged capacitor then blocks any further d.c. current flow. If the changeover switch is then connected to a.c. the lamp will illuminate at full brilliance because the capacitor will charge and discharge continuously at the supply frequency. Current is *apparently* flowing through the capacitor because electrons are moving to and fro in the wires joining the capacitor plates to the a.c. supply.

Coupling and decoupling capacitors

Capacitors can be used to separate a.c. and d.c. in an electronic circuit. If the output from circuit A, shown in Fig. 3.76(a), contains both a.c. and d.c. but only an a.c. input is required for circuit B then a *coupling* capacitor is connected between them. This blocks the d.c. while offering a low reactance to the a.c. component. Alternatively, if it is required that only d.c. be connected to circuit B, shown in Fig. 3.76(b), a *decoupling* capacitor can be connected in parallel with circuit B. This will provide a low reactance path for the a.c. component of the

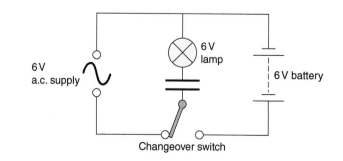

Figure 3.75 Test circuit showing capacitors in action.

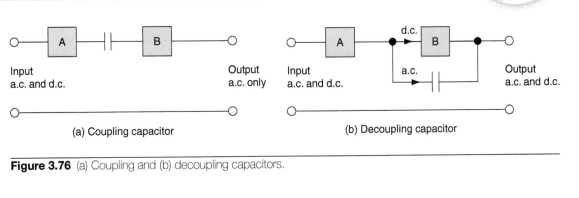

Input
a.c. and d.c.

Output
a.c. only

Input
a.c. and d.c.

d.c.

a.c.

Output
a.c. and d.c.

(a) Coupling capacitor

(b) Decoupling capacitor

Figure 3.76 (a) Coupling and (b) decoupling capacitors.

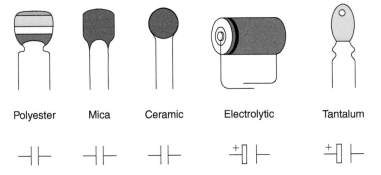

| Polyester | Mica | Ceramic | Electrolytic | Tantalum |

Figure 3.77 Capacitors and their symbols used in electronic circuits.

supply and only d.c. will be presented to the input of B. This technique is used to *filter out* unwanted a.c. in, for example, d.c. power supplies.

Types of capacitor

There are two broad categories of capacitor: non-polarized and polarized. The non-polarized type can be connected either way round, but polarized capacitors *must* be connected to the polarity indicated; otherwise a short-circuit and consequent destruction of the capacitor will result. There are many different types of capacitor, each one being distinguished by the type of dielectric used in its construction. Figure 3.77 shows some of the capacitors used in electronics.

Polyester capacitors

Polyester capacitors are an example of the plastic film capacitor. Polypropylene, polycarbonate and polystyrene capacitors are other types of plastic film capacitor. The capacitor value may be marked on the plastic film, or the capacitor colour code given in Table 3.5 may be used. This dielectric material gives a compact capacitor with good electrical and temperature characteristics. They are used in many electronic circuits, but are not suitable for high-frequency use.

Mica capacitors

Mica capacitors have excellent stability and are accurate to $\pm 1\%$ of the marked value. Since costs usually increase with increased accuracy, they tend to be more expensive than plastic film capacitors. They are used where high stability is required, for example, in tuned circuits and filters.

Ceramic capacitors

Ceramic capacitors are mainly used in high-frequency circuits subjected to wide temperature variations. They have high stability and low loss.

Table 3.7 Colour code for plastic film capacitors (values in picofarads)

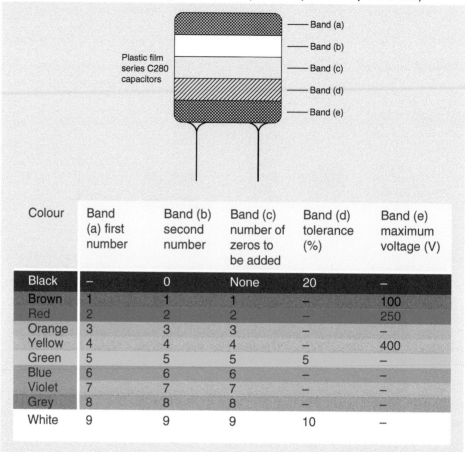

Colour	Band (a) first number	Band (b) second number	Band (c) number of zeros to be added	Band (d) tolerance (%)	Band (e) maximum voltage (V)
Black	–	0	None	20	–
Brown	1	1	1	–	100
Red	2	2	2	–	250
Orange	3	3	3	–	–
Yellow	4	4	4	–	400
Green	5	5	5	5	–
Blue	6	6	6	–	–
Violet	7	7	7	–	–
Grey	8	8	8	–	–
White	9	9	9	10	–

Electrolytic capacitors

Electrolytic capacitors are used where a large value of capacitance coupled with a small physical size is required. They are constructed on the 'Swiss roll' principle as are the paper dielectric capacitors used for power-factor correction in electrical installation circuits. The electrolytic capacitors' high capacitance for very small volume is derived from the extreme thinness of the dielectric coupled with a high dielectric strength. Electrolytic capacitors have a size gain of approximately 100 times over the equivalent non-electrolytic type. Their main disadvantage is that they are polarized and must be connected to the correct polarity in a circuit. Their large capacity makes them ideal as smoothing capacitors in power supplies.

Tantalum capacitors

Tantalum capacitors are a new type of electrolytic capacitor using tantalum and tantalum oxide to give a further capacitance/size advantage. They look like a 'raindrop' or 'blob' with two leads protruding from the bottom. The polarity and values may be marked on the capacitor, or a colour code may be used. The voltage ratings available tend to be low, as with all electrolytic capacitors. They are also extremely vulnerable to reverse voltages in excess of 0.3 V. This means that even when testing with an ohmmeter, extreme care must be taken to ensure correct polarity.

Variable capacitors

Variable capacitors are constructed so that one set of metal plates moves relative to another set of fixed metal plates as shown in Fig. 3.76. The plates are separated by air or sheet mica, which acts as a dielectric. Air dielectric variable capacitors are used to tune radio receivers to a chosen station, and small variable capacitors called *trimmers* or *presets* are used to make fine, infrequent adjustments to the capacitance of a circuit.

When choosing a capacitor for a particular application, three factors must be considered: value, working voltage and leakage current.

Energy stored in a capacitor and time constants

Following a period of charge, the capacitor will store a small amount of energy as an electrostatic charge which, we will see later, can be made to do work. The energy stored (symbol W) in a capacitor is expressed in joules and given by the formula:

$$\text{Energy} = W = 1/2CV^2 \text{ (J)}$$

where C is the capacitance of the capacitor and V is the applied voltage.

Example 1

A 60 μF capacitor is used for power-factor correction in a fluorescent luminaire. Calculate the energy stored in the capacitor when it is connected to the 230 V mains supply.

$$\text{Energy} = W = 1/2CV^2 \text{ (J)}$$
$$W = 1/2 \times 60 \times 10^{-6}\text{ F} \times (230\text{V})^2$$
$$W = 3.17\text{J}$$

Example 2

The energy stored in a certain capacitor when connected across a 400 V supply is 0.3 J.

Calculate (a) the capacitance and (b) the charge on the capacitor.

For (a),

$$W = 1/2CV^2 \text{ (J)}$$

Transposing,

$$W = \tfrac{1}{2}CV^2 \text{(J)}$$

$$C = \frac{2W}{V^2} \text{ (F)}$$

$$\therefore C = \frac{2 \times 0.3\text{J}}{(400\text{V})} \text{ (F)}$$

$$C = 3.75\,\mu\text{F}$$

For (b), the charge is given by:

$$Q = CV \, (C)$$
$$\therefore Q = 3.75 \times 10^{-6} F \times 400V$$
$$Q = 1500 \, \mu C$$

CR CIRCUITS

As we have discussed earlier in this chapter, connecting a voltage to the plates of a capacitor causes it to charge up to the potential of the supply. This involves electrons moving around the circuit to create the necessary charge conditions and, therefore, this action does not occur instantly, but takes some time, depending upon the size of the capacitor and the resistance of the circuit. Such circuits are called capacitor–resistor (CR) circuits, and have many applications in electronics as timers and triggers and for controlling the time base sweeps of a cathode ray oscilloscope.

Figure 3.78 shows the circuit diagram for a simple CR circuit and the graphs drawn from the meter readings. It can be seen that:

(a) initially the current has a maximum value and decreases slowly to zero as the capacitor charges, and

(b) initially the capacitor voltage rises rapidly but then slows down, increasing gradually until the capacitor voltage is equal to the supply voltage when fully charged.

The mathematical name for the shape of these curves is an exponential curve and, therefore, we say that the capacitor voltage is growing exponentially while the current is decaying exponentially during the charging period. The rate at which the capacitor charges is dependent upon the size of the capacitor and resistor. The bigger the values of C and R, the longer it will take to charge the capacitor. The time taken to charge a capacitor by a constant current is given by the **time constant** of the circuit which is expressed mathematically as $T = CR$, where T is the time in seconds.

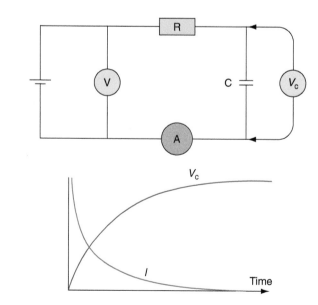

Figure 3.78 A CR circuit.

Example 1

A 60 µF capacitor is connected in series with a 20 kΩ resistor across a 12 V supply. Determine the time constant of this circuit.

$$T = CR \text{(s)}$$
$$\therefore T = 60 \times 10^{-6}\text{F} \times 20 \times 10^{3}\Omega$$
$$T = 1.2\text{s}$$

We have already seen that in practice the capacitor is not charged by a constant current but, in fact, charges exponentially. However, it can be shown by experiment that in one time constant the capacitor will have reached about 63% of its fi nal steady value, taking about fi ve times the time constant to become fully charged. Therefore, in 1.2 seconds the 60 µF capacitor of Example 1 will have reached about 63% of 12 V and after 5 T, that is 6 seconds, will be fully charged at 12 V.

GRAPHICAL DERIVATION OF CR CIRCUIT

The exponential charging and discharging curves of the CR circuit described in Example 1 may also be drawn to scale by following the procedure described below and as shown in Fig. 3.79.

1. We have calculated the time constant for the circuit (T) and found it to be 1.2 seconds.
2. We know that the maximum voltage of the fully charged capacitor will be 12 V because the supply voltage is 12 V.
3. To draw the graph we must first select suitable scales: 0–12 on the voltage axis would be appropriate for this example and 0–6 seconds on the time axis because we know that the capacitor must be fully charged in five time constants.
4. Next draw a horizontal dotted line along the point of maximum voltage, 12 V in this example.
5. Along the time axis measure off one time constant (T), distance OA in Fig. 3.79. This corresponds to 1.2 seconds because in this example T is equal to 1.2 seconds.
6. Draw the vertical dotted line AB.
7. Next, draw a full line OB; this is the start of the charging curve.
8. Select a point C, somewhere convenient and close to O along line OB.
9. Draw a horizontal line CD equal to the length of the time constant (T).
10. Draw the dotted vertical line DE.
11. Draw the line CE, the second line of our charging curve.
12. Select another point G close to C along line CE and repeat the procedures 9 to 12 to draw lines GF, JH and so on as shown in Fig. 3.79.
13. Finally, join together with a smooth curving line the points OCGJ, etc., and we have the exponential growth curve of the voltage across the capacitor.

Switching off the supply and discharging the capacitor through the 20 kΩ resistor will produce the exponential decay of the voltage across the capacitor which will be a mirror image of the growth curve. The decay curve can be derived graphically in the same way as the growth curve and is shown in Fig. 3.80.

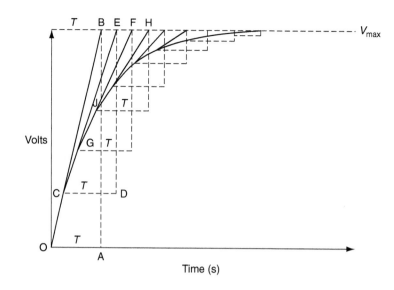

Figure 3.79 Graphical derivation of CR growth curve.

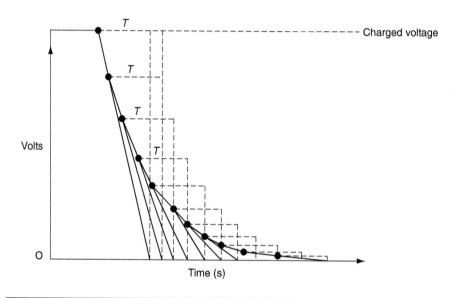

Figure 3.80 Graphical derivation of CR decay curve.

SELECTING A CAPACITOR

There are two broad categories of capacitor, the non-polarized and the polarized.

The non-polarized type is often found in electrical installation work for power-factor correction. A paper-dielectric capacitor is non-polarized and can be connected either way round.

The polarized type must be connected to the polarity indicated otherwise the capacitor will explode. Electrolytic capacitors are polarized and are used where a large value of capacitance is required in a relatively small package. We therefore fi nd polarized capacitors in electronic equipment such as smoothing or stabilized supplies, emergency lighting and alarm systems, so be careful when working on these systems. When choosing a capacitor for a particular application, three factors must be considered: value, voltage and leakage current.

The unit of capacitance is the *farad* (symbol F), to commemorate the name of the English scientist Michael Faraday. However, for practical purposes the farad is much too large, and in electrical installation work and electronics we use fractions of a farad as follows:

$$1 \text{ microfarad} = 1\,\mu F = 1 \times 10^{-6}\,F$$
$$1 \text{ nanofarad} = 1nF = 1 \times 10^{-9}\,F$$
$$1 \text{ picofarad} = 1pF = 1 \times 10^{-12}\,F$$

The power-factor correction capacitor used in a domestic fluorescent luminaire would typically have a value of $8\,\mu F$ at a working voltage of 400 V. In an electronic filter circuit a typical capacitor value might be 100 pF at 63 V.

One microfarad is one million times greater than one picofarad. It may be useful to remember that:

$$1000 \text{ pF} = 1nF$$
$$1000 \text{ nF} = 1\mu F$$

The working voltage of a capacitor is the *maximum* voltage that can be applied between the plates of the capacitor without breaking down the dielectric insulating material. This is a d.c. rating and, therefore, a capacitor with a 200 V rating must only be connected across a maximum of 200 V d.c. Since a.c. voltages are usually given as r.m.s. values, a 200 V a.c. supply would have a maximum value of about 283 V, which would damage the 200 V capacitor. When connecting a capacitor to the 230 V mains supply we must choose a working voltage of about 400 V because 230 V r.m.s. is approximately 325 V maximum. The 'factor of safety' is small and, therefore, the working voltage of the capacitor must not be exceeded.

An ideal capacitor which is isolated will remain charged forever, but in practice no dielectric insulating material is perfect, and the charge will slowly *leak* between the plates, gradually discharging the capacitor. The loss of charge by leakage through it should be very small for a practical capacitor.

Capacitor colour code

The actual value of a capacitor can be identified by using the colour codes given in Table 3.5 in the same way that the resistor colour code was applied to resistors.

Example 1

A plastic film capacitor is colour coded, from top to bottom, brown, black, yellow, black, red. Determine the value of the capacitor, its tolerance and working voltage.

From Table 3.5 we obtain the following:

Band (a) – brown has a value 1
Band (b) – black has a value 0
Band (c) – yellow indicates multiply by 10,000
Band (d) – black indicates 20%
Band (e) – red indicates 250 V

The capacitor has a value of 1,00,000 pF or $0.1\,\mu F$ with a tolerance of 20% and a maximum working voltage of 250 V.

Example 2

Determine the value, tolerance and working voltage of a polyester capacitor colour-coded, from top to bottom, yellow, violet, yellow, white, yellow.

From Table 3.5 we obtain the following:

Band (a) – yellow has a value 4
Band (b) – violet has a value 7
Band (c) – yellow indicates multiply by 10,000
Band (d) – white indicates 10%
Band (e) – yellow indicates 400 V

The capacitor has a value of 4,70,000 pF or 0.47 µF with a tolerance of 10% and a maximum working voltage of 400 V.

Example 3

A plastic film capacitor has the following coloured bands from its top down to the connecting leads: blue, grey, orange, black, brown. Determine the value, tolerance and voltage of this capacitor.

From Table 3.5 we obtain the following:

Band (a) – blue has a value 6
Band (b) – grey has a value 8
Band (c) – orange indicates multiply by 1000
Band (d) – black indicates 20%
Band (e) – brown indicates 100 V

The capacitor has a value of 68,000 pF or 68 nF with a tolerance of 20% and a maximum working voltage of 100 V.

Capacitance value codes

Where the numerical value of the capacitor includes a decimal point, it is standard practice to use the prefix for the multiplication factor in place of the decimal point. This is the same practice as we used earlier for resistors.

The abbreviation µ means microfarad, n means nanofarad and p means picofarad. Therefore, a 1.8 pF capacitor would be abbreviated to 1 p8, a 10 pF capacitor to 10 p, a 150 pF capacitor to 150 p or n15, a 2200 pF capacitor to 2n2 and a 10,000 pF capacitor to 10 n.

$$1000 \text{ pF} = 1\text{nF} = 0.001 \text{ µF}$$

Packaging electronic components

When we talk about packaging electronic components we are not referring to the parcel or box which contains the components for storage and delivery, but to the type of encapsulation in which the tiny semiconductor material is contained. Figure 3.81 shows three different package outlines for just one type of discrete component: the transistor. Identification of the pin connections for different packages is given within the text as each separate or discrete

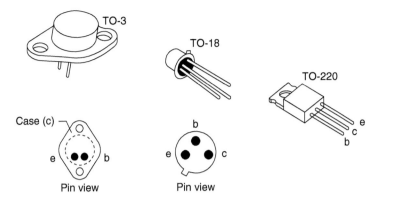

Figure 3.81 Three different package outlines for transistors.

component is considered, particularly later in this chapter when we discuss semiconductor devices.

Obtaining information and components

Electricians use electrical wholesalers and suppliers to purchase electrical cable, equipment and accessories. Similar facilities are available in most towns and cities for the purchase of electronic components and equipment. There are also a number of national suppliers who employ representatives who will call at your workshop to offer technical advice and take your order. Some of these national companies also offer a 24-hour telephone order and mail order service, such as RS at www.rswww.com or telephone 08457 201 201. Their full-colour, fully illustrated catalogues also contain an enormous amount of technical information. For local suppliers you must consult your local phone book and *Yellow Pages*.

Semiconductor materials

Modern electronic devices use the semiconductor properties of materials such as silicon or germanium. The atoms of pure silicon or germanium are arranged in a lattice structure, as shown in Fig. 3.82. The outer electron orbits contain four electrons known as *valence* electrons. These electrons are all linked to other valence electrons from adjacent atoms, forming a covalent bond. There are no free electrons in pure silicon or germanium and, therefore, no conduction can take place unless the bonds are broken and the lattice framework is destroyed.

To make conduction possible without destroying the crystal it is necessary to replace a four-valent atom with a three- or five-valent atom. This process is known as *doping*.

If a three-valent atom is added to silicon or germanium a hole is left in the lattice framework. Since the material has lost a negative charge, the material becomes positive and is known as a p-type material (p for positive).

If a five-valent atom is added to silicon or germanium, only four of the valence electrons can form a bond and one electron becomes mobile or free to carry charge. Since the material has gained a negative charge it is known as an n-type material (n for negative).

Bringing together a p-type and n-type material allows current to flow in one direction only through the p–n junction. Such a junction is called a diode, since it is the semiconductor equivalent of the vacuum diode valve used by Fleming to rectify radio signals in 1904.

Semiconductor diode

A semiconductor or junction diode consists of a p-type and n-type material formed in the same piece of silicon or germanium. The p-type material forms the anode and the n-type the cathode, as shown in Fig. 3.83. If the anode is made positive with respect to the cathode, the junction will have very little resistance and current will flow. This is referred to as forward bias. However, if reverse bias is applied, that is, the anode is made negative with respect to the cathode, the junction resistance is high and no current can flow, as shown in Fig. 3.84. The characteristics for a forward and reverse bias p–n junction are given in Fig. 3.85.

It can be seen that a small voltage is required to forward bias the junction before a current can flow. This is approximately 0.6V for silicon and 0.2V for germanium. The reverse bias potential of silicon is about 1200V and for germanium about 30V. If the reverse bias voltage is exceeded the diode will break down and current will flow in both directions. Similarly, the diode will break down if the current rating is exceeded, because excessive heat will be generated. Manufacturers' information therefore gives maximum voltage and current ratings for individual diodes which must not be exceeded. However, it is possible to connect a number of standard diodes in series or parallel, thereby sharing current or voltage, as shown in Fig. 3.86, so that the manufacturers' maximum values are not exceeded by the circuit.

Diode testing

The p–n junction of the diode has a low resistance in one direction and a very high resistance in the reverse direction.

Connecting an ohmmeter, with the red positive lead to the anode of the junction diode and the black negative lead to the cathode, would give a very low reading. Reversing the lead connections would give a high resistance reading in a 'good' component.

Zener diode

A Zener diode is a silicon junction diode but with a different characteristic than the semiconductor diode considered previously. It is a special diode with a predetermined reverse breakdown voltage, the mechanism for which was discovered by Carl Zener in 1934. Its symbol and general appearance are shown

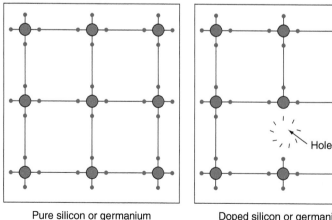

Pure silicon or germanium

Doped silicon or germanium
p-type material

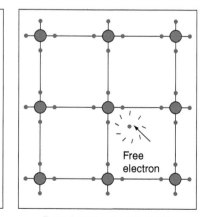

Doped silicon or germanium
n-type material

Figure 3.82 Semiconductor material.

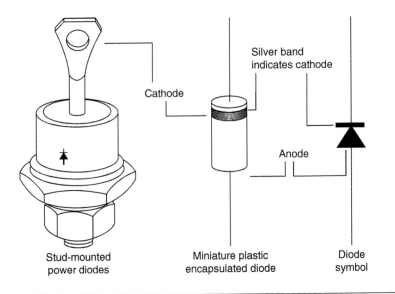

Cathode

Silver band
indicates cathode

Anode

Stud-mounted
power diodes

Miniature plastic
encapsulated diode

Diode
symbol

Figure 3.83 Symbol for and appearance of semiconductor diodes.

in Fig. 3.87. In its forward bias mode, that is, when the anode is positive and the cathode negative, the Zener diode will conduct at about 0.6 V, just like an ordinary diode, but it is in the reverse mode that the Zener diode is normally used. When connected with the anode made negative and the cathode positive, the reverse current is zero until the reverse voltage reaches a predetermined value, when the diode switches on, as shown by the characteristics given in Fig. 3.88. This is called the Zener voltage or reference voltage. Zener diodes are manufactured in a range of preferred values, for example, 2.7, 4.7, 5.1, 6.2, 6.8, 9.1, 10, 11, 12 V, etc., up to 200 V at various ratings. The diode may be damaged by overheating if the current is not limited by a series resistor, but when this is connected, the voltage across the diode remains constant. It is this property of the Zener diode which makes it useful for stabilizing power supplies.

If a test circuit is constructed as shown in Fig. 3.89, the Zener action can be observed. When the supply is less than the Zener voltage (5.1 V in this case) no current will flow and the output voltage will be equal to the input voltage. When the supply is equal to or greater than the Zener voltage, the diode will conduct and any excess voltage will appear across the 680 Ω resistor, resulting in a very stable voltage at the output. When connecting this and other electronic circuits you must take care to connect the polarity of the Zener diode as shown in the diagram. Note that current must flow through the diode to enable it to stabilize.

Light-emitting diode

The light-emitting diode (LED) is a p–n junction especially manufactured from a semiconducting material which emits light when a current of about 10 mA flows through the junction.

No light is emitted when the junction is reverse biased and if this exceeds about 5 V the LED may be damaged.

The general appearance and circuit symbol are shown in Fig. 3.90.

The LED will emit light if the voltage across it is about 2 V. If a voltage greater than 2 V is to be used then a resistor must be connected in series with the LED.

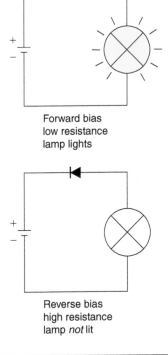

Forward bias
low resistance
lamp lights

Reverse bias
high resistance
lamp *not* lit

Figure 3.84 Forward and reverse bias of a diode.

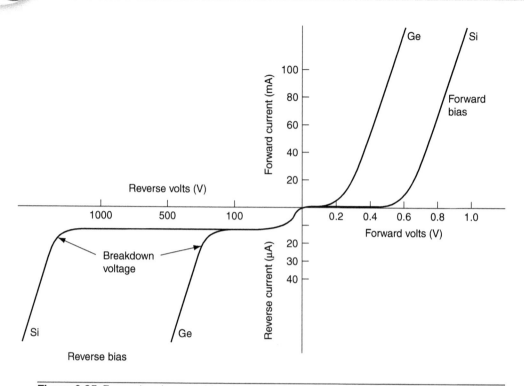

Figure 3.85 Forward and reverse bias characteristic of silicon and germanium.

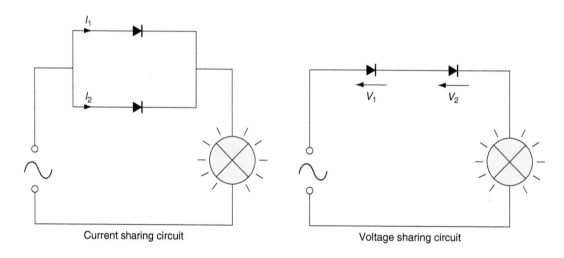

Figure 3.86 Using two diodes to reduce the current or voltage applied to a diode.

To calculate the value of the series resistor we must ask ourselves what we know about LEDs. We know that the diode requires a forward voltage of about 2 V and a current of about 10 mA must flow through the junction to give sufficient light. The value of the series resistor R will, therefore, be given by:

$$R = \frac{\text{Supply voltage} - 2\,\text{V}}{10\,\text{mA}}\ \Omega$$

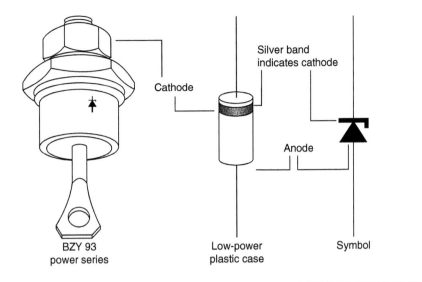

Cathode

Silver band
indicates cathode

Anode

BZY 93
power series

Low-power
plastic case

Symbol

Figure 3.87 Symbol for and appearance of Zener diodes.

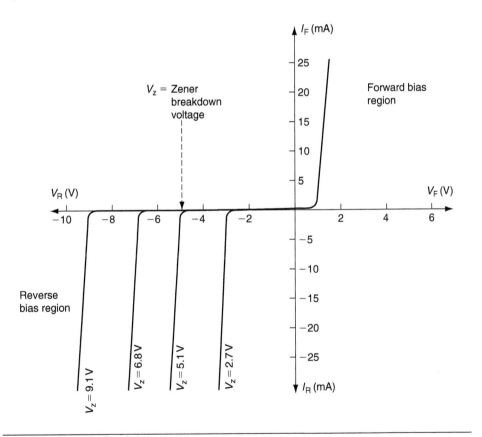

Figure 3.88 Zener diode characteristics.

Example

Calculate the value of the series resistor required when an LED is to be used to show the presence of a 12 V supply.

$$R = \frac{12\,V - 2\,V}{10\,mA}\ \Omega$$

$$R = \frac{10\,V}{10\,mA} = 1\,k\Omega$$

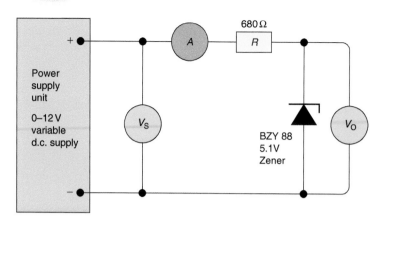

P.S.U. supply volts (V_S)	Current (A)	Output volts (V_O)
1		
2		
3		
4		
5		
6		
7		
8		
9		
10		
11		
12		

Figure 3.89 Experiment to demonstrate the operation of a Zener diode.

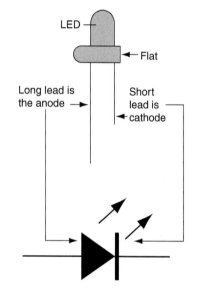

Figure 3.90 Symbol for and general appearance of an LED.

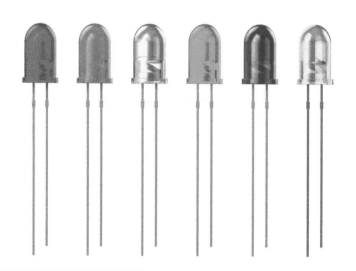

Figure 3.91 LED signal lamps.

The circuit is, therefore, as shown in Fig. 3.92.

LEDs are available in red, yellow and green and, when used with a series resistor, may replace a filament lamp. They use less current than a filament lamp, are smaller, do not become hot and last indefinitely. A filament lamp, however, is brighter and emits white light. LEDs are often used as indicator lamps, to indicate the presence of a voltage. They do not, however, indicate the *precise* amount of voltage present at that point.

Try this

LEDs

Make a list in the margin of examples where you have seen LEDs being used.

Another application of the LED is the seven-segment display used as a numerical indicator in calculators, digital watches and measuring instruments. Seven LEDs are arranged as a figure of eight so that when various segments are illuminated, the numbers 0–9 are displayed as shown in Fig. 3.93. Recent scientific discoveries have enabled this humble signal lamp to be developed into the super efficient, broad spectrum LED lamp we know today. LED lamps were discussed earlier in this chapter.

Light-dependent resistor

Almost all materials change their resistance with a change in temperature. Light energy falling on a suitable semiconductor material also causes a change in resistance. The semiconductor material of a light-dependent resistor (LDR) is encapsulated as shown in Fig. 3.94 together with the circuit symbol. The resistance of an LDR in total darkness is about $10\,M\Omega$, in normal room lighting about $5\,k\Omega$ and in bright sunlight about $100\,\Omega$. They can carry tens of milliamperes, an amount which is sufficient to operate a relay. The LDR uses this characteristic to switch on automatically street lighting and security alarms.

Photodiode

The photodiode is a normal junction diode with a transparent window through which light can enter. The circuit symbol and general appearance are shown in Fig. 3.95. It is operated in reverse bias mode and the leakage current increases in proportion to the amount of light falling on the junction. This is due to the light

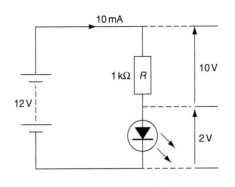

Figure 3.92 Circuit diagram for LED example.

Figure 3.93 LED used in seven-segment display.

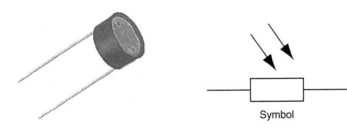

Symbol

Figure 3.94 Symbol and appearance of an LDR.

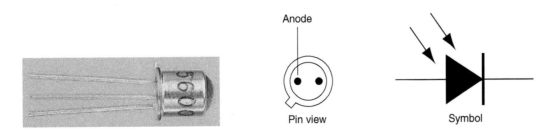

Anode

Pin view

Symbol

Figure 3.95 Symbol for pin connections and appearance of a photodiode.

energy breaking bonds in the crystal lattice of the semiconductor material to produce holes and electrons.

Photodiodes will only carry microamperes of current but can operate much more quickly than LDRs and are used as 'fast' counters when the light intensity is changing rapidly.

Thermistor

The thermistor is a thermal resistor, a semiconductor device whose resisance varies with temperature. Its circuit symbol and general appearance are shown in Fig. 3.96. They can be supplied in many shapes and are used for the measurement and control of temperature up to their maximum useful temperature limit of about 300°C. They are very sensitive and because the bead of semiconductor material can be made very small, they can measure temperature in the most inaccessible places with very fast response times. Thermistors are embedded in high-voltage underground transmission cables in order to monitor the temperature of the cable. Information about the temperature of a cable allows engineers to load the cables more efficiently. A particular cable can carry a larger load in winter, for example, when heat from the cable is being dissipated more efficiently. A thermistor is also used to monitor the water temperature of a motor car.

Transistors

The transistor has become the most important building block in electronics. It is the modern, miniature, semiconductor equivalent of the thermionic valve and was invented in 1947 by Bardeen, Shockley and Brattain at the Bell Telephone Laboratories in the United States. Transistors are packaged as separate or *discrete* components, as shown in Fig. 3.97.

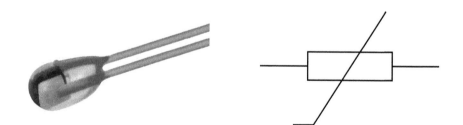

Figure 3.96 Symbol for and appearance of a thermistor.

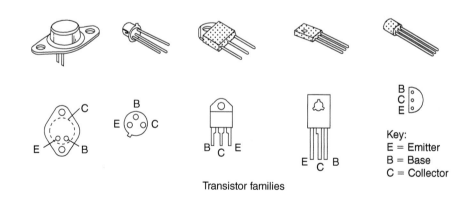

Transistor families

Key:
E = Emitter
B = Base
C = Collector

Figure 3.97 The appearance and pin connections of the transistor family.

There are two basic types of transistor: the *bipolar* or junction transistor and the *field-effect* transistor (FET).

The FET has some characteristics which make it a better choice in electronic switches and amplifiers. It uses less power and has a higher resistance and frequency response. It takes up less space than a bipolar transistor and, therefore, more of them can be packed together on a given area of silicon chip. It is, therefore, the FET which is used when many transistors are integrated on to a small area of silicon chip as in the IC that will be discussed later.

When packaged as a discrete component the FET looks much the same as the bipolar transistor. Its circuit symbol and connections are given in the Appendix. However, it is the bipolar transistor which is much more widely used in electronic circuits as a discrete component.

The bipolar transistor

The bipolar transistor consists of three pieces of semiconductor material sandwiched together, as shown in Fig. 3.98. The structure of this transistor makes it a three-terminal device having a base, collector and emitter terminal. By varying the current flowing into the base connection a much larger current flowing between collector and emitter can be controlled. Apart from the supply connections, the n-p-n and p-n-p types are essentially the same but the n-p-n type is more common.

A transistor is generally considered a current-operated device. There are two possible current paths through the transistor circuit, shown in Fig. 3.99: the base–emitter path when the switch is closed; and the collector–emitter path. Initially, the positive battery supply is connected to the n-type material of the collector, the junction is reverse biased and, therefore, no current will flow. Closing the switch will forward bias the base–emitter junction and current flowing through this junction causes current to flow across the collector–emitter junction and the signal lamp will light.

A small base current can cause a much larger collector current to flow. This is called the *current gain* of the transistor, and is typically about 100. When I say a much larger collector current, I mean a large current in electronic terms, up to about half an ampere.

We can, therefore, regard the transistor as operating in two ways: as a switch because the base current turns on and controls the collector current; and as a current amplifier because the collector current is greater than the base current.

We could also consider the transistor to be operating in a similar way to a relay. However, transistors have many advantages over electrically operated switches such as relays. They are very small, reliable, have no moving parts and, in particular, they can switch millions of times a second without arcing occurring at the contacts.

Transistor testing

A transistor can be thought of as two diodes connected together and, therefore, a transistor can be tested using an ohmmeter in the same way as was described for the diode.

Assuming that the red lead of the ohmmeter is positive, the transistor can be tested in accordance with Table 3.8.

When many transistors are to be tested, a simple test circuit can be assembled as shown in Fig. 3.100.

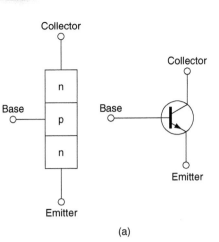
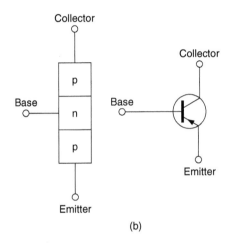
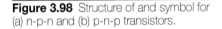

Figure 3.98 Structure of and symbol for (a) n-p-n and (b) p-n-p transistors.

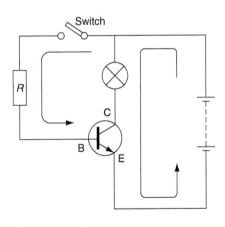

Figure 3.99 Operation of the transistor.

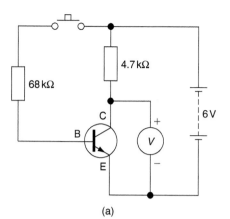

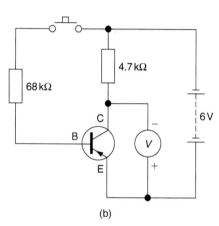

Figure 3.100 Transistor test circuits (a) n-p-n transistor test; (b) p-n-p transistor test.

Table 3.8 Transistor testing using an ohmmeter

A 'good' n-p-n transistor will give the following readings:
Red to base and black to collector = low resistance Red to base and black to emitter = low resistance
Reversed connections on the above terminals will result in a high resistance reading, as will connections of either polarity between the collector and emitter terminals.
A 'good' p-n-p transistor will give the following readings:
Black to base and red to collector = low resistance Black to base and red to emitter = low resistance
Reversed connections on the above terminals will result in a high resistance reading, as will connections of either polarity between the collector and emitter terminals.

With the circuit connected, as shown in Fig. 3.100, a 'good' transistor will give readings on the voltmeter of 6 V with the switch open and about 0.5 V when the switch is made. The voltmeter used for the test should have a high internal resistance, about ten times greater than the value of the resistor being tested – in this case 4.7 kΩ – and this is usually indicated on the back of a multi-range meter or in the manufacturers' information supplied with a new meter.

Integrated circuits

ICs were first developed in the 1960s. They are densely populated miniature electronic circuits made up of hundreds and sometimes thousands of microscopically small transistors, resistors, diodes and capacitors, all connected together on a single chip of silicon no bigger than a baby's fingernail. When assembled in a single package, as shown in Fig. 3.101, we call the device an IC.

There are two broad groups of IC: digital ICs and linear ICs. Digital ICs contain simple switching-type circuits used for logic control and calculators; linear ICs incorporate amplifier-type circuits which can respond to audio and radio frequency signals. The most versatile linear IC is the operational amplifier which has applications in electronics, instrumentation and control.

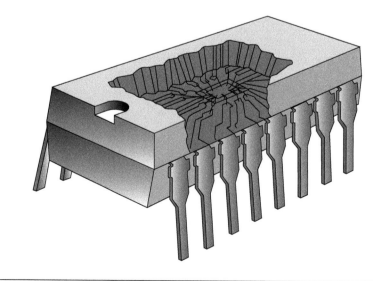

Figure 3.101 Exploded view of an IC.

The IC is an electronic revolution. ICs are more reliable, cheaper and smaller than the same circuit made from discrete or separate transistors, and electronically superior. One IC behaves differently from another because of the arrangement of the transistors within the IC.

Manufacturers' data sheets describe the characteristics of the different ICs, which have a reference number stamped on the top.

When building circuits, it is necessary to be able to identify the IC pin connection by number. The number 1 pin of any IC is indicated by a dot pressed into the encapsulation; it is also the pin to the left of the cutout (Fig. 3.102). Since the packaging of ICs has two rows of pins they are called DIL (dual in line) packaged ICs and their appearance is shown in Fig. 3.103.

ICs are sometimes connected into DIL sockets and at other times are soldered directly into the circuit. The testing of ICs is beyond the scope of a practising electrician, and when they are suspected of being faulty an identical or equivalent replacement should be connected into the circuit, ensuring that it is inserted the correct way round, which is indicated by the position of pin number 1 as described above.

The thyristor

The *thyristor* was previously known as a 'silicon controlled rectifier' since it is a rectifier which controls the power to a load. It consists of four pieces of semiconductor material sandwiched together and connected to three terminals, as shown in Fig. 3.104.

The word thyristor is derived from the Greek word *thyra* meaning door, because the thyristor behaves like a door. It can be open or shut, allowing or preventing current flow through the device. The door is opened – we say the thyristor is triggered – to a conducting state by applying a pulse voltage to the gate connection. Once the thyristor is in the conducting state, the gate loses all control over the devices. The only way to bring the thyristor back to a non-conducting state is to reduce the voltage across the anode and cathode to zero or apply reverse voltage across the anode and cathode.

We can understand the operation of a thyristor by considering the circuit shown in Fig. 3.105. This circuit can also be used to test suspected faulty components.

When SWB only is closed the lamp will not light, but when SWA is also closed, the lamp lights to full brilliance. The lamp will remain illuminated even when SWA

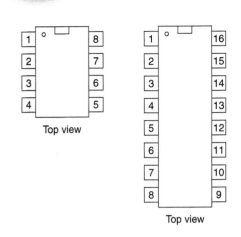

Figure 3.102 IC pin identification.

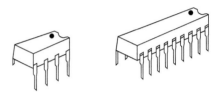

Figure 3.103 DIL packaged ICs.

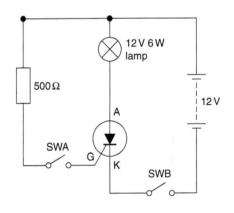

Figure 3.105 Thyristor test circuit.

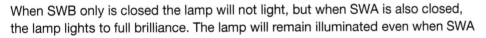

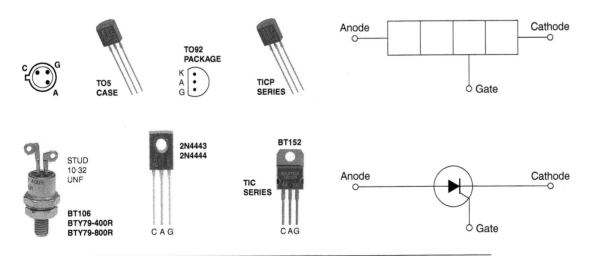

Figure 3.104 Symbol for and structure and appearance of a thyristor.

Table 3.9 Thyristor testing using an ohmmeter

A 'good' thyristor will give the following readings:
Black to cathode and red on gate = low resistance
Red to cathode and black on gate = a higher resistance value
The value of the second reading will depend upon the thyristor, and may vary from only slightly greater to very much greater
Connecting the test instrument leads from cathode to anode will result in a very high resistance reading, whatever polarity is used

is opened. This shows that the thyristor is operating correctly. Once a voltage has been applied to the gate the thyristor becomes forward conducting, like a diode, and the gate loses control.

A thyristor may also be tested using an ohmmeter as described in Table 3.9, assuming that the red lead of the ohmmeter is positive.

The thyristor has no moving parts and operates without arcing. It can operate at extremely high speeds, and the currents used to operate the gate are very small. The most common application for the thyristor is to control the power supply to a load, for example, lighting dimmers and motor speed control.

The power available to an a.c. load can be controlled by allowing current to be supplied to the load during only a part of each cycle. This can be achieved by supplying a gate pulse automatically at a chosen point in each cycle, as shown in Fig. 3.106. Power is reduced by triggering the gate later in the cycle.

The thyristor is only a half-wave device (like a diode) allowing control of only half the available power in an a.c. circuit. This is very uneconomical, and a further development of this device has been the triac which is considered next.

The triac

The triac was developed following the practical problems experienced in connecting two thyristors in parallel to obtain full-wave control, and in providing two separate gate pulses to trigger the two devices.

The triac is a single device containing a back-to-back, two-directional thyristor which is triggered on both halves of each cycle of the a.c. supply by the same gate signal. The power available to the load can, therefore, be varied between zero and full load. Its symbol and general appearance are shown in Fig. 3.107.

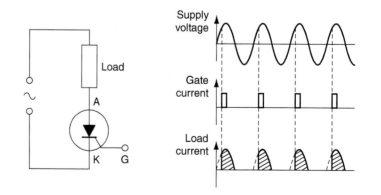

Figure 3.106 Waveforms to show the control effect of a thyristor.

Figure 3.107 Appearance of a triac.

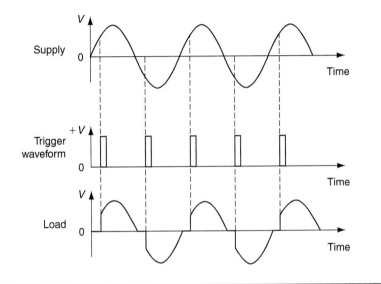

Figure 3.108 Waveforms to show the control effect of a triac.

Power to the load is reduced by triggering the gate later in the cycle, as shown by the waveforms of Fig. 3.108.

The triac is a three-terminal device, just like the thyristor, but the terms anode and cathode have no meaning for a triac. Instead, they are called main terminal one (MT_1) and main terminal two (MT_2). The device is triggered by applying a small pulse to the gate (G). A gate current of 50 mA is sufficient to trigger a triac switching up to 100 A. They are used for many commercial applications where control of a.c. power is required, for example, motor speed control and lamp dimming.

The diac

The diac is a two-terminal device containing a two-directional Zener diode. It is used mainly as a trigger device for the thyristor and triac. The symbol is shown in Fig. 3.109.

The device turns on when some predetermined voltage level is reached, say, 30 V, and, therefore, it can be used to trigger the gate of a triac or thyristor each time the input waveform reaches this predetermined value. Since the device contains back-to-back Zener diodes it triggers on both the positive and negative half-cycles.

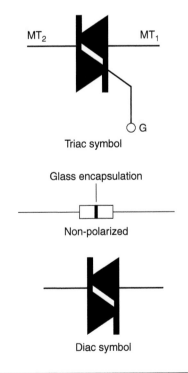

Figure 3.109 Symbol for and appearance of a diac used in triac firing circuits.

Voltage divider

Earlier in this chapter we considered the distribution of voltage across resistors connected in series. We found that the supply voltage was divided between the series resistors in proportion to the size of the resistor. If two identical resistors were connected in series across a 12 V supply, as shown in Fig. 3.110(a), both

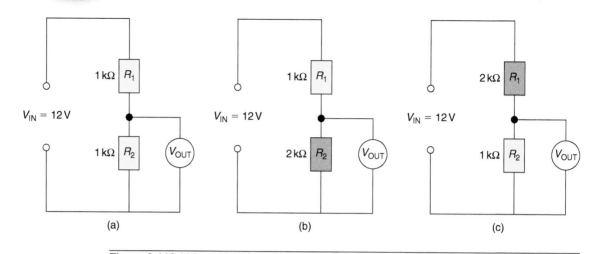

Figure 3.110 Voltage divider circuit.

common sense and a simple calculation would confirm that 6 V would be measured across the output. In the circuit shown in Fig. 3.110(b), the 1 and 2 kΩ resistors divide the input voltage into three equal parts. One part, 4 V, will appear across the 1 kΩ resistor and two parts, 8 V, will appear across the 2 kΩ resistor. In Fig. 3.110(c) the situation is reversed and, therefore, the voltmeter will read 4 V. The division of the voltage is proportional to the ratio of the two resistors and, therefore, we call this simple circuit a *voltage divider* or *potential divider*. The values of the resistors R_1 and R_2 determine the output voltage as follows:

$$V_{OUT} = V_{IN} \times \frac{R_2}{R_1 + R_2} \text{(V)}$$

For the circuit shown in Fig. 3.109(b)

$$V_{OUT} = 12 \text{ V} \times \frac{2 \text{ k}\Omega}{1 \text{k}\Omega + 2 \text{ k}\Omega} = 8 \text{ V}$$

For the circuit shown in Fig. 3.109(c)

$$V_{OUT} = 12 \text{ V} \times \frac{1 \text{ k}\Omega}{2 \text{ k}\Omega + 1 \text{ k}\Omega} = 4 \text{ V}$$

Example 1

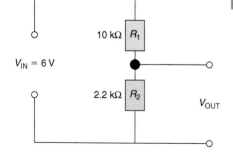

Figure 3.111 Voltage divider circuit for Example 1.

For the circuit shown in Fig. 3.111, calculate the output voltage.

$$V_{OUT} = 6 \text{ V} \times \frac{2.2 \text{ k}\Omega}{10 \text{ k}\Omega + 2.2 \text{ k}\Omega} = 1.08 \text{ V}$$

Example 2

For the circuit shown in Fig. 3.112(a), calculate the output voltage.

We must first calculate the equivalent resistance of the parallel branch:

$$\frac{1}{R_T} = \frac{1}{R_1} + \frac{1}{R_2}$$

$$\frac{1}{R_T} = \frac{1}{10\,k\Omega} + \frac{1}{10\,k\Omega} = \frac{1+1}{10\,k\Omega} = \frac{2}{10\,k\Omega}$$

$$R_T = \frac{10\,k\Omega}{2} = 5\,k\Omega$$

The circuit may now be considered as shown in Fig. 3.111(b):

$$V_{OUT} = 6\,V \times \frac{10\,k\Omega}{5\,k\Omega + 10\,k\Omega} = 4\,V$$

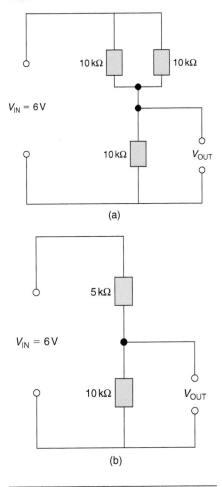

(a)

(b)

Figure 3.112 (a) Voltage divider circuit for Example 2; (b) equivalent circuit for Example 2.

Voltage dividers are used in electronic circuits to produce a reference voltage which is suitable for operating transistors and ICs. The volume control in a radio or the brightness control of a cathode-ray oscilloscope requires a continuously variable voltage divider and this can be achieved by connecting a variable resistor or potentiometer, as shown in Fig. 3.113. With the wiper arm making a connection at the bottom of the resistor, the output would be zero. When connection is made at the centre, the voltage would be 6 V, and at the top of the resistor the voltage would be 12 V. The voltage is continuously variable between 0 and 12 V simply by moving the wiper arm of a suitable variable resistor such as those shown in Fig. 3.113.

When a voltmeter is connected to a voltage divider it 'loads' the circuit, causing the output voltage to fall below the calculated value. To avoid this, the resistance of the voltmeter should be at least ten times as great as the value of the resistor across which it is connected. For example, the voltmeter connected across the voltage divider shown in Fig. 3.110(b) must be greater than 20 kΩ, and across 3.110(c) greater than 10 kΩ. This problem of loading the circuit occurs when taking voltage readings in electronic circuits and therefore a high impedance voltmeter should always be used to avoid instrument errors.

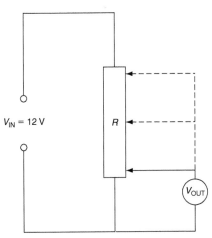

Figure 3.113 Constantly variable voltage divider circuit.

Rectification of a.c.

When a d.c. supply is required, batteries or a rectified a.c. supply can be provided. Batteries have the advantage of portability, but a battery supply is more expensive than using the a.c. mains supply suitably rectified. **Rectification** is the conversion of an a.c. supply into a unidirectional or d.c. supply. This is one of the many applications for a diode which will conduct in one direction only; that is, when the anode is positive with respect to the cathode.

Half-wave rectification

The circuit is connected as shown in Fig. 3.114. During the first half-cycle the anode is positive with respect to the cathode and, therefore, the diode will conduct. When the supply goes negative during the second half-cycle, the anode

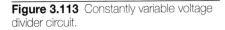

Definition

Rectification is the conversion of an a.c. supply into a unidirectional or d.c. supply.

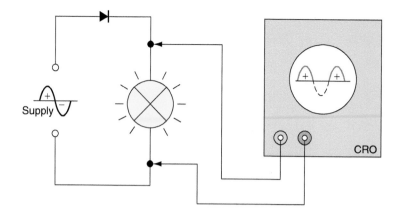

Figure 3.114 Half-wave rectification.

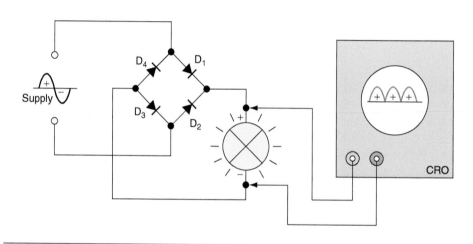

Figure 3.115 Full-wave rectification using a bridge circuit.

is negative with respect to the cathode and, therefore, the diode will not allow current to flow. Only the positive half of the waveform will be available at the load and the lamp will light at reduced brightness.

Full-wave rectification

Figure 3.115 shows an improved rectifier circuit which makes use of the whole a.c. waveform and is, therefore, known as a full-wave rectifier. When the four diodes are assembled in this diamond-shaped configuration, the circuit is also known as a *bridge rectifier*. During the first half-cycle diodes D_1 and D_3 conduct, and diodes D_2 and D_4 conduct during the second half-cycle. The lamp will light to full brightness.

Full-wave and half-wave rectification can be displayed on the screen of a CRO and will appear as shown in Figs. 3.114 and 3.115.

Smoothing

The circuits of Figure 3.116 convert an alternating waveform into a waveform which never goes negative, but they cannot be called continuous d.c. because they contain a large alternating component. Such a waveform is too bumpy to be used to supply electronic equipment but may be used for battery charging. To be useful in electronic circuits the output must be smoothed. The simplest way to smooth an output is to connect a large-value capacitor across the output terminals, as shown in Fig. 3.116.

When the output from the rectifier is increasing, as shown by the dotted lines of Fig. 3.117, the capacitor charges up. During the second quarter of the cycle, when the output from the rectifier is falling to zero, the capacitor discharges into the load. The output voltage falls until the output from the rectifier once again charges the capacitor. The capacitor connected to the full-wave rectifier circuit is charged up twice as often as the capacitor connected to the half-wave circuit and, therefore, the output ripple on the full-wave circuit is smaller, giving better smoothing. Increasing the current drawn from the supply increases the size of the ripple. Increasing the size of the capacitor reduces the amount of ripple.

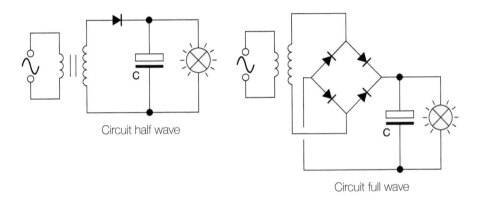

Figure 3.116 Rectified a.c. with smoothing capacitor connected.

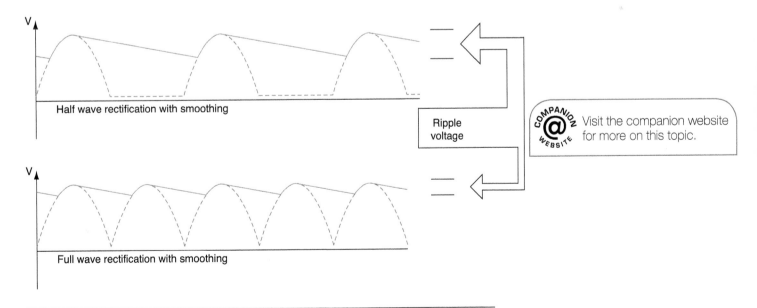

Visit the companion website for more on this topic.

Figure 3.117 Output waveforms with smoothing showing reduced ripple with full wave.

Try this

Battery Charger

- Do you have a battery charger for your car battery?
- What type of circuit do you think is inside?
- Carefully look inside and identify the components.

Low-pass filter

The ripple voltage of the rectified and smoothed circuit shown in Fig. 3.116 can be further reduced by adding a low-pass filter, as shown in Fig. 3.118. A low-pass filter allows low frequencies to pass while blocking higher frequencies. Direct current has a frequency of zero hertz, while the ripple voltage of a full-wave rectifier has a frequency of 100 Hz. Connecting the low-pass filter will allow the d.c. to pass while blocking the ripple voltage, resulting in a smoother output voltage.

The low-pass filter shown in Fig. 3.118 does, however, increase the output resistance, which encourages the voltage to fall as the load current increases. This can be reduced if the resistor is replaced by a choke, which has a high impedance to the ripple voltage but a low resistance, which reduces the output ripple without increasing the output resistance.

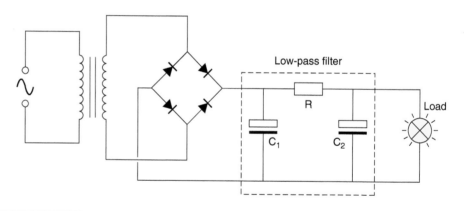

Figure 3.118 Rectified a.c. with low-pass filter connected.

Stabilized power supplies

The power supplies required for electronic circuits must be ripple-free, stabilized and have good regulation; that is, the voltage must not change in value over the whole load range. A number of stabilizing circuits are available which, when connected across the output of the circuit shown in Fig. 3.119, give a constant or stabilized voltage output. These circuits use the characteristics of the Zener diode which was described by the experiment in Fig. 3.89.

Figure 3.119 shows an d.c. supply which has been rectified, smoothed and stabilized. You could build and test this circuit at college if your lecturers agree.

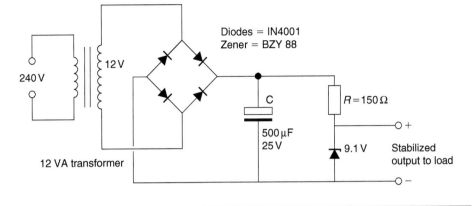

Figure 3.119 Stabilized d.c. supply.

Electronic systems

So far in this chapter we have looked at the basic electronic components. Let us now look at some electronic systems which are made up of those individual components.

Fire alarm circuits (BS 5839 and BS EN 54-2: 1998)

Through one or more of the various statutory Acts, all public buildings are required to provide an effective means of giving a warning of fire so that life and property may be protected. An effective system is one which gives a warning of fire while sufficient time remains for the fire to be put out and any occupants to leave the building.

Fire alarm circuits are wired as either normally open or normally closed. In a **normally open circuit**, the alarm call points are connected in parallel with each other so that when any alarm point is initiated the circuit is completed and the sounder gives a warning of fire. The arrangement is shown in Fig. 3.120. It is essential for some parts of the wiring system to continue operating even when attacked by fire. For this reason the master control and sounders should be wired

> **Definition**
>
> *Fire alarm circuits* are wired as either normally open or normally closed. In a *normally open circuit*, the alarm call points are connected in parallel with each other so that when any alarm point is initiated the circuit is completed and the sounder gives a warning of fire.

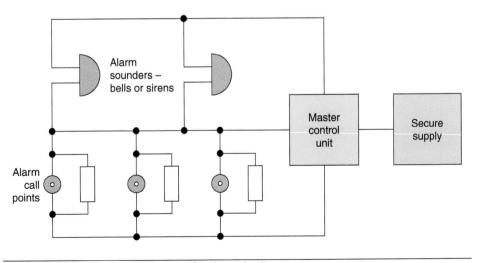

Figure 3.120 A simple normally open fire alarm circuit.

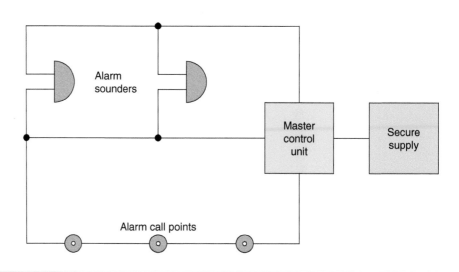

Figure 3.121 A simple normally closed fire alarm circuit.

Figure 3.122 Breakglass manual call point.

in MI or FP 200 cable. The alarm call points of a normally open system must also be wired in MI or FP 200 cable, unless a monitored system is used. In its simplest form this system requires a high-value resistor to be connected across the call-point contacts, which permits a small current to circulate and operate an indicator, declaring the circuit healthy. With a monitored system, PVC insulated cables may be used to wire the alarm call points.

In a **normally closed circuit**, the alarm call points are connected in series to normally closed contacts as shown in Fig. 3.121. When the alarm is initiated, or if a break occurs in the wiring, the alarm is activated. The sounders and master control unit must be wired in MI or FP 200 cable, but the call points may be wired in PVC insulated cable since this circuit will always 'fail safe'.

Alarm call points

Manually operated alarm call points should be provided in all parts of a building where people may be present, and should be located so that no one need walk for more than 30 m from any position within the premises in order to give an alarm. A breakglass manual call point is shown in Fig. 3.122. They should be located on exit routes and, in particular, on the floor landings of staircases and exits to the street. They should be fixed at a height of 1.4 m above the floor at easily accessible, well-illuminated and conspicuous positions.

Automatic detection of fire is possible with heat and smoke detectors. These are usually installed on the ceilings and at the tops of stairwells of buildings because heat and smoke rise. Smoke detectors tend to give a faster response than heat detectors, but whether manual or automatic call points are used should be determined by their suitability for the particular installation. They should be able to discriminate between a fire and the normal environment in which they are to be installed.

Sounders

The positions and numbers of **sounders** should be such that the alarm can be distinctly heard above the background noise in every part of the premises. The sounders should produce a minimum of 65 dB, or 5 dB above any ambient sound which might persist for more than 30 seconds. If the sounders are to arouse sleeping persons then the minimum sound level should be increased to 75 dB at the bedhead. Bells, hooters or sirens may be used but in any one installation they must all be of the same type. Examples of sounders are shown in Fig. 3.123. Normal speech is about 5 dB.

Fire alarm design considerations

Since all fire alarm installations must comply with the relevant statutory regulations, good practice recommends that contact be made with the local fire prevention officer at the design stage in order to identify any particular local regulations and obtain the necessary certification.

Larger buildings must be divided into zones so that the location of the fire can be quickly identified by the emergency services. The zones can be indicated on an indicator board situated in, for example, a supervisor's office or the main reception area.

In selecting the zones, the following rules must be considered:

1 Each zone should not have a floor area in excess of 2000 m^2.
2 Each zone should be confined to one storey, except where the total floor area of the building does not exceed 300 m^2.
3 Staircases and very small buildings should be treated as one zone.
4 Each zone should be a single fire compartment. This means that the walls, ceilings and floors are capable of containing the smoke and fire.

At least one fire alarm sounder will be required in each zone, but all sounders in the building must operate when the alarm is activated.

The main sounders may be silenced by an authorized person, once the general public have been evacuated from the building, but the current must be diverted to a supervisory buzzer which cannot be silenced until the system has been restored to its normal operational state.

A fire alarm installation may be linked to the local fire brigade's control room by the telecommunication network, if the permission of the fire authority and local telecommunication office is obtained.

The electricity supply to the fire alarm installation must be secure in the most serious conditions. In practice the most reliable supply is the mains supply, backed up by a 'standby' battery supply in case of mains failure. The supply should be exclusive to the fire alarm installation, fed from a separate switch fuse, painted red and labelled 'Fire Alarm – Do Not Switch Off'. Standby battery supplies should be capable of maintaining the system in full normal operation for at least 24 hours and, at the end of that time, be capable of sounding the alarm for at least 30 minutes.

Fire alarm circuits are Band I circuits and consequently cables forming part of a fire alarm installation must be physically segregated from all Band II circuits unless they are insulated for the highest voltage (IET Regulations 528.1 and 560.7.1).

Figure 3.123 Typical fire alarm sounders.

Intruder alarms

The installation of security alarm systems in the United Kingdom is already a multi-million-pound business and yet it is also a relatively new industry. As society becomes increasingly aware of crime prevention, it is evident that the market for security systems will expand.

Not all homes are equally at risk, but all homes have something of value to a thief. Properties in cities are at highest risk, followed by homes in towns and villages, and at least risk are homes in rural areas. A nearby motorway junction can, however, greatly increase the risk factor. Flats and maisonettes are the most vulnerable, with other types of property at roughly equal risk. Most intruders are young, fit and foolhardy opportunists. They ideally want to get in and away quickly but, if they can work unseen, they may take a lot of trouble to gain access to a property by, for example, removing the glass from a window.

Most intruders are looking for portable and easily saleable items such as mobile phones, television sets, home computers, jewellery, cameras, silverware, money, cheque books or credit cards. The Home Office has stated that only 7% of homes are sufficiently protected against intruders, although 75% of householders believe they are secure. Taking the simplest precautions will reduce the risk, while installing a security system can greatly reduce the risk of a successful burglary.

Security lighting

Security lighting is the first line of defence in the fight against crime. A recent study carried out by Middlesex University has shown that in two London boroughs the crime figures were reduced by improving the lighting levels. Police forces agree that homes and public buildings which are externally well illuminated are a much less attractive target for the thief.

Security lighting installed on the outside of the home may be activated by external detectors. These detectors sense the presence of a person outside the protected property and additional lighting is switched on. This will deter most potential intruders while also acting as courtesy lighting for visitors (Fig. 3.124).

Passive infra-red detectors

Passive infra-red (PIR) detector units allow a householder to switch on lighting units automatically whenever the area covered is approached by a moving body whose thermal radiation differs from the background. This type of detector is ideal for driveways or dark areas around the protected property. It also saves energy because the lamps are only switched on when someone approaches the protected area. The major contribution to security lighting comes from the 'unexpected' high-level illumination of an area when an intruder least expects it. This surprise factor often encourages the potential intruder to 'try next door'.

PIR detectors are designed to sense heat changes in the field of view dictated by the lens system. The field of view can be as wide as 180°, as shown by the diagram in Fig. 3.125. Many of the 'better' detectors use a split lens system so that a number of beams have to be broken before the detector switches on the security lighting. This capability overcomes the problem of false alarms, and a typical PIR is shown in Fig. 3.126.

PIR detectors are often used to switch LED or tungsten halogen floodlights because, of all available luminaires, LED or tungsten halogen offers instant high-level illumination. Light fittings must be installed out of reach of an intruder in order to prevent sabotage of the security lighting system.

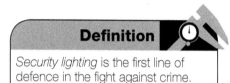

Definition

Security lighting is the first line of defence in the fight against crime.

Figure 3.124 Security lighting reduces crime.

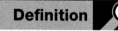

Definition

PIR detector units allow a householder to switch on lighting units automatically whenever the area covered is approached by a moving body whose thermal radiation differs from the background.

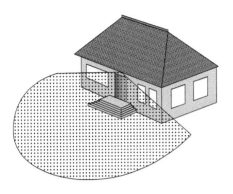

Figure 3.125 PIR detector and field of detection.

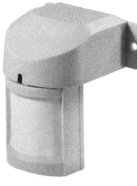

Figure 3.126 A typical PIR detector.

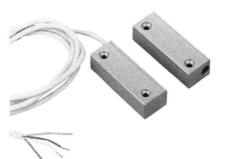

Figure 3.127 Proximity switches for perimeter protection.

Intruder alarm systems

Alarm systems are now increasingly considered to be an essential feature of home security for all types of homes and not just property in high-risk areas. An **intruder alarm system** serves as a deterrent to a potential thief and often reduces home insurance premiums. In the event of a burglary they alert the occupants, neighbours and officials to a possible criminal act and generate fear and uncertainty in the mind of the intruder which encourages a more rapid departure. Intruder alarm systems can be broadly divided into three categories – those which give perimeter protection, space protection or trap protection. A system can comprise one or a mixture of all three categories.

A **perimeter protection system** places alarm sensors on all external doors and windows so that an intruder can be detected as he or she attempts to gain access to the protected property. This involves fitting proximity switches to all external doors and windows.

A **movement or heat detector** placed in a room will detect the presence of anyone entering or leaving that room. PIR detectors and ultrasonic detectors give space protection. Space protection does have the disadvantage of being triggered by domestic pets but it is simpler and, therefore, cheaper to install. Perimeter protection involves a much more extensive and, therefore, expensive installation, but is easier to live with.

Trap protection places alarm sensors on internal doors and pressure pad switches under carpets on through routes between, for example, the main living area and the master bedroom. If an intruder gains access to one room he cannot move from it without triggering the alarm.

Proximity switches

These are designed for the discreet protection of doors and windows. They are made from moulded plastic and are about the size of a chewing-gum packet, as shown in Fig. 3.127. One moulding contains a reed switch, the other a magnet, and when they are placed close together the magnet maintains the contacts of the reed switch in either an open or closed position. Opening the door or window separates the two mouldings and the switch is activated, triggering the alarm.

PIR detectors

These are activated by a moving body which is warmer than the surroundings. The PIR shown in Fig. 3.128 has a range of 12 m and a detection zone of 110° when mounted between 1.8 and 2 m high.

Figure 3.128 PIR intruder alarm detector.

Figure 3.129 Intruder alarm sounder.

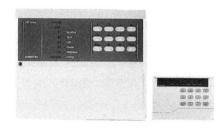

Figure 3.130 Intruder alarm control panel with remote keypad.

Intruder alarm sounders

Alarm sounders give an audible warning of a possible criminal act. Bells or sirens enclosed in a waterproof enclosure, such as those shown in Fig. 3.129, are suitable. It is usual to connect two sounders on an intruder alarm installation, one inside to make the intruder apprehensive and anxious, hopefully encouraging a rapid departure from the premises, and one outside. The outside sounder should be displayed prominently, since the installation of an alarm system is thought to deter the casual intruder and a ringing alarm encourages neighbours and officials to investigate a possible criminal act.

Control panel

The control panel, such as that shown in Fig. 3.130, is at the centre of the intruder alarm system. All external sensors and warning devices radiate from the control panel. The system is switched on or off at the control panel using a switch or coded buttons. To avoid triggering the alarm as you enter or leave the premises, there are exit and entry delay times to allow movement between the control panel and the door.

Supply

The supply to the intruder alarm system must be secure and this is usually achieved by an a.c. mains supply and battery back-up. Nickel–cadmium rechargeable cells are usually mounted in the sounder housing box.

Design considerations

It is estimated that there is now a 5% chance of being burgled, but the installation of a security system does deter a potential intruder. Every home in Britain will almost certainly contain electrical goods, money or valuables of value to an intruder. Installing an intruder alarm system tells the potential intruder that you intend to make his job difficult, which in most cases encourages him to look for easier pickings.

The type and extent of the intruder alarm installation, and therefore the cost, will depend upon many factors including the type and position of the building, the contents of the building, the insurance risk involved and the peace of mind offered by an alarm system to the owner or occupier of the building.

The designer must ensure that an intruder cannot sabotage the alarm system by cutting the wires or pulling the alarm box from the wall. Most systems will trigger if the wires are cut and sounders should be mounted in any easy-to-see but difficult-to-reach position.

Intruder alarm circuits are Band I circuits and should, therefore, be segregated from mains supply cables which are designated as Band II circuits or insulated to the highest voltage present if run in a common enclosure with Band II cables (IET Regulation 528.1).

Closed circuit television

Closed circuit television (CCTV) is now an integral part of many security systems. CCTV systems range from a single monitor with just one camera dedicated to monitoring perhaps a hotel car park, through to systems with many internal and external cameras connected to several locations for monitoring perhaps a shopping precinct.

CCTV cameras are also required to operate in total darkness when floodlighting is impractical. This is possible by using infra-red lighting which renders the scene under observation visible to the camera while to the human eye it appears to be in total darkness.

Cameras may be fixed or movable under remote control, such as those used for motorway traffic monitoring. Typically an external camera would be enclosed in a weatherproof housing such as the one shown in Fig. 3.131. Using remote control, the camera can be panned, tilted or focused and have its viewing screen washed and wiped.

Pictures from several cameras can be multiplexed on to a single co-axial video cable, together with all the signals required for the remote control of the camera.

A permanent record of the CCTV pictures can be stored and replayed by incorporating a video tape recorder into the system, as is the practice in most banks and building societies.

Security cameras should be robustly fixed and cable runs designed so that they cannot be sabotaged by a potential intruder.

Figure 3.131 CCTV camera.

Emergency lighting (BS 5266 and BS EN 1838)

Emergency lighting should be planned, installed and maintained to the highest standards of reliability and integrity, so that it will operate satisfactorily when called into action, no matter how infrequently this may be.

Emergency lighting is not required in private homes because the occupants are familiar with their surroundings, but in public buildings people are in unfamiliar surroundings. In an emergency people do not always act rationally, but well-illuminated and easily identified exit routes can help to reduce panic.

Emergency lighting is provided for two reasons; to illuminate escape routes, called 'escape' lighting; and to enable a process or activity to continue after a normal lights failure, called 'standby' lighting.

Escape lighting is usually required by local and national statutory authorities under legislative powers. The escape lighting scheme should be planned so that identifiable features and obstructions are visible in the lower levels of illumination which may prevail during an emergency. Exit routes should be clearly indicated by signs and illuminated to a uniform level, avoiding bright and dark areas.

Standby lighting is required in hospital operating theatres and in industry, where an operation or process once started must continue, even if the mains lighting fails. Standby lighting may also be required for security reasons. The cash points in public buildings may need to be illuminated at all times to discourage acts of theft occurring during a mains lighting failure.

Definition

Emergency lighting is not required in private homes because the occupants are familiar with their surroundings, but in public buildings people are in unfamiliar surroundings. In an emergency people do not always act rationally, but well-illuminated and easily identified exit routes can help to reduce panic.

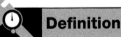
Definition

Emergency lighting is provided for two reasons: to illuminate escape routes, called 'escape' lighting; and to enable a process or activity to continue after a normal lights failure, called 'standby' lighting.

Emergency supplies

Since an emergency occurring in a building may cause the mains supply to fail, the emergency lighting should be supplied from a source which is independent from the main supply. In most premises the alternative power supply would be from batteries, but generators may also be used. Generators can have a large capacity and duration, but a major disadvantage is the delay of time while the generator runs up to speed and takes over the load. In some premises a delay of more than five seconds is considered unacceptable, and in these cases a battery supply is required to supply the load until the generator can take over.

The emergency lighting supply must have an adequate capacity and rating for the specified duration of time (IET Regulation 313.2). BS 5266 and BS EN 1838 states that after a battery is discharged by being called into operation for its specified duration of time, it should be capable of once again operating for the specified duration of time following a recharge period of not longer than 24 hours. The duration of time for which the emergency lighting should operate will be specified by a statutory authority but is normally one to three hours. The British Standard states that escape lighting should operate for a minimum of 1 hour. Standby lighting operation time will depend upon financial considerations and the importance of continuing the process or activity.

There are two possible modes of operation for emergency lighting installations: maintained and non-maintained.

Maintained emergency lighting

In a maintained system the emergency lamps are continuously lit using the normal supply when this is available, and change over to an alternative supply when the mains supply fails. The advantage of this system is that the lamps are continuously proven healthy and any failure is immediately obvious. It is a wise precaution to fit a supervisory buzzer or LED indicator in the emergency supply to prevent accidental discharge of the batteries, since it is not otherwise obvious which supply is being used.

Maintained emergency lighting is normally installed in theatres, cinemas, discotheques and places of entertainment where the normal lighting may be dimmed or extinguished while the building is occupied. The emergency supply for this type of installation can also be supplied from a central battery, the emergency lamps being wired in parallel from the low-voltage supply as shown in Fig. 3.132. Escape sign lighting units used in commercial facilities should be wired in the maintained mode.

Non-maintained emergency lighting

In a non-maintained system the emergency lamps are only illuminated if the normal mains supply fails. Failure of the main supply de-energizes a solenoid and

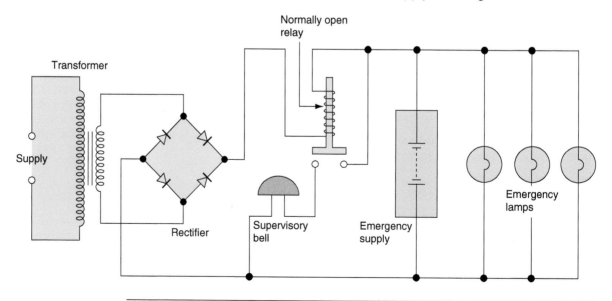

Figure 3.132 Maintained emergency lighting.

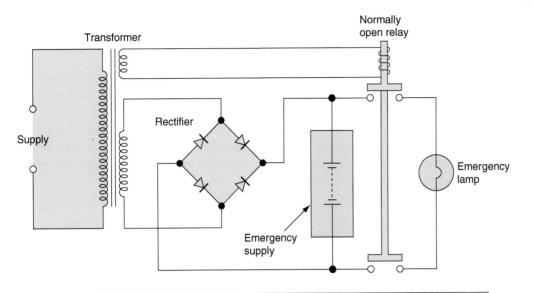

Figure 3.133 Non-maintained emergency lighting.

a relay connects the emergency lamps to a battery supply, which is maintained in a state of readiness by a trickle charge from the normal mains supply. When the normal supply is restored, the relay solenoid is energized, breaking the relay contacts, which disconnects the emergency lamps, and the charger recharges the battery. Figure 3.133 illustrates this arrangement.

The disadvantage with this type of installation is that broken lamps are not detected until they are called into operation in an emergency, unless regularly maintained. The emergency supply is usually provided by a battery contained within the luminaire, together with the charger and relay, making the unit self-contained. Self-contained units are cheaper and easier to install than a central battery system, but the central battery can have a greater capacity and duration, and permit a range of emergency lighting luminaires to be installed.

Maintenance

The contractor installing the emergency lighting should provide a test facility which is simple to operate and secure against unauthorized interference, usually a simple key switch. The emergency lighting installation must be segregated completely from any other wiring, so that a fault on the main electrical installation cannot damage the emergency lighting installation (IET Regulation 528.1). Figure 3.134 shows a trunking which provides for segregation of circuits.

The batteries used for the emergency supply should be suitable for this purpose. Motor vehicle batteries are not suitable for emergency lighting applications, except in the starter system of motor-driven generators. The fuel supply to a

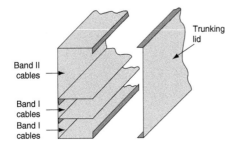

Figure 3.134 Segregation of cables in trunking.

motor-driven generator should be checked. The battery room of a central battery system must be well ventilated and, in the case of a motor-driven generator, adequately heated to ensure rapid starting in cold weather.

The British Standard recommends that the full load should be carried by the emergency supply for at least one hour in every six months. After testing, the emergency system must be carefully restored to its normal operative state. A record should be kept of each item of equipment and the date of each test by a qualified or responsible person. It may be necessary to produce the record as evidence of satisfactory compliance with statutory legislation to a duly authorized person.

Self-contained units are suitable for small installations of up to about 12 units. The batteries contained within these units should be replaced about every five years, or as recommended by the manufacturer, and be connected to the a.c. mains supply through a 'test' switch.

Telephone socket outlets

The installation of telecommunications equipment could, for many years, only be undertaken by British Telecom engineers, but today an electrical contractor may now supply and install telecommunications equipment.

On new premises the electrical contractor may install sockets and the associated wiring to the point of intended line entry, but the connection of the incoming line to the installed master socket must only be made by the telephone company's engineer.

On existing installations, additional secondary sockets may be installed to provide an extended plug-in facility, as shown in Fig. 3.135. Any number of secondary sockets may be connected in parallel, but the number of telephones which may be connected at any one time is restricted.

Each telephone or extension bell is marked with a ringing equivalence number (REN) on the underside. Each exchange line has a maximum capacity of REN 4 and, therefore, the total REN values of all the connected telephones must not exceed four if they are to work correctly.

An extension bell may be connected to the installation by connecting the two bell wires to terminals 3 and 5 of a telephone socket. The extension bell must be of the high impedance type having an REN rating. All equipment connected to a BT exchange line must display the green circle of approval.

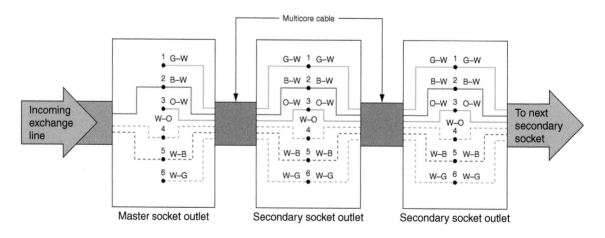

Figure 3.135 Telephone circuit outlet connection diagram.

Table 3.10 Telephone cable identification

Code	Base colour	Stripe
G–W	Green	White
B–W	Blue	White
O–W	Orange	White
W–O	White	Orange
W–B	White	Blue
W–G	White	Green

Table 3.11 Telephone socket terminal identification. Terminals 1 and 6 are frequently unused, and therefore four-core cable may normally be installed. Terminal 4, on the incoming exchange line, is only used on a PBX line for earth recall

Socket terminal	Circuit
1	Spare
2	Speech circuit
3	Bell circuit
4	Earth recall
5	Speech circuit
6	Spare

The multi-core cable used for wiring extension socket outlets should be of a type intended for use with telephone circuits, which will normally be between 0.4 and 0.68 mm in cross-section. Telephone cable conductors are identified in Table 3.10 and the individual terminals in Table 3.11. The conductors should be connected as shown in Fig. 3.135. Telecommunications cables are Band I circuits and must be segregated from Band II circuits containing mains cables (IET Regulation 528.1).

When you have completed the questions, check out the answers at the back of the book.

Note: more than one multiple-choice answer may be correct.

1   The opposition to current flow in an a.c. resistive circuit is called:
   a.   resistance
   b.   inductance
   c.   reactance
   d.   impedance.

2   The opposition to current flow in an a.c. capacitive or inductive current is called:
   a.   resistance
   b.   inductance
   c.   reactance
   d.   impedance.

3   The total opposition to current flow in any a.c. circuit is called:
   a.   resistance
   b.   inductance
   c.   reactance
   d.   impedance.

4   A straight line having definite length and direction that represents to scale a quantity such as current voltage or impedance is called:
   a.   a series a.c. circuit
   b.   capacitive reactance
   c.   a phasor, as in a phasor diagram
   d.   the impedance triangle.

5   An a.c. series circuit has an inductive reactance of $4\,\Omega$ and a resistance of $3\,\Omega$. The impedance of this circuit will be:
   a. $5\,\Omega$
   b. $7\,\Omega$
   c. $12\,\Omega$
   d. $25\,\Omega$.

6   An a.c. series circuit has a capacitive reactance of $12\,\Omega$ and a resistance of $9\,\Omega$. The impedance of this current will be:
   a. $3\,\Omega$
   b. $15\,\Omega$
   c. $20\,\Omega$
   d. $108\,\Omega$.

7   The inductive reactance of a 100 mH coil when connected to a 50 Hz supply
    will be:
    a.   5 Ω
    b.   20 Ω
    c.   31.42 Ω
    d.   31.42 kΩ.

8   The capacitive reactance of a 100 µF capacitor when connected to a 50 Hz
    supply will be:
    a.   5 Ω
    b.   20 Ω
    c.   31.8 Ω
    d.   31.8 kΩ.

9   A circuit with *bad* power factor causes:
    a.   a fall in the supply voltage
    b.   an increase in the supply voltage
    c.   more current to be taken from the supply
    d.   less current to be taken from the supply.

10 One application for a series d.c. motor is:
    a.   an electric train
    b.   a microwave oven
    c.   a central heating pump
    d.   an electric drill.

11 One application for an a.c. induction motor is:
    a.   an electric train
    b.   a microwave oven
    c.   a central heating pump
    d.   an electric drill.

12 One application for a shaded pole a.c. motor is:
    a.   an electric train
    b.   a microwave oven
    c.   a central heating pump
    d.   an electric drill.

13 An electromagnetic switch operated by a solenoid is one definition of:
    a.   a transformer
    b.   an a.c. motor
    c.   a relay
    d.   an inductive coil.

14 An electronic circuit resistor is colour coded green, blue, brown, gold. It has
    a value of:
    a.   56 Ω 6 10%
    b.   65 Ω 6 5%
    c.   560 Ω 6 5%
    d.   650 Ω 6 10%.

15 An electronic device which will allow current to flow through it in one direction only is a:
   a.   light-dependent resistor (LDR)
   b.   light-emitting diode (LED)
   c.   semiconductor diode
   d.   thermistor.

16 An electronic device whose resistance varies with temperature is a:
   a.   light-dependent resistor (LDR)
   b.   light-emitting diode (LED)
   c.   semiconductor diode
   d.   thermistor.

17 An electronic device which emits red, green, or yellow light when a current of about 10 mA flows through it is a:
   a.   light-dependent resistor (LDR)
   b.   light-emitting diode (LED)
   c.   semiconductor diode
   d.   thermistor.

18 An electronic device whose resistance changes as a result of light energy falling upon it is a:
   a.   light-dependent resistor (LDR)
   b.   light-emitting diode (LED)
   c.   semiconductor diode
   d.   thermistor.

19 A street lamp has a luminous intensity of 2000 cd and is suspended 5 m above the ground. The illuminance on the pavement below the lamp will be:
   a.   40 lx
   b.   80 lx
   c.   400 lx
   d.   800 lx.

20 An electric motor is connected to a motor starter to:
   a.   improve the efficiency of the motor
   b.   provide the initial starting torque for the motor
   c.   reduce heavy starting currents
   d.   provide overload and no-volt protection.

21 The light output of a high pressure mercury vapour lamp is a:
   a.   blue white but cold light with fairly good colour rendering
   b.   very yellow light with poor colour rendering
   c.   warm yellow light with good colour rendering
   d.   cold white light with very good colour rendering properties, making it suitable for colour photography.

22 The light output of a high-pressure sodium lamp is a:

   a.   blue white but cold light with fairly good colour rendering

   b.   very yellow light with poor colour rendering

   c.   warm yellow light with good colour rendering

   d.   cold white light with very good colour rendering properties, making it suitable for colour photography.

23 The light output of a metal halide lamp is a:

   a.   blue white but cold light with fairly good colour rendering

   b.   very yellow light with poor colour rendering

   c.   warm yellow light with good colour rendering

   d.   cold white light with very good colour rendering properties, making it suitable for colour photography.

24 The light output of a low pressure sodium lamp is a:

   a.   blue white but cold light with fairly good colour rendering

   b.   very yellow light with poor colour rendering

   c.   warm yellow light with good colour rendering

   d.   cold white light with very good colour rendering properties, making it suitable for colour photography.

25 An electrical immersion water heating element gives up its heat energy to the water by a process called:

   a.   conduction

   b.   convection

   c.   radiation

   d.   dissimilar metals.

26 Sketch the construction of a simple alternator and label all the parts.

27 State how an e.m.f. is induced in an alternator. Sketch and name the shape of the generated e.m.f.

28 Calculate or state the average r.m.s. and maximum value of the domestic a.c. mains supply and show these values on a sketch of the mains supply.

29 Use a sketch with notes of explanation to describe 'good' and 'bad' power factor.

30 State how power factor correction is achieved on:

   a.   a fluorescent light fitting

   b.   an electric motor.

31 Use a sketch to help you describe the meaning of the words:

   a.   inductance

   b.   mutual inductance.

32 Use a sketch with notes of explanation to describe how a force is applied to a conductor in a magnetic circuit and how this principle is applied to an electric motor.

33 Use a sketch with notes of explanation to show how a turning force is applied to the rotor and, therefore, the drive shaft of an electric motor.

34 Give three applications for each of the following types of motor:
   a. a d.c. series motors
   b. an a.c. induction motor
   c. an a.c. split-phase motor
   d. an a.c. shaded pole motor.

35 Sketch an electronic circuit which will give full-wave rectification of the a.c. mains supply. Assume a 12 V output from a 230 V supply.

36 Briefly describe how a fuse operates or blows when a fault occurs in a circuit protected by a fuse.

37 Briefly describe how an MCB gives protection to a circuit under overload conditions.

38 Briefly describe how an MCB gives short-circuit protection to a circuit.

39 Briefly describe how an RCD operates under fault conditions. You may find a sketch helpful to your explanation.

40 Briefly describe how an RCBO gives protection to a circuit. Why is this device superior to an MCB or RCD?

41 Briefly describe what we mean by the stroboscopic effect when considering a lighting scheme for an industrial environment.

42 With the aid of a sketch, briefly describe the principle of operation of a room thermostat. Where would you install this device in a building?

43 With the aid of a sketch, briefly describe the principle of operation of an immersion heater thermostat.

44 Sketch the circuit diagram for a switch-start fluorescent luminare.

45 Describe with the aid of a sketch:
   a. an LED signal lamp and
   b. an LED energy efficient lamp in a GU10 enclosure.
   State an application for each lamp.

Electrical installations: Fault diagnosis and rectification

Unit 303 of the City and Guilds 2365-03 syllabus

Learning outcomes – when you have completed this chapter you should:

- understand how electrical fault diagnosis is reported;
- understand how electrical faults are diagnosed;
- understand the process of fault rectification;
- be able to diagnose faults on electrical systems.

This chapter has free associated content, including animations and instructional videos, to support your learning.

When you see the logo, visit the website below to access this material: www.routledge.com/cw/linsley

Recording faults

If an electrical installation fault is identified as a result of a periodic inspection for the purpose of completing an Electrical Installation Condition Report, then it should be documented as detailed in Appendix 6 of the IET Regulations.

The 18th Edition of the IET Regulations has introduced changes to the Periodic Inspection Report. This has now become the Electrical Installation Condition Report and incorporates changes to the classification coding system and a new inspection schedule that you will find in Appendix 6 of the IET Regulations. The inspection schedule for a domestic installation includes over 60 items which must be inspected and the electrician carrying out the inspection must comment on the condition of each item using the coding system described below. **The 18th Edition of the IET Wiring Regulations now requires that an inspection of accessible roof spaces be carried out where electrical equipment is present in that roof space.**

The objective in producing this new Electrical Installation Condition Report inspection schedule is to provide the electrician carrying out the inspection with guidelines, so that the inspection is done in a structured and consistent way and for the client to better understand the result of the inspection.

The classification codes to be used on the condition report inspection schedule are:

✓ meaning an acceptable condition. The item inspected has been classified as acceptable.
C1 meaning an unacceptable condition. The item inspected has been classified as unacceptable. Immediate danger is present and the safety of those using the installation is at risk. For example, live parts are directly accessible. *Immediate remedial action is required.*
C2 meaning an unacceptable condition. The item inspected has been classified as unacceptable. Potential danger is present and the safety of those using the installation may be at risk. For example, the absence of the main protective bonding conductors. *Urgent remedial action is required.*
C3 meaning improvement is recommended. The item inspected is not dangerous for continued use but the inspector recommends that improvements be made. For example, the installation does not have RCDs installed for additional protection. *Improvements are recommended.*
FI meaning Further Investigation is required. This investigation has uncovered a deficiency. The item inspected has been classified as unacceptable. The deficiency cannot be identified at this time and further investigation is required. *Urgent remedial action is required.*

Let us assume that while carrying out a visual inspection, the electrician observes a badly damaged socket outlet. On close inspection he realizes that it is possible to touch a live part through the cracked plate. This fault will attract a C1 code because there is immediate danger.

Alternatively, if on close inspection the electrician realizes that it is possible to touch an earth connection, then it will attract a C2 code: there is potential danger here because the earth wire may become live under fault conditions.

The presence of a C1 or C2 code will result in the overall condition of the electrical installation being reported as unsatisfactory – it has failed to meet the standard. The inspector should report their findings to the duty holder or person responsible for the installation immediately, both verbally and in writing, that a risk of injury exists. If possible, immediately dangerous situations, given a C1 code, should be made safe, isolated or rectified on discovery.

A code C3 in itself would not warrant an unsatisfactory result. It is simply stating that while the installation is not compliant with the latest edition of the IET Regulations, it is compliant with a previous edition, and the item is not unsafe and does not necessarily require upgrading.

N/V meaning not verified. Although a particular item on the schedule is relevant to the installation, the inspector was unable to verify its condition.

LIM meaning limitation. A particular item on the schedule is relevant to the installation but there were certain limitations in being able to check the condition.

N/A meaning not applicable. The particular item on the schedule is not relevant to the installation being inspected.

The Electrical Installation Condition Report should only be used for an existing building. The report should include:

- schedules of both the inspection and test results in section A;
- the reasons for producing the report such as a change of occupancy or landlord's maintenance should be identified in section B;
- the extent and limitations of the report should be stated in section D;
- the inspector producing the final report should give a summary of the condition of the installation in terms of the safety of the installation in section E, stating how any C1, C2 and C3 codes given for items in the schedule may affect the overall condition of the installation being classified as unsatisfactory;
- the recommended date of the next Electrical Installation Condition Report should be given in section F.

See Section G of the ON Site Guide. Further information on the new coding system can be downloaded free from the Electricity Safety Council's Best Practice Guide 4.

Fault diagnosis

To diagnose and find faults in electrical installations and equipment is probably one of the most difficult tasks undertaken by an electrician. The knowledge of fault finding and the diagnosis of faults can never be completely 'learned' because no two fault situations are exactly the same. As the systems we install become more complex, then the faults developed on these systems become more complicated to solve. To be successful the individual must have a thorough knowledge of the installation or piece of equipment and have a broad range of the skills and competences associated with the electrical industries.

The ideal person will tackle the problem using a reasoned and logical approach, recognize his own limitations and seek help and guidance where necessary.

The tests recommended by the IET Regulations can be used as a diagnostic tool but the safe working practices described by the Electricity at Work Regulations and elsewhere must always be observed during the fault-finding procedures.

If possible, fault finding should be planned ahead to avoid inconvenience to other workers and to avoid disruption of the normal working routine. **However, a faulty piece of equipment or a fault in the installation is not normally a planned event and usually occurs at the most inconvenient time.** The diagnosis and rectification of a fault is therefore often carried out in very stressful circumstances.

Symptoms of an electrical fault

The basic symptoms of an electrical fault may be described in one or a combination of the following ways:

1 There is a complete loss of power.
2 There is partial or localized loss of power.
3 The installation or piece of equipment is failing because of the following:

- an individual component is failing;
- the whole plant or piece of equipment is failing;
- the insulation resistance is low;
- the overload or protective devices operate frequently;
- electromagnetic relays will not latch, giving an indication of undervoltage.

Causes of electrical faults

A fault is not a natural occurrence; it is an unplanned event which occurs unexpectedly. The fault in an electrical installation or piece of equipment may be caused by:

- negligence – that is, lack of proper care and attention;
- misuse – that is, not using the equipment properly or correctly;
- abuse – that is, deliberate ill-treatment of the equipment.

If the installation was properly designed in the first instance to perform the tasks required of it by the user, then the *negligence, misuse or abuse* must be the fault of the user. However, if the installation does not perform the tasks required of it by the user, then the negligence is due to the electrical contractor not designing the installation to meet the needs of the user.

Negligence on the part of the user may be due to insufficient maintenance or lack of general care and attention, such as not repairing broken equipment or removing covers or enclosures which were designed to prevent the ingress of dust or moisture.

Misuse of an installation or piece of equipment may occur because the installation is being asked to do more than it was originally designed to do, due to the expansion of a company, for example. Circuits are sometimes overloaded

Definition

A *fault* is not a natural occurrence; it is an unplanned event which occurs unexpectedly.

because a company grows and a greater demand is placed on the existing installation by the introduction of new or additional machinery and equipment.

Where do electrical faults occur?

1   Faults occur in wiring systems, but not usually along the length of the cable, unless it has been damaged by a recent event such as an object being driven through it or a JCB digger pulling up an underground cable. Cable faults usually occur at each end, where the human hand has been at work at the point of cable interconnections. This might result in broken conductors, trapped conductors or loose connections in joint boxes, accessories or luminaires.

All cable connections must be made mechanically and electrically secure. They must also remain accessible for future inspection, testing and maintenance (IET Regulation 526.3). The only exceptions to this rule are when:

- underground cables are connected in a compound-filled or encapsulated joint;
- floor-warming or ceiling-warming heating systems are connected to a cold tail;
- a joint is made by welding, brazing, soldering or a compression tool;
- a joint is made in a maintenance-free accessory complying with BS 5733 and marked with the symbol MF, as shown in Fig. 6.32 of Chapter 6.

Since they are accessible, cable interconnections are an obvious point of investigation when searching out the cause of a fault.

2   Faults also occur at cable terminations. The IET Regulations require that a cable termination of any kind must securely anchor all conductors to reduce mechanical stresses on the terminal connections. All conductors of flexible cords must be terminated within the terminal connection; otherwise the current carrying capacity of the conductor is reduced, which may cause local heating. Flexible cords and fine multiwire conductors are delicate – has the terminal screw been over-tightened, thus breaking the connection as the conductors flex or vibrate? Cables and flexible cords must be suitable for the temperature to be encountered at the point of termination or must be provided with additional insulation sleeves to make them suitable for the surrounding temperatures (IET Regulations 522.2 and 526.9).

3   Faults also occur at accessories such as switches, sockets, control gear, motor contactors or at the point of connection with electronic equipment. The source of a possible fault is again at the point of human contact with the electrical system and again the connections must be checked as described in the first two points above. Contacts that make and break a circuit are another source of wear and possible failure, so switches and motor contactors may fail after extensive use. Socket outlets that have been used extensively and loaded to capacity, in, say, kitchens, are another source of fault due to overheating or loose connections. Electronic equipment can be damaged by the standard tests described in the IET Regulations and must, therefore, be disconnected before testing begins.

4   Faults occur on instrumentation panels either as a result of a faulty instrument or as a result of a faulty monitoring probe connected to the instrument. Many panel instruments are standard sizes connected to CTs or VTs and this is another source of possible faults of the types described in points 1 to 3.

5   Faults occur in protective devices for the reasons given in points 1 to 3 above but also because they may have been badly selected for the job in hand and

Figure 4.1 Using a voltage tester.

Key fact

Whatever method is used to make the connection in conductors, the connection must be both electrically and mechanically sound if we are to avoid high resistance joints, corrosion and erosion at the point of termination.

Safety first

Cable fault
- faults do occur in wiring systems,
- but not usually along the cable length;
- faults usually occur at each end,
- where the human hand has been at work making connections.

do not offer adequate protection, or selectivity as described in Chapter 6, Fig. 6.40 of this book.

6   Faults often occur in luminaires (light fittings) because the lamp has expired. Discharge lighting (fluorescent fittings) also require a 'starter' to be in good condition, although many fluorescent luminaires these days use starter-less electronic control gear. The points made in 1 to 3 about cable and flexible cord connections are also relevant to luminaire faults.

7   Faults occur when terminating flexible cords and fine multiwire conductors as a result of the flexible cable being of a smaller cross-section than the load demands, because it is not adequately anchored to reduce mechanical stresses on the connection or because the flexible cord is not suitable for the ambient temperature to be encountered at the point of connection. When terminating flexible cords, the insulation should be carefully removed without cutting out any flexible cord strands of wire because this effectively reduces the cross-section of the conductor. The conductor strands should be twisted together and then doubled over, if possible, and terminated in the appropriate connection, as shown in Fig. 4.2. The connection screws should be opened fully so that they will not snag the flexible cord as it is eased into the connection. The insulation should go up to, but not into, the termination. The terminal screws should then be tightened. When terminating very fine conductors, see also IET Regulation 526.9 which gives us the following advice:

- To avoid separation or spreading of individual wires, suitable terminals must be used or the conductor ends treated; for example, by enclosing the individual wires of multiwire in a brass ferrule or claw washer as shown in Figure 4.2.
- Soldering or tinning of the whole conductor end of multiwire is not permitted if screw terminals are used.
- Soldered or tinned conductor ends are not permissible at connection and junction points which may be subject in service to relative movement or vibration.

8   Faults occur in electrical components, equipment and accessories such as motors, starters, switch gear, control gear, distribution panels, switches,

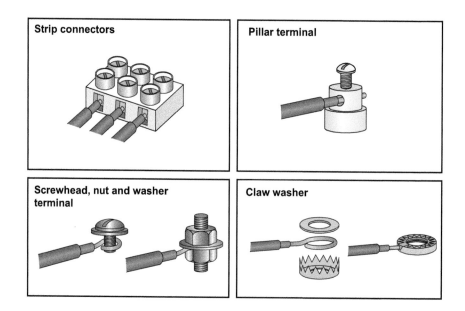

Figure 4.2 Methods of connecting small conductors.

sockets and luminaires because these all have points at which electrical connections are made. It is unusual for an electrical component to become faulty when it is relatively new because it will have been manufactured and tested to comply with the appropriate British Standard. Through overuse or misuse components and equipment do become faulty but most faults are caused by poor installation techniques.

9 The 18th Edition of the IET Regulations at 532.6 and 421.1.7 has introduced a new topic, Arc Fault Detection Devices. (AFDD) These are compulsory in the EU Countries but only recommended in this country by the IET and the British standards. They are recognised as giving additional protection against fires caused by arc faults. Arc faults can be formed by cable insulation defects, damage to cables by impact or penetration by nails and screws, and loose terminal connections. AFDDs detect faults which MCBs and RCDs cannot detect. They are installed at the origin of the final circuit to be protected.

Modern electrical installations using new materials can now last longer than 50 years. Therefore, they must be properly installed by skilled (electrically) or instructed (electrically) persons. Good design, good workmanship and the use of proper materials are essential if the installation is to comply with the relevant regulations (IET Regulations 133.1 to 134.1).

Fault finding

Before an electrician can begin to diagnose the cause of a fault he must:

- have a thorough knowledge and understanding of the electrical installation or electrical equipment;
- collect information about the fault and the events occurring at or about the time of the fault from the people who were in the area at the time;
- begin to predict the probable cause of the fault using his own and other people's skills and expertise;
- test some of the predictions using a logical approach to identify the cause of the fault.

Most importantly, electricians must use their detailed knowledge of electrical circuits and equipment learned through training and experience and then apply this knowledge to look for a solution to the fault.

Let us, therefore, now briefly consider some of the basic wiring circuits.

Lighting circuits

Table A1 in Appendix A of the *On Site Guide* deals with the assumed current demand of points, and states that for lighting outlets we should assume a current equivalent to a minimum of 100 W per lampholder. This means that for a domestic lighting circuit rated at 5A, a maximum of 11 lighting outlets could be connected to each circuit. In practice, it is usual to divide the fixed lighting outlets into a convenient number of circuits of seven or eight outlets each. In this way the whole installation is not plunged into darkness if one lighting circuit fails (IET Regulation 314.1).

Lighting circuits are usually wired in 1.0 or 1.5 mm cable using either a loop-in or joint-box method of installation. The loop-in method is universally employed with conduit installations or when access from above or below is prohibited after installation, as is the case with some industrial installations or blocks of flats. In this method the only joints are at the switches or lighting points, the live conductors being looped from switch to switch and the neutrals from one lighting point to another.

The use of junction boxes with fixed brass terminals is the method often adopted in domestic installations, since the joint boxes can be made accessible but are out of site in the loft area and under floorboards.

The live conductors must be broken at the switch position in order to comply with the IET Regulations (612.6). A ceiling rose may only be connected to installations operating at 250 V maximum and must only accommodate one flexible cord unless it is specially designed to take more than one (IET Regulations 559.6.1.1 to 559.6.1.3). Lampholders suspended from flexible cords must be capable of suspending the mass of the luminaire fixed to the lampholder (559.6.1.5).

The type of circuit used will depend upon the installation conditions and the customer's requirements. One light controlled by one switch is called one-way switch control. A room with two access doors might benefit from a two-way switch control so that the lights may be switched on or off at either position. A long staircase with more than two switches controlling the same lights would require intermediate switching.

One-way, two-way or intermediate switches can be obtained as plate switches for wall mounting or ceiling-mounted cord switches. Cord switches can provide a convenient method of control in bedrooms or bathrooms and for independently controlling an office luminaire.

Socket outlet circuits

Where portable equipment is to be used, it should be connected by a plug to a conveniently accessible socket outlet (IET Regulation 553.1.7). Pressing the plug into a socket outlet connects the appliance to the source of supply. **Socket outlets** therefore provide an easy and convenient method of connecting portable electrical appliances to a source of supply.

Socket outlets can be obtained in 15, 13, 5 and 2A ratings, but the 13A flat pin type complying with BS 1363 is the most popular for domestic installations in the United Kingdom. Each 13A plug contains a cartridge fuse to give maximum potential protection to the flexible cord and the appliances which it serves.

Definition
Socket outlets provide an easy and convenient method of connecting portable electrical appliances to a source of supply.

Figure 4.3 Socket outlet.

Socket outlets may be wired on a ring or radial circuit and in order that every appliance can be fed from an adjacent and conveniently accessible socket outlet, the number of sockets is unlimited provided that the floor area covered by the circuit does not exceed that given in Appendix 15 of the IET Regulations and Table H7 in Appendix H of the On Site Guide.

In a radial circuit each socket outlet is fed from the previous one. Live is connected to live, neutral to neutral and earth to earth at each socket outlet. The fuse and cable sizes are given in Appendix 15 but circuits may also be expressed with a block diagram, as shown in Fig. 4.4. The number of permitted socket outlets is unlimited but each radial circuit must not exceed the floor area stated and the known or estimated load.

Where two or more circuits are installed in the same premises, the socket outlets and permanently connected equipment should be reasonably shared out among the circuits, so that the total load is balanced.

When designing ring or radial circuits special consideration should be given to the loading in kitchens which may require separate circuits. This is because the maximum demand of current-using equipment in kitchens may exceed the rating of the circuit cable and protection devices.

Ring and radial circuits may be used for domestic or other premises where the maximum demand of the current-using equipment is estimated not to exceed the rating of the protective devices for the chosen circuit.

Ring circuits are very similar to radial circuits in that each socket outlet is fed from the previous one, but in ring circuits the last socket is wired back to the source of supply. Each ring final circuit conductor must be looped into every socket outlet or joint box which forms the ring and must be electrically continuous throughout its length. The number of permitted socket outlets is unlimited but each ring circuit must not cover more than $100\,\text{m}^2$ of floor area.

The circuit details are given in Appendix 15 of the IET Regulations but may also be expressed by the block diagram given in Fig. 4.5.

Additional protection by 30 mA residual current device (RCD) is now required in addition to overcurrent protection for all socket outlet circuits that are to be used by ordinary persons and intended for general use.

This additional protection is provided in case basic protection or fault protection fails or if the user of the installation is careless (IET Regulations 411.3.3, 415.1.1 and 522.6.102).

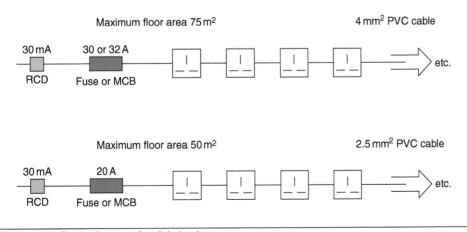

Figure 4.4 Block diagram of radial circuits.

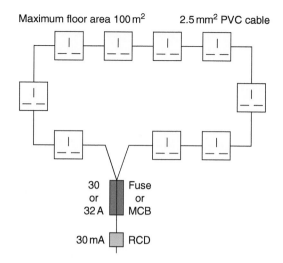

Figure 4.5 Block diagram of ring circuits.

Note: An ordinary person is one who is neither an electrically skilled nor an instructed person.

Designing out faults

The designer of the installation cannot entirely design out the possibility of a fault occurring but he can design in 'damage limitation' should a fault occur.

For example, designing in two, three or four lighting and power circuits will reduce the damaging effect of any one circuit failing because not all lighting and power will be lost as a result of a fault. Limiting faults to only one of many circuits is good practice because it limits the disruption caused by a fault. IET Regulation 314 tells us to divide an installation into circuits as necessary so as to:

1 avoid danger and minimize inconvenience in the event of a fault occurring;
2 facilitate safe operation, inspection testing and maintenance;
3 reduce unwanted RCD tripping, sometimes called 'nuisance tripping'.

Requirements for successful electrical fault finding

The steps involved in successfully finding a fault can be summarized as follows:

1 Gather *information* by talking to people and looking at relevant sources of information such as manufacturer's data, circuit diagrams, charts and schedules.
2 *Analyse* the evidence and use standard tests and a visual inspection to predict the cause of the fault.
3 *Interpret* test results and diagnose the cause of the fault.
4 *Rectify* the fault.
5 *Carry out* functional tests to verify that the installation or piece of equipment is working correctly and that the fault has been rectified.

Requirements for safe working procedures

The following five safe working procedures must be applied before undertaking the fault diagnosis:

1 The circuits must be isolated using a 'safe isolation procedure', such as that described below.
2 All test equipment must be 'approved' and calibrated and connected to the test circuits by recommended test probes as described by the Health and Safety Executive (HSE) Guidance Note GS 38 and shown in Fig. 4.6. The test equipment used must also be 'proved' on a known supply or by means of a proving unit such as that shown in Fig. 4.7.
3 Isolation devices must be 'secured' in the 'off' position as shown in Fig. 4.8. The key is retained by the person working on the isolated equipment.
4 Warning notices must be posted.
5 All relevant safety and functional tests must be completed before restoring the supply.

Live testing

The Electricity at Work Regulations tell us that it is 'preferable' that supplies be made dead before work commences (Regulation 4(3)). However, they do acknowledge that some work, such as fault finding and testing, may require the electrical equipment to remain energized. Therefore, if the fault finding and testing can only be successfully carried out 'live', then the person carrying out the fault diagnosis must:

* be trained so that he understands the equipment and the potential hazards of working live and can, therefore, be deemed to be 'skilled (electrically)' to carry out the activity;
* only use approved test equipment;
* set up barriers and warning notices so that the work activity does not create a situation dangerous to others.

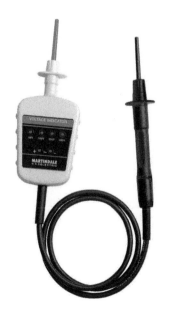

Figure 4.6 Typical voltage indicator.

Figure 4.8 Secure isolation of a supply.

Figure 4.7 Voltage proving unit.

Note that while live testing may be required in order to find the fault, live repair work must not be carried out. The individual circuit or item of equipment must first be isolated.

Secure electrical isolation

Electric shock occurs when a person becomes part of the electrical circuit. The level or intensity of the shock will depend upon many factors, such as age, fitness and the circumstances in which the shock is received. The lethal level is approximately 50 mA, above which muscles contract, the heart flutters and breathing stops. A shock above the 50 mA level is therefore fatal unless the person is quickly separated from the supply. Below 50 mA only an unpleasant tingling sensation may be experienced or you may be thrown across a room or shocked enough to fall from a roof or ladder, but the resulting fall may lead to serious injury.

To prevent people from receiving an electric shock accidentally, all circuits contain protective devices. All exposed metal is earthed; fuses and miniature circuit-breakers (MCBs) are designed to trip under fault conditions, and residual current devices (RCDs) are designed to trip below the fatal level.

Construction workers and particularly electricians do receive electric shocks, usually as a result of carelessness or unforeseen circumstances. As an

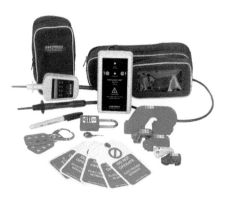

Figure 4.9 Lock-out kit.

electrician working on electrical equipment you must always make sure that the equipment is switched off or electrically isolated before commencing work. Every circuit must be provided with a means of **isolation** (IET Regulation 132.15). When working on portable equipment or desktop units it is often simply a matter of unplugging the equipment from the adjacent supply. Larger pieces of equipment and electrical machines may require isolating at the local isolator switch before work commences. To deter anyone from re-connecting the supply while work is being carried out on equipment, a sign 'Danger – Electrician at Work' should be displayed on the isolator and the isolation 'secured' with a small padlock or the fuses removed so that no one can re-connect while work is being carried out on that piece of equipment. The Electricity at Work Regulations 1989 are very specific at Regulation 12(1) that we must ensure the disconnection and separation of electrical equipment from every source of supply and that this disconnection and separation is secure. Where a test instrument or voltage indicator is used to prove the supply dead, Regulation 4(3) of the Electricity at Work Regulations 1989 recommends that the following procedure is adopted.

1 First connect the test device such as that shown in Fig. 4.6 to the supply which is to be isolated. The test device should indicate main's voltage.

2 Next, isolate the supply and observe that the test device now reads zero volts.

3 Then connect the same test device to a known live supply or proving unit such as that shown in Fig. 4.7 to 'prove' that the tester is still working correctly.

4 Finally secure the isolation and place warning signs; only then should work commence.

The test device being used by the electrician must incorporate safe test leads which comply with the Health and Safety Executive Guidance Note 38 on electrical test equipment. These leads should incorporate barriers to prevent the user from touching live terminals when testing and incorporating a protective resistor and be well insulated and robust, such as those shown in Fig. 4.11.

To isolate a piece of equipment or individual circuit successfully, competently, safely and in accordance with all the relevant regulations, we must follow a procedure such as that given by the flow diagram of Fig. 4.12. Start at the top

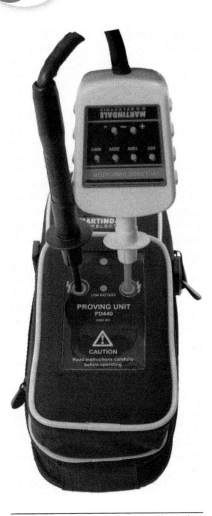

Figure 4.10 Voltage tester and proving unit.

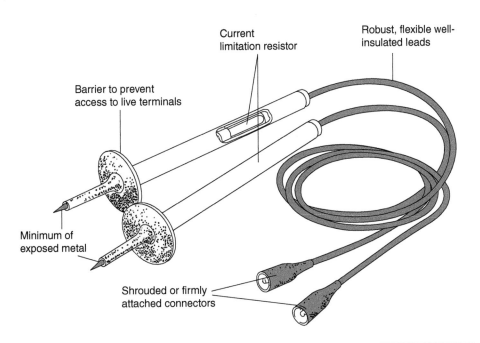

Figure 4.11 Recommended type of test probe and leads.

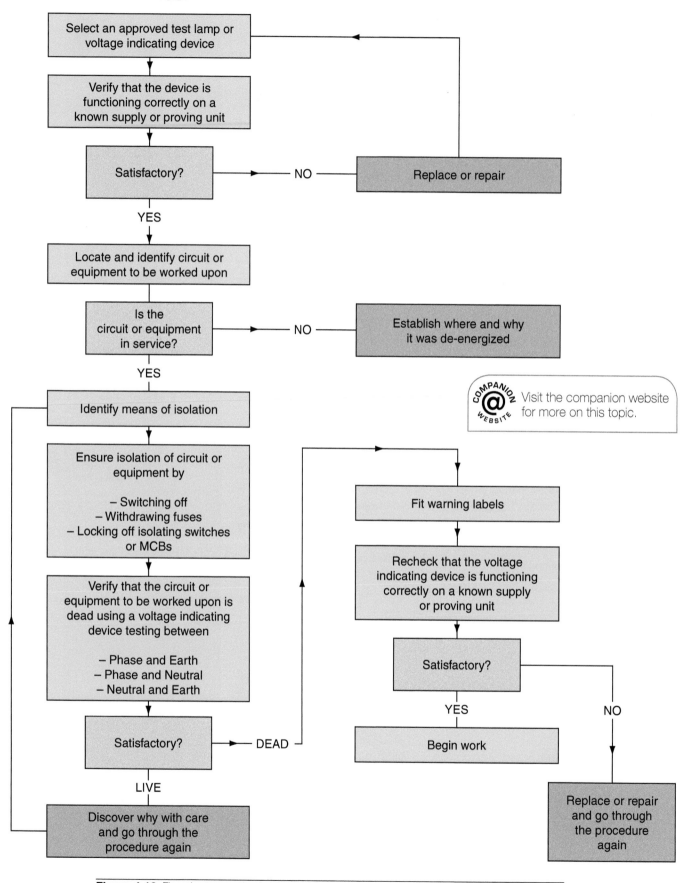

Figure 4.12 Flowchart or method statement for a secure isolation procedure.

and work down the flow diagram. When the heavy outlined amber boxes are reached, pause and ask yourself whether everything is satisfactory up to this point. If the answer is 'yes', move on. If the answer is 'no', go back as indicated by the diagram.

Permit-to-work system

The **permit-to-work procedure** is a type of 'safe system to work' procedure used in specialized and potentially dangerous plant process situations. The procedure was developed for the chemical industry, but the principle is equally applicable to the management of complex risk in other industries or situations. For example:

- working on part of an assembly line process where goods move through a complex, continuous process from one machine to another (e.g. the food industry);
- repairs to railway tracks, tippers and conveyors;
- working in confined spaces (e.g. vats and storage containers);
- working on or near overhead crane tracks;
- working underground or in deep trenches;
- working on pipelines;
- working near live equipment or unguarded machinery;
- roof work;
- working in hazardous atmospheres (e.g. the petroleum industry);
- working near or with corrosive or toxic substances.

All the above situations are high-risk working situations that should be avoided unless you have received special training and will probably require the completion of a permit-to-work. Permits-to-work must adhere to the following eight principles:

1 Wherever possible the hazard should be eliminated so that the work can be done safely without a permit-to-work.

2 The Site Manager has overall responsibility for the permit-to-work even though he may delegate the responsibility for its issue.

3 The permit must be recognized as the master instruction, which, until it is cancelled, overrides all other instructions.

4 The permit applies to everyone on-site, other trades and subcontractors.

5 The permit must give detailed information; for example: (i) which piece of plant has been isolated and the steps by which this has been achieved; (ii) what work is to be carried out; (iii) the time at which the permit comes into effect.

6 The permit remains in force until the work is completed and is cancelled by the person who issued it.

7 No other work is authorized. If the planned work must be changed, the existing permit must be cancelled and a new one issued.

8 Responsibility for the plant must be clearly defined at all stages because the equipment that is taken out of service is released to those who are to carry out the work.

The people doing the work, the people to whom the permit is given, take on the responsibility of following and maintaining the safeguards set out in the permit, which will define what is to be done (no other work is permitted) and the time scale in which it is to be carried out.

Definition

The *permit-to-work procedure* is a type of 'safe system to work' procedure used in specialized and potentially dangerous plant process situations.

The permit-to-work system must help communication between everyone involved in the process or type of work. Employers must train staff in the use of such permits and, ideally, training should be designed by the company issuing the permit, so that sufficient emphasis can be given to particular hazards present and the precautions which will be required to be taken. For further details see Permit to Work @www.hse.gov.uk.

Faulty equipment: To repair or replace?

Having successfully diagnosed the cause of the fault we have to decide if we are to repair or replace the faulty component or piece of equipment.

In many cases the answer will be straightforward and obvious, but in some circumstances the solution will need to be discussed with the customer. Some of the issues which may be discussed are as follows:

- What is the cost of replacement? Will the replacement cost be prohibitive? Is it possible to replace only some of the components? Will the labour costs of the repair be more expensive than a replacement? Do you have the skills necessary to carry out the repair? Would the repaired piece of equipment be as reliable as a replacement?
- Is a suitable replacement available within an acceptable time? These days, manufacturers carry small stocks to keep costs down.
- Can the circuit or system be shut down to facilitate a repair or replacement?
- Can alternative or temporary supplies and services be provided while replacements or repairs are carried out?

Selecting test equipment

The HSE has published a guidance note (GS 38) which advises electricians and other electrically competent people on the selection of suitable test probes, voltage-indicating devices and measuring instruments. This is because they consider suitably constructed test equipment to be as vital for personal safety as the training and practical skills of the electrician. In the past, unsatisfactory test probes and voltage indicators have frequently been the cause of accidents, and therefore all test probes must now incorporate the following features:

1 The probes must have finger barriers or be shaped so that the hand or fingers cannot make contact with the live conductors under test.
2 The probe tip must not protrude more than 2 mm, and preferably only 1 mm, be spring-loaded and screened.
3 The lead must be adequately insulated and coloured so that one lead is readily distinguished from the other.
4 The lead must be flexible and sufficiently robust.
5 The lead must be long enough to serve its purpose but not too long.
6 The lead must not have accessible exposed conductors even if it becomes detached from the probe or from the instrument.
7 Where the leads are to be used in conjunction with a voltage detector they must be protected by a current limiting resistor.

A suitable probe and lead is shown in Fig. 4.11.

GS 38 also tells us that where the test is being made simply to establish the presence or absence of a voltage, the preferred method is to use a proprietary test lamp or voltage indicator which is suitable for the working voltage, rather than a multimeter. Accident history has shown that incorrectly set multimeters or makeshift devices for voltage detection have frequently caused accidents. Figure 4.6 shows a suitable voltage indicator. Test lamps and voltage indicators are not fail-safe, and therefore GS 38 recommends that they should be regularly proved, preferably before and after use, as described previously in the flowchart for a safe isolation procedure.

The IET Regulations (BS 7671) also specify the test voltage or current required to carry out particular tests satisfactorily. All testing must, therefore, be carried out using an 'approved' test instrument if the test results are to be valid. The test instrument must also carry a calibration certificate; otherwise the recorded results may be void. Calibration certificates usually last for a year. Test instruments must, therefore, be tested and recalibrated each year by an approved supplier. This will maintain the accuracy of the instrument to an acceptable level, usually within 2% of the true value.

Modern digital test instruments are reasonably robust, but to maintain them in good working order they must be treated with care. An approved test instrument costs as much as a good-quality camera; it should, therefore, receive the same care and consideration.

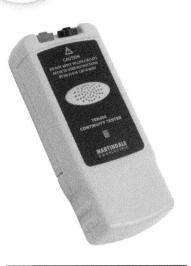

Figure 4.13 Continuity tester.

Continuity tester

To measure accurately the resistance of the conductors in an electrical installation, we must use an instrument which is capable of producing an open-circuit voltage of between 4 and 24 V a.c. or d.c., and delivering a short-circuit current of not less than 200 mA (IET Regulation 643.2.1). The functions of continuity testing and insulation resistance testing are usually combined in one test instrument.

Insulation resistance tester

The test instrument must be capable of detecting insulation leakage between live conductors and between live conductors and earth. To do this and comply with IET Regulation 643.3 the test instrument must be capable of producing a test voltage of 250, 500 or 1,000 V and delivering an output current of not less than 1 mA at its normal voltage.

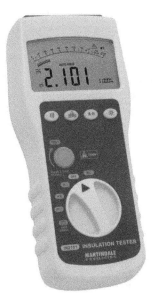

Figure 4.14 Insulation and continuity tester.

Earth fault loop impedance tester

The test instrument must be capable of delivering fault currents as high as 25A for up to 40 ms using the supply voltage. During the test, the instrument does an Ohm's law calculation and displays the test result as a resistance reading.

RCD tester

Where circuits are protected by an RCD we must carry out a test to ensure that the device will operate very quickly under fault conditions and within the time limits set by the IET Regulations. The instrument must, therefore, simulate a fault and measure the time taken for the RCD to operate. The instrument is, therefore, calibrated to give a reading measured in milliseconds to an in-service accuracy of 10%.

If you purchase good-quality 'approved' test instruments and leads from specialist manufacturers they will meet all the regulations and standards and therefore give valid test results. However, to carry out all the tests required by the IET Regulations will require a number of test instruments and this will represent a major capital investment in the region of £1000.

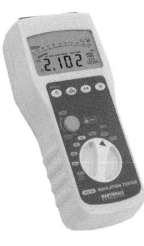

Figure 4.15 Multi voltage insulation and continuity tester.

Figure 4.16 Digital multimeter suitable for testing electrical and electronic circuits.

The specific tests required by the IET Regulations: BS 7671 are described in detail in Chapter 5 of this book under the subheading 'Inspection and testing techniques'.

Electrical installation circuits usually carry in excess of 1A and often carry hundreds of amperes. Electronic circuits operate in the milliampere or even microampere range. The test instruments used on electronic circuits must have a *high impedance* so that they do not damage the circuit when connected to take readings. All instruments cause some disturbance when connected into a circuit because they consume some power in order to provide the torque required to move the pointer. In power applications these small disturbances seldom give rise to obvious errors, but in electronic circuits a small disturbance can completely invalidate any readings taken. We must, therefore, choose our electronic test equipment with great care, as described in Chapter 4 of *Basic Electrical Installation Work 9th Edition* and shown here at Fig. 4.16.

So far in this chapter, I have been considering standard electrical installation circuits wired in conductors and cables using standard wiring systems. However, you may be asked to diagnose and repair a fault on a system that is unfamiliar to you or outside your experience and training. If this happens to you, I would suggest that you immediately tell the person ordering the work or your supervisor that it is beyond your knowledge and experience. I have said earlier that fault diagnosis can only be carried out successfully by someone with a broad range of experience and a thorough knowledge of the installation or equipment that is malfunctioning. The person ordering the work will not think you a fool for saying straightaway that the work is outside your experience. It is better to be respected for your honesty than to attempt something that is beyond you at the present time and which could create bigger problems and waste valuable repair time.

Let us now consider some situations where special precautions or additional skills and knowledge may need to be applied.

Special situations or hazardous electrical locations

All electrical installations and installed equipment must be safe to use and free from the dangers of electric shock, but some installations or locations require special consideration because of the inherent dangers of the installed conditions. The danger may arise because of the corrosive or explosive nature of the atmosphere, because the installation must be used in damp or low-temperature conditions or because there is a need to provide additional mechanical protection for the electrical system. Part 7 of the IET Regulations deals with these special installations or locations. In this section we will consider some of the installations which require special consideration.

Optical fibre cables

The introduction of fibre-optic cable systems and digital transmissions will undoubtedly affect future cabling arrangements and the work of the electrician. Networks based on the digital technology currently being used so successfully by the telecommunications industry are very likely to become the long-term standard for computer systems. Fibre-optic systems dramatically reduce the number of cables required for control and communications systems, and this will in turn reduce the physical room required for these systems. Fibre-optic cables are also immune to electrical noise when run parallel to mains cables and, therefore, the present rules of segregation and screening may change in the

Figure 4.17 Optical fibre cables.

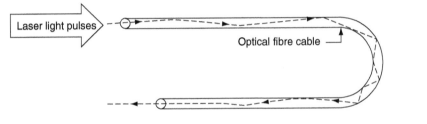

Figure 4.18 Digital pulses of laser light down an optical fibre cable.

Visit the companion website for more on this topic.

future. There is no spark risk if the cable is accidentally cut and, therefore, such circuits are intrinsically safe.

Optical fibre cables are communication cables made from optical-quality plastic, the same material from which spectacle lenses are manufactured. The energy is transferred down the cable as digital pulses of laser light as against current flowing down a copper conductor in electrical installation terms. The light pulses stay within the fibre-optic cable because of a scientific principle known as 'total internal refraction' which means that the laser light bounces down the cable and when it strikes the outer wall it is always deflected inwards and, therefore, does not escape out of the cable, as shown in Fig. 4.18.

The cables are very small because the optical quality of the conductor is very high and signals can be transmitted over great distances. They are cheap to produce and lightweight because these new cables are made from high-quality plastic and not high-quality copper. Single-sheathed cables are often called 'simplex' cables and twin-sheathed cables 'duplex'; that is, two simplex cables together in one sheath. Multicore cables are available containing up to 24 single fibres.

Fibre-optic cables look like steel wire armour (SWA) cables (but of course are lighter) and should be installed in the same way and given the same level of protection as SWA cables. Avoid tight-radius bends if possible and kinks at all costs. Cables are terminated in special joint boxes which ensure cable ends are cleanly cut and butted together to ensure the continuity of the light pulses. Fibre-optic cables are Band I circuits when used for data transmission and must therefore be segregated from other mains cables to satisfy the IET Regulations.

The testing of fibre-optic cables requires that special instruments be used to measure the light attenuation (i.e. light loss) down the cable. Finally, when working with fibre-optic cables, electricians should avoid direct eye contact with the low-energy laser light transmitted down the conductors.

Electrostatic discharge

Static electricity is a voltage charge which builds up to many thousands of volts between two surfaces when they rub together. A dangerous situation occurs when the static charge has built up to a potential capable of striking an arc through the airgap separating the two surfaces.

Static charges build up in a thunderstorm. A lightning strike is the discharge of the thunder cloud, which may have built up to a voltage of 100 MV, to the general mass of earth which is at 0 V. Lightning discharge currents are of the order of 20 kA, hence the need for lightning conductors on vulnerable buildings in order to discharge the energy safely.

Static charge builds up between any two insulating surfaces or between an insulating surface and a conducting surface, but it is not apparent between two conducting surfaces.

A motor car moving through the air builds up a static charge which sometimes gives the occupants a minor shock as they step out and touch the door handle.

Static electricity also builds up in modern offices and similar carpeted areas. The combination of synthetic carpets, man-made footwear materials and dry, air-conditioned buildings contribute to the creation of static electrical charges building up on people moving about these buildings. Individuals only become aware of the charge if they touch earthed metalwork, such as a stair banister rail, before the static electricity has been dissipated. The effect is a sensation of momentary shock.

The precautions against this problem include using floor coverings that have been 'treated' to increase their conductivity or that contain a proportion of natural fibres that have the same effect. The wearing of leather-soled footwear also reduces the likelihood of a static charge persisting, as does increasing the humidity of the air in the building.

A nylon overall and nylon bed sheets build up static charge which is the cause of the 'crackle' when you shake them. Many flammable liquids have the same properties as insulators, and therefore liquids, gases, powders and paints moving through pipes build up a static charge.

Petrol pumps, operating theatre oxygen masks and car spray booths are particularly at risk because a spark in these situations may ignite the flammable liquid, powder or gas.

So how do we protect ourselves against the risks associated with static electrical charges? I said earlier that a buildup of static charge is not apparent between two conducting surfaces, and this gives a clue to the solution. Bonding surfaces together with protective equipotential bonding conductors prevents a buildup of static electricity between the surfaces. If we use large-diameter pipes, we reduce the flow rates of liquids and powders and, therefore, we reduce the buildup of static charge. Hospitals use cotton sheets and uniforms, and use protective equipotential bonding extensively in operating theatres. Rubber, which contains

Definition

Static electricity is a voltage charge which builds up to many thousands of volts between two surfaces when they rub together.

Definition

Static charge builds up between any two insulating surfaces or between an insulating surface and a conducting surface, but it is not apparent between two conducting surfaces.

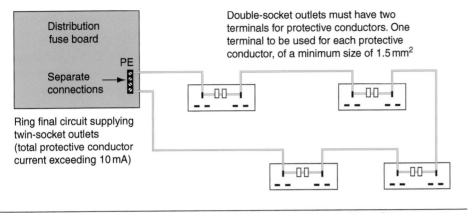

Figure 4.19 Recommended method of connecting IT equipment to socket outlets.

a proportion of graphite, is used to manufacture anti-static trolley wheels and surgeons' boots. Rubber constructed in this manner enables any buildup of static charge to 'leak' away. Increasing humidity also reduces static charge because the water droplets carry away the static charge, thus removing the hazard.

Avoiding shutdown of IT equipment

Every modern office now contains computers, and many systems are linked together or networked. Most computer systems are sensitive to variations or distortions in the mains supply and many computers incorporate filters which produce high-protective conductor currents of around 2 or 3 mA. This is clearly not a fault current, but is typical of the current which flows in the circuit protective conductor of IT equipment under normal operating conditions. IET Regulations 543.7 deal with the earthing requirements for the installation of equipment having high-protective conductor currents. IET Guidance Note 7 recommends that IT equipment should be connected to double sockets as shown in Fig. 4.19.

Surge protection

A transient overvoltage or surge is a voltage spike of very short duration. It may be caused by a lightning strike or a switching action on the system. It sends a large voltage spike for a few microseconds down the mains supply which is sufficient to damage sensitive electronic equipment. Supplies to computer circuits must be 'clean' and 'secure'. Mainframe computers and computer networks are sensitive to mains distortion or interference, which is referred to as 'noise'. Noise is mostly caused by switching an inductive circuit which causes a transient spike, or by brush gear making contact with the commutator segments of an electric motor. These distortions in the mains supply can cause computers to 'crash' or provoke errors and are shown in Fig. 4.20.

To avoid this, a 'clean' supply is required for the computer network. This can be provided by taking the ring or radial circuits for the computer supplies from a point as close as possible to the intake position of the electrical supply to the building. A clean earth can also be taken from this point, which is usually one core of the cable and not the armour of an SWA cable, and distributed around the final wiring circuit. Alternatively, the computer supply can be cleaned by means of a filter such as that shown in Fig. 4.21 or by installing surge protection devices.

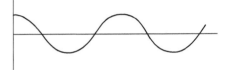

Clean supply

Spikes, caused by an over-voltage transient surging through the mains

'Noise': unwanted electrical signals picked up by power lines or supply cords

Figure 4.20 Distortions in the a.c. mains supply.

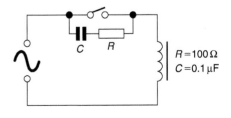

$R = 100\,\Omega$
$C = 0.1\,\mu F$

Figure 4.21 A simple noise suppressor.

A surge protection device is a device intended to limit transient overvoltages, and to divert damaging surge currents away from sensitive equipment.

Section 534 of the IET Regulations contains the requirements for the installation of surge protective devices (SPDs) to limit transient overvoltages where required by section 443 of the IET Regulations or where specified by the designer.

Secure supplies

The mains electrical supply in the United Kingdom is extremely reliable and secure. However, the loss of supply to a mainframe computer or computer network for even a second can cause the system to 'crash', and hours, or even days, of work can be lost.

Definition

A UPS is essentially a battery supply electronically modified to provide a clean and secure a.c. supply.

One solution to this problem is to protect 'precious' software systems with an uninterruptible power supply (UPS). A **UPS** is essentially a battery supply electronically modified to provide a clean and secure a.c. supply. The UPS is plugged into the mains supply and the computer systems are plugged into the UPS.

A UPS to protect a small network of, say, six PCs is physically about the size of one PC hard drive and is usually placed under or at the side of an operator's desk.

It is best to dedicate a ring or radial circuit to the UPS and either to connect the computer equipment permanently or to use non-standard outlets to discourage the unauthorized use and overloading of these special supplies by, for example, kettles.

Finally, remember that most premises these days contain some computer equipment and systems. Electricians intending to isolate supplies for testing or modification should *first check and then check again* before they finally isolate the supply in order to avoid loss or damage to computer systems.

Damage to electronic devices by 'overvoltage'

The use of electronic circuits in all types of electrical equipment has increased considerably over recent years. Electronic circuits and components can now be found in leisure goods, domestic appliances, motor starting and control circuits, discharge lighting, emergency lighting, alarm circuits and special-effects lighting systems. All electronic circuits are low-voltage circuits carrying very small currents.

Electrical installation circuits usually carry in excess of 1A and often carry hundreds of amperes. Electronic circuits operate in the milliampere or even microampere range. The test instruments used on electronic circuits must have a *high impedance* so that they do not damage the circuit when connected to take readings.

The use of an insulation resistance test as described by the IET Regulations (described in Chapter 5 of this book), must be avoided with any electronic equipment. The working voltage of this instrument can cause total devastation to modern electronic equipment. When carrying out an insulation resistance test as part of the prescribed series of tests for an electrical installation, all electronic equipment must first be disconnected or damage will result.

Any resistance measurements made on electronic circuits must be achieved with a battery-operated ohmmeter with a high impedance to avoid damaging the electronic components as shown earlier in this chapter at Fig. 4.16.

Risks associated with high-frequency or large capacitive circuits

Induction heating processes use high-frequency power to provide very focused heating in industrial processes.

The induction heater consists of a coil of large cross-section. The work-piece or object to be heated is usually made of ferrous metal and is placed inside the coil. When the supply is switched on, eddy currents are induced into the work-piece and it heats up very quickly so that little heat is lost to conduction and convection.

The frequency and size of the current in the coil determines where the heat is concentrated in the work-piece:

* the higher the current, the greater the surface penetration;
* the longer the current is applied, the deeper the penetration;
* the higher the frequency, the less the depth of heat penetration.

For shallow penetration, high-frequency, high-current, short-time application is typically used for tool tempering. Other applications are brazing and soldering industrial and domestic gas boiler parts.

When these machines are not working they look very harmless but when they are working they operate very quietly and there is no indication of the intense heat that they are capable of producing. Domestic and commercial microwave ovens operate at high frequency. The combination of risks of high frequency and intense heating means that before any maintenance, repair work or testing is carried out, the machine must first be securely isolated and no one should work on these machines unless they have received additional training to enable them to do so safely.

Industrial wiring systems are very inductive because they contain many inductive machines and circuits, such as electric motors, transformers, welding plants and discharge lighting. The inductive nature of the industrial load causes the current to lag behind the voltage and creates a bad power factor. Power factor is the percentage of current in an alternating current circuit that can be used as energy for the intended purpose. A power factor of, say, 0.7 indicates that 70% of the current supplied is usefully employed by the industrial equipment.

An inductive circuit, such as that produced by an electric motor, induces an electromagnetic force which opposes the applied voltage and causes the current waveform to lag the voltage waveform. Magnetic energy is stored up in the load during one half cycle and returned to the circuit in the next half cycle. If a capacitive circuit is employed, the current leads the voltage since the capacitor stores energy as the current rises and discharges it as the current falls. So here we have the idea of a solution to the problem of a bad power factor created by inductive industrial loads. Power factor and power factor improvement is discussed below.

The **power factor** at which consumers take their electricity from the local electricity supply authority is outside the control of the supply authority. The power factor of the consumer is governed entirely by the electrical plant and equipment that is installed and operated within the consumer's buildings. Domestic consumers do not have a bad power factor because they use very little inductive equipment. Most of the domestic load is neutral and at unity power factor.

Definition

Induction heating processes use high-frequency power to provide very focused heating in industrial processes.

Definition

The *power factor* of the consumer is governed entirely by the electrical plant and equipment that is installed and operated within the consumer's buildings.

Electricity supply authorities discourage the use of equipment and installations with a low power factor because they absorb part of the capacity of the generating plant and the distribution network to no useful effect. They, therefore, penalize industrial consumers with a bad power factor through a maximum demand tariff, metered at the consumer's intake position. If the power factor falls below a datum level of between 0.85 and 0.9 then extra charges are incurred. In this way industrial consumers are encouraged to improve their power factor.

Power factor improvement of most industrial loads is achieved by connecting capacitors to either:

- individual items of equipment; or
- banks of capacitors may be connected to the main busbars of the installation at the intake position.

The method used will depend upon the utilization of the installed equipment by the industrial or commercial consumer. If the load is constant then banks of capacitors at the mains intake position would be indicated. If the load is variable then power factor correction equipment could be installed adjacent to the machine or piece of equipment concerned.

Power factor correction by capacitors is the most popular method because of the following:

- They require no maintenance.
- Capacitors are flexible and additional units may be installed as an installation or system is extended.
- Capacitors may be installed adjacent to individual pieces of equipment or at the mains intake position. Equipment may be placed on the floor or fixed high up and out of the way.

Capacitors store charge and must be disconnected before the installation or equipment is tested in accordance with Section 6 of the IET Regulations BS 7671.

Small power factor correction capacitors, as used in discharge lighting, often incorporate a high-value resistor connected across the mains terminals. This discharges the capacitor safely when not in use. Banks of larger capacity capacitors may require discharging to make them safe when not in use. To discharge a capacitor safely and responsibly it must be discharged slowly over a period in excess of five 'time-constants' through a suitable discharge resistor. Capacitors are discussed in this book in Chapter 3.

Presence of storage batteries

Since an emergency occurring in a building may cause the mains supply to fail, the **emergency lighting** should be supplied from a source which is independent from the main supply. A battery's ability to provide its output instantly makes it a very satisfactory source of standby power. In most commercial, industrial and public service buildings housing essential services, the alternative power supply would be from batteries, but generators may also be used. Generators can have a large capacity and duration, but a major disadvantage is the delay of time while the generator runs up to speed and takes over the load. In some premises a delay of more than five seconds is considered unacceptable, and in these cases a battery supply is required to supply the load until the generator can take over.

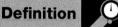

Definition

Power factor improvement of most industrial loads is achieved by connecting capacitors to either:
- individual items of equipment; or
- banks of capacitors.

Definition

Since an emergency occurring in a building may cause the mains supply to fail, the *emergency lighting* should be supplied from a source which is independent from the main supply.

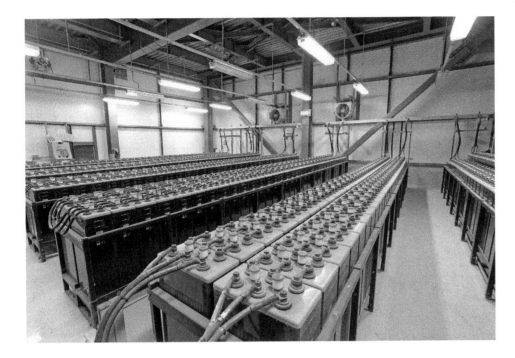

The emergency lighting supply must have an adequate capacity and rating for the specified duration of time (IET Regulation 313.2). BS 5266 and BS EN 1838 states that after a battery is discharged by being called into operation for its specified duration of time, it should be capable of once again operating for the specified duration of time following a recharge period of no longer than 24 hours. The duration of time for which the emergency lighting should operate will be specified by a statutory authority but is normally one to three hours. The British Standard states that escape lighting should operate for a minimum of one hour. Standby lighting operation time will depend upon financial considerations and the importance of continuing the process or activity within the premises after the mains supply has failed.

The contractor installing the emergency lighting should provide a test facility which is simple to operate and secure against unauthorized interference. The emergency lighting installation must be segregated completely from any other wiring, so that a fault on the main electrical installation cannot damage the emergency lighting installation (IET Regulation 528.1).

The batteries used for the emergency supply should be suitable for this purpose. Motor vehicle batteries are not suitable for emergency lighting applications, except in the starter system of motor-driven generators. The fuel supply to a motor-driven generator should be checked. The battery room of a central battery system must be well ventilated and, in the case of a motor-driven generator, adequately heated to ensure rapid starting in cold weather.

The British Standard recommends that the full load should be carried by the emergency supply for at least one hour in every six months. After testing, the emergency system must be carefully restored to its normal operative state. A record should be kept of each item of equipment and the date of each test by a qualified or responsible person. It may be necessary to produce the record as evidence of satisfactory compliance with statutory legislation to a duly authorized person.

Self-contained units are suitable for small installations of up to about 12 units. The batteries contained within these units should be replaced about every five years, or as recommended by the manufacturer.

Storage batteries are secondary cells. A secondary cell has the advantage of being rechargeable. If the cell is connected to a suitable electrical supply, electrical energy is stored on the plates of the cell as chemical energy. When the cell is connected to a load, the chemical energy is converted to electrical energy.

A lead-acid cell is a secondary cell. Each cell delivers about 2 V, and when six cells are connected in series a 12 V battery is formed.

A lead-acid battery is constructed of lead plates which are deeply ribbed to give maximum surface area for a given weight of plate. The plates are assembled in groups, with insulating separators between them. The separators are made of a porous insulating material, such as wood or ebonite, and the whole assembly is immersed in a dilute sulphuric acid solution in a plastic container.

The capacity of a cell to store charge is a measure of the total quantity of electricity which it can cause to be displaced around a circuit after being fully charged. It is stated in ampere-hours, abbreviation Ah, and calculated at the 10-hour rate which is the steady load current which would completely discharge the battery in 10 hours. Therefore, a 50 Ah battery will provide a steady current of 5A for 10 hours.

Maintenance of lead-acid batteries

- The plates of the battery must always be covered by dilute sulphuric acid. If the level falls, it must be topped up with distilled water.
- Battery connections must always be tight and should be covered with a thin coat of petroleum jelly.
- The specific gravity or relative density of the battery gives the best indication of its state of charge. A discharged cell will have a specific gravity of 1.150, which will rise to 1.280 when fully charged. The specific gravity of a cell can be tested with a hydrometer.
- To maintain a battery in good condition it should be regularly trickle-charged. A rapid charge or discharge encourages the plates to buckle, and may cause permanent damage. Most batteries used for standby supplies today are equipped with constant voltage chargers. The principle of these is that after the battery has been discharged by it being called into operation, the terminal voltage will be depressed and this enables a relatively large current (1–5A) to flow from the charger to recharge the battery. As the battery becomes more fully charged its voltage will rise until it reaches the constant voltage level where the current output from the charger will drop until it is just sufficient to balance the battery's internal losses. The main advantage of this system is that the battery controls the amount of charge it receives and is therefore automatically maintained in a fully charged condition without human intervention and without the use of any elaborate control circuitry.
- The room used to charge the emergency supply storage batteries must be well ventilated because the charged cell gives off hydrogen and oxygen, which are explosive in the correct proportions.

Working alone

Some working situations are so potentially hazardous that not only must PPE be worn but you must also never work alone and safe working procedures must be in place before your work begins to reduce the risk.

It is unsafe to work in isolation in the following situations:

- when working above ground;
- when working below ground;
- when working in confined spaces;
- when working close to unguarded machinery;
- when a fire risk exists;
- when working close to toxic or corrosive substances such as battery acid.

Working above ground

The new Work at Height Regulations 2005 tells us that a person is at height if that person could be injured by falling from it. The Regulations require the following:

- We should avoid working at height if at all possible.
- No work should be done at height which can be done on the ground. For example, equipment can be assembled on the ground and then taken up to height, perhaps for fixing.
- Ensure the work at height is properly planned.
- Take account of any risk assessments carried out under Regulation 3 of the Management of Health and Safety at Work Regulations.

Working below ground

Working below ground might be working in a cellar or an unventilated basement with only one entrance/exit. There is a risk that this entrance/exit might become blocked by materials, fumes or fire. When working in trenches there is always the risk of the sides collapsing if they are not adequately supported by temporary steel sheets. There is also the risk of falling objects, so always:

- wear a hard hat;
- never go into an unsupported excavation;
- erect barriers around the excavation;
- provide good ladder access;
- ensure the work is properly planned;
- take account of the risk assessment before starting work.

Working in confined spaces

When working in confined spaces there is always the risk that you may become trapped or overcome by a lack of oxygen, or by gas, fumes, heat or an accumulation of dust. Examples of confined spaces are:

- storage tanks and silos on farms;
- enclosed sewer and pumping stations;
- furnaces;
- ductwork.

In my experience, electricians spend a lot of time on their knees in confined spaces because many electrical cable systems run out of sight away from the public areas of a building.

The Confined Spaces Regulations 1997 require that:

* a risk assessment is carried out before work commences;
* if there is a serious risk of injury in entering the confined space then the work should be done on the outside of the vessel;
* follow a safe working procedure such as a 'permit-to-work procedure' which was discussed earlier in this chapter, and put adequate emergency arrangements in place before work commences.

Working near unguarded machinery

There is an obvious risk in working close to unguarded machinery and, indeed, most machinery will be guarded but in some production processes and with overhead travelling cranes, this is not always possible. To reduce the risks associated with these hazards:

* have the machinery stopped and isolated before your work activity begins;
* put temporary barriers in place;
* make sure the machine operator knows that you are working on the equipment;
* identify the location of emergency stop buttons;
* take account of the risk assessment before work commences.

A risk of fire

When working in locations containing stored flammable materials such as petrol, paraffin, diesel or bottled gas, there is always the risk of fire. To minimize the risk:

* take account of the risk assessment before work commences;
* keep the area well ventilated;
* locate the fire extinguishers;
* secure your exit from the area;
* locate the nearest fire alarm point;
* follow a safe working procedure and put adequate emergency arrangements in place before work commences.

Working in hazardous areas

The British Standards concerned with hazardous areas were first published in the 1920s and were concerned with the connection of electrical apparatus in the mining industry. Since those early days many national and international standards, as well as codes of practice, have been published to inform the manufacture, installation and maintenance of electrical equipment in all hazardous areas.

The relevant British Standards for Electrical Apparatus for Potentially Explosive Atmospheres are BS 5345, BS EN 60079 and BS EN 50014: 1998.

They define a **hazardous area** as 'any place in which an explosive atmosphere may occur in such quantity as to require special precautions to protect the safety of workers'. Clearly these regulations affect the petroleum industry, but they also apply to petrol filling stations.

Safety first

Working alone

* Never work alone in:
  - confined spaces
  - storage tanks
  - enclosed ductwork.

Definition

They define a hazardous area as 'any place in which an explosive atmosphere may occur in such quantity as to require special precautions to protect the safety of workers'.

Most flammable liquids only form an explosive mixture between certain concentration limits. Above and below this level of concentration the mix will not explode. The lowest temperature at which sufficient vapour is given off from a flammable substance to form an explosive gas–air mixture is called the **flashpoint**. A liquid which is safe at normal temperatures will require special consideration if heated to flashpoint. An area in which an explosive gas–air mixture is present is called a hazardous area, as defined by the British Standards, and any electrical apparatus or equipment within a hazardous area must be classified as flameproof.

Flameproof electrical equipment is constructed so that it can withstand an internal explosion of the gas for which it is certified, and prevent any spark or flame resulting from that explosion from leaking out and igniting the surrounding atmosphere. This is achieved by manufacturing flameproof equipment to a robust standard of construction. All access and connection points have wide machined flanges which damp the flame in its passage across the flange. Flanged surfaces are firmly bolted together with many recessed bolts, as shown in Fig. 4.22. Wiring systems within a hazardous area must be to flameproof fittings using an appropriate method, such as:

- PVC cables encased in solid drawn heavy-gauge screwed steel conduit terminated at approved enclosures having wide flanges and bolted covers.
- Mineral insulated cables terminated into accessories with approved flameproof glands. These have a longer gland thread than normal MICC glands. Where the cable is laid underground, it must be protected by a PVC sheath and laid at a depth of not less than 500 mm.
- PVC armoured cables terminated into accessories with approved flameproof glands or any other wiring system which is approved by the British Standard. All certified flameproof enclosures will be marked Ex, indicating that they are suitable for potentially explosive situations, or EEx, where equipment is certified to the harmonized European Standard. All the equipment used in a flameproof installation must carry the appropriate markings, as shown in Fig. 4.23, if the integrity of the wiring system is to be maintained.

Flammable and explosive installations are to be found in the petroleum and chemical industries, which are classified as group II industries. Mining is classified as group I and receives special consideration from the Mining Regulations because of the extreme hazards of working underground. Petrol filling pumps must be wired and controlled by flameproof equipment to meet

Definition

The lowest temperature at which sufficient vapour is given off from a flammable substance to form an explosive gas–air mixture is called the *flashpoint*.

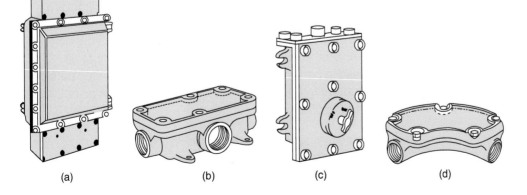

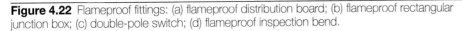

Figure 4.22 Flameproof fittings: (a) flameproof distribution board; (b) flameproof rectangular junction box; (c) double-pole switch; (d) flameproof inspection bend.

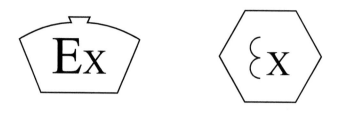

Figure 4.23 Flameproof equipment markings.

the requirements of the Petroleum Regulation Act 1928 and 1936 and any local licensing laws concerning the keeping and dispensing of petroleum spirit.

Hazardous area classification

The British Standard divides the risk associated with inflammable gases and vapours into three classes or zones.

- Zone 0 is the most hazardous, and is defined as a zone or area in which an explosive gas–air mixture is *continuously present* or present for long periods. ('Long periods' is usually taken to mean that the gas–air mixture will be present for longer than 1000 hours per year.)
- Zone 1 is an area in which an explosive gas–air mixture is *likely to occur* in normal operation. (This is usually taken to mean that the gas–air mixture will be present for up to 1000 hours per year.)
- Zone 2 is an area in which an explosive gas–air mixture is *not likely* to occur in normal operation and if it does occur it will exist for a very short time. (This is usually taken to mean that the gas–air mixture will be present for less than 10 hours per year.)

If an area is not classified as zone 0, 1 or 2, then it is deemed to be non-hazardous, so that normal industrial electrical equipment may be used.

The electrical equipment used in zone 2 will contain a minimum amount of protection. For example, normal sockets and switches cannot be installed in a zone 2 area, but oil-filled radiators may be installed if they are directly connected and controlled from outside the area. Electrical equipment in this area should be marked Ex'o' for oil-immersed or Ex'p' for powder-filled.

In zone 1 all electrical equipment must be flameproof, as shown in Fig. 4.22, and marked Ex'd' to indicate a flameproof enclosure.

Ordinary electrical equipment cannot be installed in zone 0, even when it is flameproof protected. However, many chemical and oil-processing plants are entirely dependent upon instrumentation and data transmission for their safe operation. Therefore, very low-power instrumentation and data-transmission circuits can be used in special circumstances, but the equipment must be *intrinsically safe*, and used in conjunction with a 'safety barrier' installed outside the hazardous area. Intrinsically safe equipment must be marked Ex'ia' or Ex's', specially certified for use in zone 0.

Definition

An *intrinsically safe circuit* is one in which no spark or thermal effect is capable of causing ignition of a given explosive atmosphere.

Intrinsic safety

By definition, an **intrinsically safe circuit** is one in which no spark or thermal effect is capable of causing ignition of a given explosive atmosphere. The

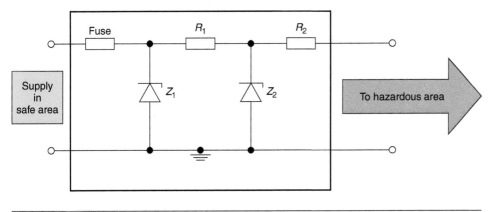

Figure 4.24 Zener safety barrier.

intrinsic safety of the equipment in a hazardous area is assured by incorporating a Zener diode safety barrier into the control circuit such as that shown in Fig. 4.24. In normal operation, the voltage across a Zener diode is too low for it to conduct, but if a fault occurs, the voltage across Z_1 and Z_2 will rise, switching them on and blowing the protective fuse. Z_2 is included in the circuit as a 'backup' in case the first Zener diode fails.

An intrinsically safe system, suitable for use in zone 0, is one in which *all* the equipment, apparatus and interconnecting wires and circuits are intrinsically safe.

Index of Protection (IP) BS EN 60529

IET Regulation 416.2.1 tells us that where barriers and enclosures have been installed to prevent direct contact with live parts, they must afford a degree of protection not less than IP2X and IPXXB, but what does this mean?

The Index of Protection is a code which gives us a means of specifying the suitability of equipment for the environmental conditions in which it will be used. The tests to be carried out for the various degrees of protection are given in the British and European Standard BS EN 60529.

The code is written as IP (Index of Protection) followed by two numbers XX. The first number gives the degree of protection against the penetration of solid objects into the enclosure. The second number gives the degree of protection against water penetration. For example, a piece of equipment classified as IP45 will have barriers installed which prevent a 1mm diameter rigid steel bar from making contact with live parts and be protected against the ingress of water from jets of water applied from any direction. Where a degree of protection is not specified, the number is replaced by an 'X' which simply means that the degree of protection is not specified although some protection may be afforded. The 'X' is used instead of '0' since '0' would indicate that no protection was given. The index of protection codes is shown in Fig. 4.25.

Appendix 5 of the IET Regulations identifies the required IP classification for electrical equipment being used in hazardous conditions and requiring water protection as follows:

- IPX1 or IPX2 where water vapour occasionally condenses on electrical equipment;
- IPX3 where sprayed water forms a continuous film on the floor;
- IPX4 where equipment may be subjected to splashed water, e.g. construction sites;

- IPX5 where hosed water is regularly used, e.g. car washing;
- IPX6 for seashore locations, e.g. marinas and piers;
- IPX7 for locations which may become flooded, immersing equipment in water;
- IPX8 where electrical equipment is permanently immersed in water, e.g. swimming pools.

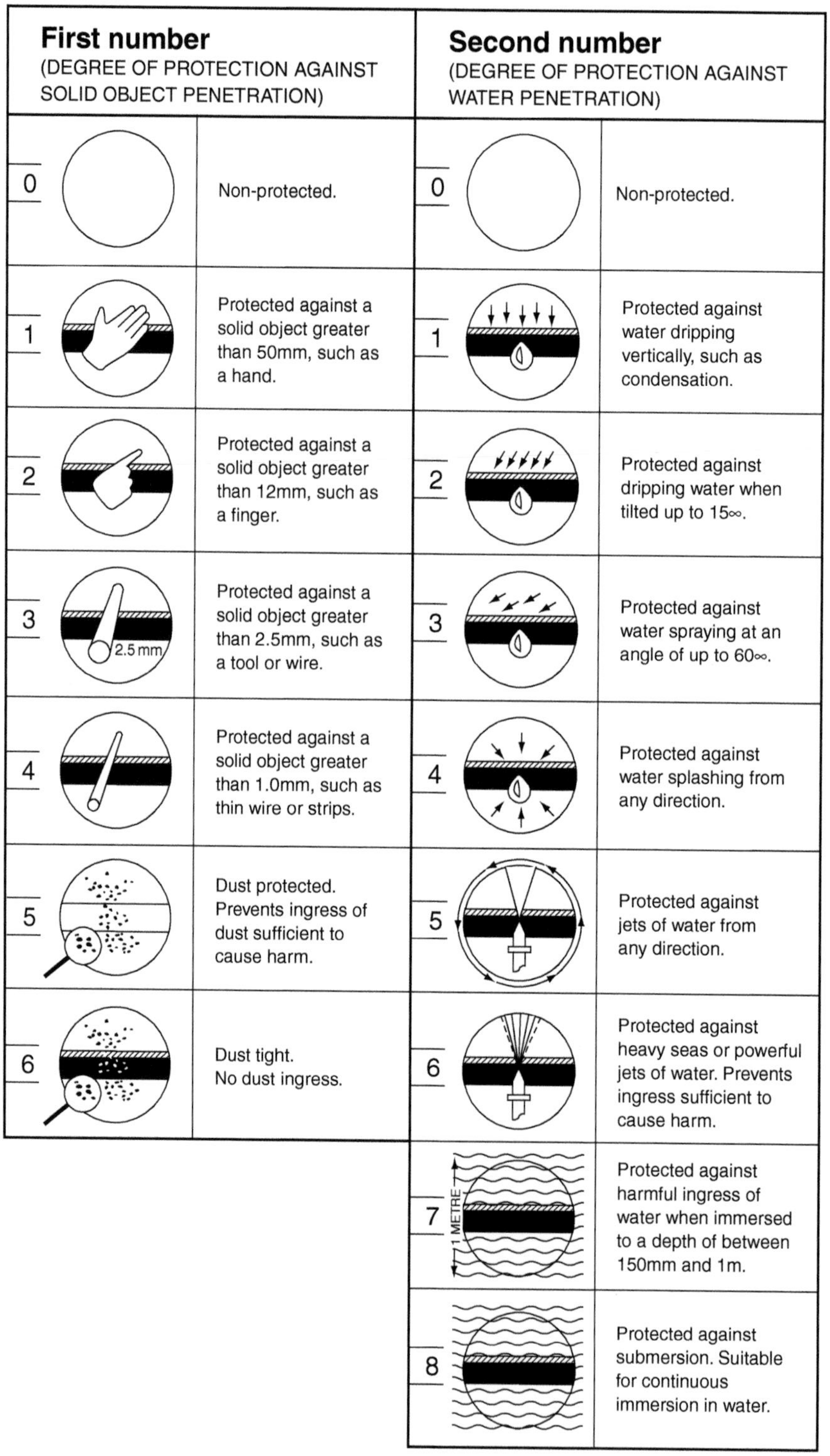

First number (DEGREE OF PROTECTION AGAINST SOLID OBJECT PENETRATION)		Second number (DEGREE OF PROTECTION AGAINST WATER PENETRATION)	
0	Non-protected.	0	Non-protected.
1	Protected against a solid object greater than 50mm, such as a hand.	1	Protected against water dripping vertically, such as condensation.
2	Protected against a solid object greater than 12mm, such as a finger.	2	Protected against dripping water when tilted up to 15∞.
3	Protected against a solid object greater than 2.5mm, such as a tool or wire.	3	Protected against water spraying at an angle of up to 60∞.
4	Protected against a solid object greater than 1.0mm, such as thin wire or strips.	4	Protected against water splashing from any direction.
5	Dust protected. Prevents ingress of dust sufficient to cause harm.	5	Protected against jets of water from any direction.
6	Dust tight. No dust ingress.	6	Protected against heavy seas or powerful jets of water. Prevents ingress sufficient to cause harm.
		7	Protected against harmful ingress of water when immersed to a depth of between 150mm and 1m.
		8	Protected against submersion. Suitable for continuous immersion in water.

Figure 4.25 The British Standards Index of protection codes.

Appendix 5 of the IET Regulations also identifies the required IP classification to prevent dust and objects from penetrating electrical equipment as follows:

- IP2X to prevent penetration by solid objects as thick as a finger, approximately 12mm;
- IP3X to prevent penetration by small objects of which the smallest is 2.5mm;
- IP4X to prevent penetration by very small objects of which the smallest is 1.0mm;
- IP5X where light dust penetration would not harm the electrical equipment;
- IP6X where dust must not penetrate the equipment;
- IPXXB means total protection.

Construction and demolition site installations

The IET Regulations have given us the whole of Section 704 for the regulations which relate to Construction and Demolition site installations.

Construction and demolition sites are potentially dangerous in many ways, tripping, falling, and the constantly changing and temporary nature of the working environment create these hazards. The risk of electric shock is high because of the following factors:

- a construction site or demolition site is, by definition, a temporary state. Upon completion there will be either a building with all the necessary safety features, or a brown field site where the building previously stood.
- there is the possibility of damage to cables and equipment as a consequence of the temporary nature of the site and because the site is not always sealed in the early stages from the weather.
- mobile equipment such as electrical tools and hand lamps with trailing leads will be in use.
- other trades will use the temporary electrical installation for their own mobile equipment.
- there will be many extraneous conductive parts on site which cannot practically be bonded because of the changing nature of the construction process.

The requirements of Section 704 are not applicable to those parts of the construction and demolition site that have been provided for administration, welfare and sanitation purposes such as offices, canteen, toilet and first aid. For example, cabins situated in a fixed position or a specific part of a building set aside for that purpose. The environment in such areas is likely to be less dangerous than the construction or demolition site itself. Regulation 704.1.1.

Temporary electrical supplies provided on construction sites can save many man hours of labour by providing the energy required for fixed and portable tools and lighting which speeds up the completion of a project. However, construction sites are dangerous places and the temporary electrical supply which is installed to assist the construction process must comply with all of the relevant wiring regulations for permanent installations (IET Regulation 110.1). All equipment must be of a robust construction in order to fulfil the on-site electrical requirements while being exposed to rough handling, vehicular nudging, the wind, rain and sun. All equipment socket outlets, plugs and couplers must be of the industrial type to BS EN 60439 and BS EN 60309 and specified by IET Regulation 704.511.1, as shown in Fig. 4.26.

Table 4.1 Common IP ratings used in the electrical industry and their description

| First IP code number indicating solid object penetration | | | Second IP code number indicating water penetration protection | | | | | | | | |
IP code	Contact penetration	Foreign body Penetration	IP X0 No protection	IP X1 Protection against vertically falling dripping water	IP X2 Protection against dripping water if housing is angled up to 15°	IP X3 Protection against spray water from all directions even if angled up to 60°	IP X4 Protection against splash water from all directions	IP X5 Protection against jet water from all directions	IP X6 Protection against heavy jet water from all directions	IP X7 Protection against temporary immersion in water	IP X8 Protection against effects of long-term immersion in water
IP 0X	No contact protection	No protection against solid foreign bodies	IP0								
IP 1X	Protection against large area contact (back of hand)	Protection against solid foreign bodies greater than 50mm diameter	IP10	IP11	IP12						
IP 2X	Protection against contact with one finger	Protection against solid foreign bodies greater than 12.5mm diameter	IP20	IP21	IP21	IP34					
IP 3X	Protection against contact with tools, wires, etc. greater than 2.5mm diameter	Protection against solid foreign bodies greater than 2.5mm diameter	IP30	IP31	IP32	IP33	IP34				
IP 4X	Protection against contact with tools, wires, etc. greater than 1mm diameter	Protection against solid foreign bodies greater than 1mm diameter	IP40	IP41	IP42	IP43	IP44				
IP 5X	Protection against contact with tools, wires, etc. greater than 1mm diameter	Protection against disruptive dust deposits in the inside	IP50				IP54	IP55			
IP 6X	Protection against contact with tools, wires, etc. greater than 1mm diameter	Protection from dust	IP60					IP65	IP66	IP67	IP68

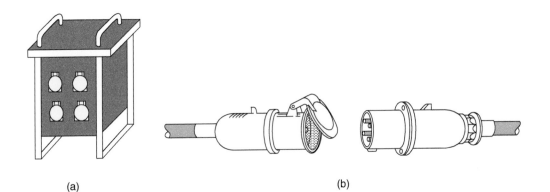

(a)   (b)

Figure 4.26 110V distribution unit and cable connector, suitable for construction site electrical supplies: (a) reduced-voltage distribution unit incorporating industrial sockets to BS EN 60309; and (b) industrial plug and connector.

Figure 4.27 Temporary electrical supply.

Where an electrician is not permanently on-site, MCBs are preferred so that overcurrent protection devices can be safely reset by an unskilled person. The British Standards Code of Practice 1017, *The Distribution of Electricity on Construction and Building Sites*, advises that protection against earth faults may be obtained by first providing a low impedance path, so that overcurrent devices can operate quickly, as described in Chapter 4, and second, by fitting an RCD in addition to the overcurrent protection device (IET Regulation 704.410.3.10). The 18th edition of the IET Regulations considers construction sites very special locations, devoting the whole of Section 704 to their requirements. A construction site installation should be tested and inspected in accordance with Part 6 of the IET Regulations every three months throughout the construction period.

The source of supply for the temporary installation may be from a petrol or diesel generating set or from the local supply company. When the local electricity company provides the supply, the incoming cable must be terminated in a non-combustible, waterproof and locked enclosure to prevent unauthorized access and provide metering arrangements. Regulation 421 and 422 and Section 704.

IET Regulations 704 and 411 tell us that reduced low voltage is strongly *preferred* for mobile equipment and tools used on construction and demolition sites.

The distribution of electrical supplies on a construction site would typically be as follows:

- 400V three phase for supplies to major items of plant having a rating above 3.75kW such as cranes and lifts. These supplies must be wired in armoured cables.
- 230V single phase for supplies to items of equipment which are robustly installed such as floodlighting towers, small hoists and site offices. These supplies must be wired in armoured cable unless they are installed inside the site offices.
- 110V single phase for supplies to all mobile hand tools and all mobile lighting equipment. The supply is usually provided by a reduced voltage distribution unit which incorporates splashproof sockets fed from a centre-tapped 110V transformer. This arrangement limits the voltage to earth to 55V, which is recognized as safe in most locations. A 110V distribution unit is shown in Fig. 4.26. Edison screw lamps are used for 110V lighting supplies so that they are not interchangeable with 230V site office lamps.

There are occasions when even a 110V supply from a centre-tapped transformer is too high; for example, supplies to inspection lamps for use inside damp or confined places. In these circumstances a safety extra-low voltage (SELV) supply would be required.

Industrial plugs have a keyway which prevents a tool from one voltage from being connected to the socket outlet of a different voltage. They are also colour coded for easy identification as follows:

- 400V – red
- 230V – blue
- 110V – yellow
- 50V – white
- 25V – violet.

The reduced low voltage system has been used on UK construction and demolition sites for more than fifty years to provide a safe system in potentially hazardous conditions. It also provides a supply that is sufficiently powerful for most site electrical equipment to function efficiently and correctly. However battery power tools are increasingly popular for small drilling cutting and grinding operations.

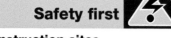

Safety first

Construction sites
- low voltage or
- battery tools must be used.

Agricultural and horticultural premises

Especially adverse installation conditions are to be encountered on agricultural and horticultural premises because of the presence of livestock, vermin, dampness, corrosive substances and mechanical damage. The 18th edition of the IET Wiring Regulations considers these installations very special locations and has devoted the whole of Section 705 to their requirements. In situations accessible to livestock the electrical equipment should be of a type which is appropriate for the external influences likely to occur, and should have at least protection IP44; that is, protection against solid objects and water splashing from any direction (IET Regulation 705.512.2, see also Fig. 4.25).

In buildings intended for livestock, all fixed wiring systems must be inaccessible to the livestock and cables liable to be attacked by vermin must be suitably protected (IET Regulation 705.513.2).

PVC cables enclosed in heavy-duty PVC conduit are suitable for installations in most agricultural buildings. All exposed and extraneous metalwork must be provided with supplementary protective bonding in areas where livestock is kept (IET Regulation 705.415.2.1). In many situations, waterproof socket outlets to BS 196 must be installed. All socket outlet circuits must be protected by an RCD complying with the appropriate British Standard and the operating current must not exceed 30 mA.

Cables buried on agricultural or horticultural land should be buried at a depth of not less than 600 mm, or 1000 mm where the ground may be cultivated, and the cable must have an armour sheath and be further protected by cable tiles. Overhead cables must be insulated and installed so that they are clear of farm machinery or placed at a minimum height of 6 m to comply with IET Regulation 705.522.

Horses and cattle have a very low body resistance, which makes them susceptible to an electric shock at voltages lower than 25 V r.m.s. The sensitivity of farm animals to electric shock means that they can be contained by an electric fence. An animal touching the fence receives a short pulse of electricity which passes through the animal to the general mass of earth and back to an earth electrode sunk near the controller, as shown in Fig. 4.28. The pulses are generated by a capacitor–resistor circuit inside the controller which may be mains or battery operated (capacitor–resistor circuits are discussed in Chapter 3). There must be no risk to any human coming into contact with the controller, which should be manufactured to BS 2632. The output voltage of the controller must not exceed 10 kV and the energy must not be greater than 5 J. The duration of the pulse must not be greater than 1.5 ms and the pulse must never have a frequency greater than one pulse per second. This shock level is very similar to that which can be experienced by touching a spark plug lead on a motor car. The energy levels are very low at 5 J. There are 3.6 million joules of energy in 1 kWh.

Earth electrodes connected to the earth terminal of an electric fence controller must be separate from the earthing system of any other circuit and should be situated outside the resistance area of any electrode used for protective earthing. The electric fence controller and the fence wire must be installed so that they do not come into contact with any power, telephone or radio systems, including poles. Agricultural and horticultural installations should be tested and inspected in accordance with Part 6 of the Wiring Regulations every three years.

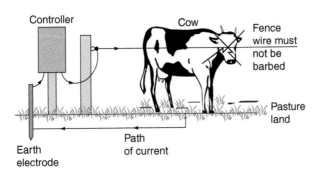

Visit the companion website for more on this topic.

Figure 4.28 Farm animal control by electric fence.

Figure 4.29 They say that the grass is greener on the other side of the fence, but one false move and this sheep will get an electric shock.

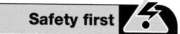

Safety first

Electric shock
- Animals and humans must be protected against electric shock.

Key fact

Caravans
- every caravan pitch must have at least one 16A industrial socket;
- each socket must have RCD and overcurrent protection;

IET Regulation 708.553.

Caravans and caravan sites

The electrical installations on caravan sites, and within caravans, must comply in all respects with the wiring regulations for buildings. All the dangers which exist in buildings are present in and around caravans, including the added dangers associated with repeated connection and disconnection of the supply and the flexing of the caravan installation in a moving vehicle. The 18th edition of the IET Regulations has devoted Section 721 to the electrical installation in caravans and motor caravans and Section 708 to caravan parks.

Touring caravans must be supplied from a 16A industrial-type socket outlet adjacent to the caravan park pitch, having a degree of protection of at least IP44, similar to that shown in Fig. 4.26. Each socket outlet must be provided with individual overcurrent protection and an individual residual current circuit-breaker with a rated tripping current of 30 mA (IET Regulations 708.55 1 to 14).

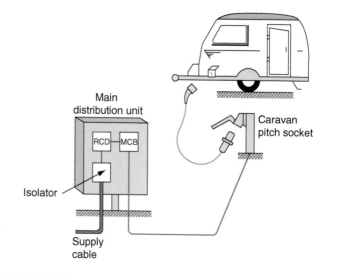

Figure 4.30 Electrical supplies to caravans.

The distance between the caravan connector and the site socket outlet must not be more than 20 m (IET Regulation 708.55.1.2). These requirements are shown in Fig. 4.30.

The supply cables must be installed outside the pitch area and be buried at a depth of at least 0.6 m (IET Regulation 708.521).

The caravan or motor caravan must be provided with a mains isolating switch and an RCD to break all live conductors (IET Regulation 721.411). An adjacent notice detailing how to connect and disconnect the supply safely must also be provided, as shown in IET Regulation 721.514. Electrical equipment must not be installed in fuel storage compartments (IET Regulation 721.528). Caravans flex when being towed, and therefore the installation must be wired in flexible or stranded conductors of at least 1.5 mm cross-section. The conductors must be supported on horizontal runs at least every 25 cm and the metalwork of the caravan and chassis must be bonded with 4.0 mm^2 cable.

The wiring of the extra-low-voltage battery supply must be run in such a way that it does not come into contact with the 230 V wiring system (IET Regulation 721.528.1).

The caravan should be connected to the pitch socket outlet by means of a flexible cable, no longer than 25 m and having a minimum cross-sectional area of 2.5 mm^2 or as detailed in Section 708 Notes and Table 721 of the IET Regulations.

Because of the mobile nature of caravans it is recommended that the electrical installation be tested and inspected at intervals considered appropriate, preferably not less than once every three years and annually if the caravan is used frequently (IET Regulation 721 Periodic Inspection Notes).

Bathroom installations

Rooms containing a fixed bath tub or shower basin are considered an area of increased shock risk and, therefore, additional regulations are specified in Section 701 of the IET Regulations. This is to reduce the risk of electric shock to people in circumstances where body resistance is lowered because of contact with water. The regulations can be summarized as follows:

- socket outlets must not be installed and no provision is made for connection of portable appliances unless the socket outlet can be fixed three metres horizontally beyond the zone 1 boundary within the bath or shower room (IET Regulation 701.512.3);
- only shaver sockets which comply with BS EN 60742, that is, those which contain an isolating transformer, may be installed in zone 2 or outside the zones in the bath or shower room (IET Regulation 701.512.3);
- all circuits serving a bath or shower room, that is, both power and lighting, circuits which are serving the location, or passing through zones 1 and 2 not serving the location, must be additionally protected by an RCD having a rated maximum operating current of 30 mA (IET Regulation 701.411.3.3);
- the 18th Edition of the IET Wiring Regulations at (701.411.3.3) has introduced an additional requirement that any circuit 'passing through' zones 1 and 2 but not serving the location must also be RCD protected;
- there are restrictions as to where appliances, switchgear and wiring accessories may be installed. See *Zones for bath and shower rooms* below;

- local supplementary protective bonding (IET Regulation 701.415.2) must be provided to all gas, water and central heating pipes in addition to metallic baths, *unless the following two requirements are both met*:

  (i) all bathroom circuits, both lighting and power, are protected by a 30 mA RCD in addition to a circuit breaker or fuse; and

  (ii) the bath or shower is located in a building with main protective bonding in place as shown in figs 6.6 to 6.8 (IET Regulation 411.3.1.2 and 701.415.2).

Note: Local supplementary protective bonding may be an additional requirement of the Local Authority regulations in, for example, licensed premises, student accommodation and rented property as shown in Fig 4.33.

Zones for bath and shower rooms

Locations that contain a bath or shower are divided into zones or separate areas, as shown in Fig. 4.31.

Zone 0 – the bath tub or shower basin itself, which can contain water and is, therefore, the most dangerous zone

Zone 1 – the next most dangerous zone in which people stand in water

Zone 2 – the next most dangerous zone in which people might be in contact with water

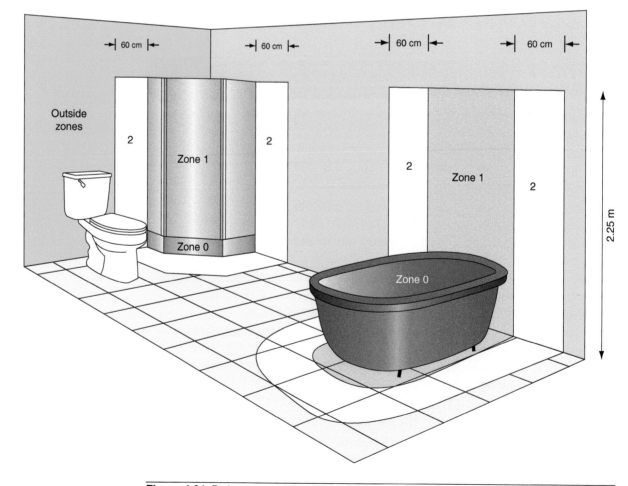

Figure 4.31 Bathroom zone dimensions. (Any window recess is zone 2.)

Outside zones – people are least likely to be in contact with water but are still in a potentially dangerous environment and the general IET Regulations apply.

- spaces under the bath which are accessible '*only with the use of a tool*' are outside zones;
- spaces under the bath which are accessible '*without the use of a tool*' are zone 1.

Electrical equipment and accessories are restricted within the zones.

Zone 0 – being the most potentially dangerous zone, for all practical purposes no electrical equipment can be installed in this zone. However, the IET Regulations permit that where SELV fixed equipment with a rated voltage not exceeding 12 V a.c. cannot be located elsewhere, it may be installed in this zone (IET Regulation 701.55). The electrical equipment must have at least IPX7 protection against total immersion in water (IET Regulation 701.512.2).

Zone 1 – water heaters, showers and shower pumps and SELV fixed equipment may be installed in zone 1. The electrical equipment must have at least IPX4 protection against water splashing from any direction. If the electrical equipment may be exposed to water jets from, for example, commercial cleaning equipment, then the electrical equipment must have IPX5 protection. (The Index of protection codes was discussed earlier and is shown in Fig. 4.25.)

Figure 4.32 Zone 2 shaver socket.

Zone 2 – luminaires and fans, and equipment from zone 1 plus shaver units to BS EN 60742 may be installed in zone 2. The electrical equipment must be suitable for installation in that zone according to the manufacturer's instructions and have at least IPX4 protection against splashing or IPX5 protection if commercial cleaning is anticipated.

Outside zones – appliances are allowed plus accessories except socket outlets unless the location containing the bath or shower is very big and the socket outlet can be installed at least 3 m horizontally beyond the zone 1 boundary (IET Regulation 701.512.3) and has additional RCD protection (IET Regulation 701.411.3.3).

If underfloor heating is installed in these areas it must have an overall earthed metallic grid or the heating cable must have an earthed metallic sheath, which is connected to the protective conductor of the supply circuit (IET Regulation 701.753).

Supplementary protective bonding

Modern plumbing methods make considerable use of non-metals (PTFE tape on joints, for example). Therefore, the metalwork of water and gas installations cannot be relied upon to be continuous throughout.

The IET Regulations describe the need to consider additional protection by supplementary protective bonding in situations where there is a high risk of electric shock (e.g. in kitchens and bathrooms) (IET Regulation 415.2).

In kitchens, supplementary protective bonding of hot and cold taps, sink tops and exposed water and gas pipes *is only required* if an earth continuity test proves that they are not already effectively and reliably connected to the main protective bonding, having negligible impedance, by the soldered pipe fittings of the installation. If the test proves unsatisfactory, the metalwork must be bonded using a single-core copper conductor with PVC green/yellow insulation, which will normally be 4 mm^2 for domestic installations but must comply with IET Regulation 543.1.1.

In rooms containing a fixed bath or shower, supplementary protective bonding conductors *must* be installed to reduce to a minimum the risk of an electric shock unless the following *two* conditions are met:

1 all bathroom circuits are protected by a fuse or MCB plus a 30 mA RCD; and
2 the bathroom is located in a building with a main protective bonding system in place (IET Regulation 701.415.2). Such a system is shown in Fig. 5.14 in this edition.

Supplementary equipotential bonding conductors in domestic premises will normally be of 4 mm^2 copper with PVC insulation to comply with IET Regulation 543.1.1 and must be connected between all exposed metalwork (e.g. between metal baths, bath and sink taps, shower fittings, metal waste pipes and radiators) as shown in Fig. 4.33.

The bonding connection must be made to a cleaned pipe, using a suitable bonding clip. Fixed at, or near, the connection must be a permanent label saying 'Safety electrical connection – do not remove' (IET Regulation 514.13.1), as shown in Fig. 4.34.

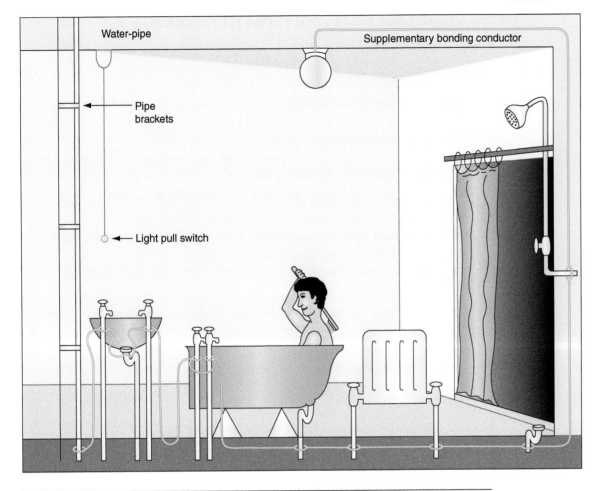

Water-pipe

Supplementary bonding conductor

Pipe
brackets

Light pull switch

Figure 4.33 Supplementary protective bonding in bathrooms to metal pipework.

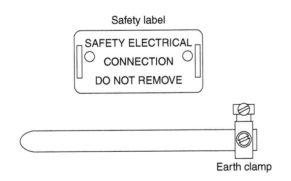

Safety label

SAFETY ELECTRICAL
CONNECTION
DO NOT REMOVE

Earth clamp

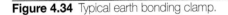

Figure 4.34 Typical earth bonding clamp.

Legrand 'rapid' earth bonding clamp.

Adjustable earth bonding clamps.

Restoring systems to working order

When the fault has been identified and repaired as described in this chapter, the circuit, system or equipment must be inspected, tested and functional checks carried out as required by IET Regulations Chapter 61.

The purpose of inspecting and testing the repaired circuit, system or equipment is to confirm the electrical integrity of the system before it is re-energized.

The tests recommended by Part 6 of the IET Regulations are:

1 Test the continuity of the protective conductors including the main protective bonding conductors and supplementary bonding conductors.
2 Test the continuity of all ring final circuit conductors.
3 Test the insulation resistance between live conductors and earth.
4 Test the polarity to verify that single pole control and protective devices are connected in the line conductor only.
5 Test the earth electrode resistance where the installation incorporates an earth electrode as a part of the earthing system.

The supply may now be connected and the following tests carried out:

6 Test the polarity using an approved test lamp or voltage indicator.
7 Test the earth fault loop impedance where the protective measures used require a knowledge of earth fault loop impedance.

These tests *where relevant* must be carried out in the order given above to comply with IET Regulation 643.1.

If any test indicates a failure, that test and any preceding test must be repeated after the fault has been rectified. This is because the earlier tests may have been influenced by the fault.

The above tests are described in Chapter 5 of this book, in Part 6 of the IET Regulations and in Guidance Note 3 published by the IET.

Functional testing (IET Regulation 643.10)

Following the carrying out of the relevant tests described above we must carry out functional testing to ensure that:

* the circuit, system or equipment works correctly;
* it works as it did before the fault occurred;
* it continues to comply with the original specification;
* it is electrically safe;
* it is mechanically safe;
* it meets all the relevant regulations, in particular the IET Regulations (BS 7671).

IET Regulation 643.10 tells us to check the effectiveness of the following assemblies to show that they are properly mounted, adjusted and installed:

* residual current devices (RCDs);
* switchgear;
* control gear;
* controls and interlocks.

Wherever RCDs are installed, a label shall be fixed near to each RCD stating 'The device must be tested 6 monthly' (IET Regulation 514.12.2). Note: In the 17th edition the test period was three monthly.

Restoration of the building structure

If the structure, or we sometimes call it the fabric, of the building has been damaged as a result of your electrical repair work, it must be made good before you hand the installation, system or equipment back to the client.

Where a wiring system passes through elements of the building construction such as floors, walls, roofs, ceilings, partitions or cavity barriers, the openings remaining after the passage of the wiring system must be sealed according to

 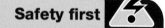

the degree of fire resistance demonstrated by the original building material (IET Regulation 527.2).

You should always make good the structure of the building using appropriate materials *before* you leave the job so that the general building structural performance and fire safety are not reduced. If additionally there is a little cosmetic plastering and decorating to be done, then who will actually carry out this work is a matter of negotiation between the client and the electrical contractor.

Disposal of waste

Having successfully diagnosed the electrical fault and carried out the necessary repairs or having completed any work in the electrical industry, we come to the final practical task: leaving the site in a safe and clean condition and the removal of any waste material. This is an important part of your company's 'good customer relationships' with the client. We also know from Chapter 3 of *Basic Electrical Installation Work 8th Edition* that we have a 'duty of care' for the waste that we produce as an electrical company (see Chapter 3, under the subheading 'Controlled Waste Regulations 1998').

We have also said many times in this book that having a good attitude towards health and safety, working conscientiously and neatly, keeping passageways clear and regularly tidying up the workplace is the sign of a good and competent craftsman. But what do you do with the rubbish that the working environment produces? Well:

- All the packaging material for electrical fittings and accessories usually goes into either your employer's skip or the skip on-site designated for that purpose.
- All the offcuts of conduit, trunking and tray also go into the skip.
- In fact, most of the general site debris will probably go into the skip and the waste disposal company will take the skip contents to a designated local council landfill area for safe disposal.
- The part coils of cable and any other reusable leftover lengths of conduit, trunking or tray will be taken back to your employer's stores area. Here it will be stored for future use and the returned quantities deducted from the costs allocated to that job.
- What goes into the skip for normal disposal into a landfill site is usually a matter of common sense. However, some substances require special consideration and disposal. We will now look at asbestos and large quantities of used fluorescent tubes which are classified as 'special waste' or 'hazardous waste'.

Asbestos is a mineral found in many rock formations. When separated it becomes a fluffy, fibrous material with many uses. It was used extensively in the construction industry during the 1960s and 1970s for roofing material, ceiling and floor tiles, fire-resistant board for doors and partitions, for thermal insulation and commercial and industrial pipe lagging.

There are three main types of asbestos:

- chysotile which is white and accounts for about 90% of the asbestos in use today;
- amosite which is brown;
- crocidolite which is blue.

Safety first

Fire

If your electrical activities cause damage to the fabric of the building then:

- the openings remaining must be sealed according to the degree of fire resistance demonstrated by the original building material.

IET Regulation 527.2.

Figure 4.34 Asbestos warning sign.

Asbestos cannot be identified by colour alone and a laboratory analysis is required to establish its type. Blue and brown are the two most dangerous forms of asbestos and have been banned from use since 1985. White asbestos was banned from use in 1999.

In the buildings where it was installed some 40 years ago, when left alone, it does not represent a health hazard, but those buildings are increasingly becoming in need of renovation and modernization. It is in the dismantling and breaking up of these asbestos materials that the health hazard increases. Asbestos is a serious health hazard if the dust is inhaled. The tiny asbestos particles find their way into delicate lung tissue and remain embedded for life, causing constant irritation and, eventually, serious lung disease.

Working with asbestos materials is not a job for anyone in the electrical industry. If asbestos is present in situations or buildings where you are expected to work, it should be removed by a specialist contractor before your work commences. Specialist contractors, who will wear fully protective suits and use breathing apparatus, are the only people who can safely and responsibly carry out the removal of asbestos. They will wrap the asbestos in thick plastic bags and store them temporarily in a covered and locked skip. This material is then disposed of in a special landfill site with other toxic industrial waste materials and the site monitored by the local authority for the foreseeable future.

There is a lot of work for electrical contractors in many parts of the country, updating and improving the lighting in government buildings and schools. This work often involves removing the old fluorescent fittings hanging on chains or fixed to beams and installing a suspended ceiling and an appropriate number of recessed modular fluorescent fittings or modular LED fittings. So what do we do with the old fittings? Well, the fittings are made of sheet steel, a couple of plastic lamp holders, a little cable, a starter and ballast. All of these materials can go into the ordinary skip. However, the fluorescent tubes contain a little mercury and fluorescent powder with toxic elements, which cannot be disposed of in the normal landfill sites. Hazardous Waste Regulations were introduced in July 2005 and under these regulations lamps and tubes are classified as hazardous. While each lamp contains only a small amount of mercury, vast numbers of lamps and tubes are disposed of in the United Kingdom every year, resulting in a significant environmental threat.

The environmentally responsible way to dispose of fluorescent lamps and tubes is to recycle them.

The process usually goes like this:

- your employer arranges for the local electrical wholesaler to deliver a plastic waste container of an appropriate size for the job;
- expired lamps and tubes are placed whole into the container, which often has a grating inside to prevent the tubes from breaking when being transported;
- when the container is full of used lamps and tubes, you telephone the electrical wholesaler and ask them to pick up the filled container and deliver it to one of the specialist recycling centres;
- your electrical company will receive a 'Duty of Care Note' and full recycling documents which ought to be filed safely as proof that the hazardous waste was recycled safely;
- the charge is approximately 50 p for each 1800 mm tube and this cost is passed on to the customer through the final account.

Safety first

Waste
- Clean up before you leave the job.
- Put waste in the correct skip.
- Recycle used lamps and tubes.
- Get rid of all waste responsibly.

Figure 4.35 Be careful what you put in the skip.

The Control of Substances Hazardous to Health (COSHH) Regulations and the Controlled Waste Regulations 1998 have encouraged specialist companies to set up businesses dealing with the responsible disposal of toxic waste material. Specialist companies have systems and procedures which meet the relevant regulation, and they will usually give an electrical company a certificate to say that they have disposed of a particular waste material responsibly. The system is called 'Waste Transfer Notes'. The notes will identify the type of waste, by whom it was taken and its final place of disposal. The person handing over the waste material to the waste disposal company will be given a copy of the notes and this must be filed in a safe place, probably in the job file or a dedicated file. It is the proof that your company has carried out its duty of care to dispose of the waste responsibly. The cost of this service is then passed on to the customer. These days, large employers and local authorities insist that waste is disposed of properly.

The Environmental Health Officer at your local Council Offices will always give advice and point you in the direction of specialist companies dealing with toxic waste disposal.

Hand over to the client

Handing over the repaired circuit, system or equipment is an important part of the fault diagnosis and repair process. You are effectively saying to the client, 'here is your circuit, system or equipment, it is now safe to use and it works as it should work'.

The client will probably be interested in the following:

- What has been done to identify and repair the fault?
- The possible reasons why the fault occurred and recommendations which will prevent a recurrence of the problem.
- A demonstration of the operation of the circuit, system or equipment to show that the fault has been fully rectified.
- Finally, the handing over of certificates of inspection and test results and manufacturer's instructions, if new equipment has been installed.

Check your understanding

When you have completed these questions, check out the answers at the back of the book.

Note: more than one multiple-choice answer may be correct.

1   To diagnose and find electrical faults the ideal person will:
   a.   first isolate the whole system
   b.   use a logical and methodical approach
   c.   carry out the relevant tests recommended by Part 6 of the IET Regulations
   d.   recognize his own limitations and seek help and guidance where necessary.

2   The 'symptoms' of an electrical fault might be:
   a.   there is a complete loss of power
   b.   nothing unusual is happening
   c.   there is a local or partial loss of power
   d.   the local isolator switch is locked off.

3   A fault is not a natural occurrence, it is not planned and occurs unexpectedly. It may be caused by:
   a.   regular maintenance
   b.   negligence
   c.   misuse
   d.   abuse.

4   The main lighting in a room having only one entrance would probably be controlled by a:
   a.   pull switch
   b.   intermediate switch
   c.   two-way switch
   d.   one-way switch.

5   The main lighting in a room having two entrances would probably be controlled by a:
   a.   pull switch
   b.   intermediate switch
   c.   two-way switch
   d.   one-way switch.

6   Electrical test equipment must always:
    a.   work on a.c. and d.c. supplies
    b.   have an in-date calibration certificate
    c.   incorporate a range selector switch
    d.   incorporate probes which comply with GS 38.

7   A final circuit feeding socket outlets with a rated current of less than 20A and used by ordinary persons for general use must always have:
    a.   overcurrent protection
    b.   30mA RCD protection
    c.   splash-proof protection
    d.   incorporate an industrial-type socket outlet.

8   The IET Regulation test for 'continuity of conductors' tests:
    a.   line conductors
    b.   neutral conductors
    c.   CPCs
    d.   protective bonding conductors.

9   The IET Regulation test for 'continuity of ring final circuit conductors' tests:
    a.   line conductors
    b.   neutral conductors
    c.   CPCs
    d.   protective bonding conductors.

10   The IET Regulation test for 'insulation resistance' tests:
    a.   the effectiveness of conductors' insulation
    b.   the earth loop impedance
    c.   the earth conductors to the earth electrode
    d.   that protective devices are connected in the line conductor.

11   The IET Regulation test for 'polarity' tests:
    a.   the earth loop impedance
    b.   the effective operation of RCDs
    c.   the effective operation of controls and switchgear
    d.   that protective devices are connected in the line conductor.

12   The IET Regulation test for 'functional testing' tests:
    a.   the earth loop impedance
    b.   the effective operation of RCDs
    c.   the effective operation of controls and switchgear
    d.   that protective devices are connected in the line conductor.

13   State three symptoms of an electrical fault.

14   State three causes of electrical faults.

15   Make a list of 10 places where faults might occur on an electrical system.

16   List four steps involved in fault finding.

17   State five requirements for safe working procedures when fault finding.

18   State four factors that might influence the decision to either repair or replace faulty equipment.

19   State four safety features you would look for before selecting test equipment.

20   State the advantages and dangers associated with optical fibre cables.

21   State briefly the meaning of 'static electricity'. What action is taken to reduce the buildup of a static charge on an electrical system?

22   State the problems which an unexpected mains failure of IT equipment would create.

23   Use a sketch to describe 'clean supplies', 'spikes' and 'noise' on IT supplies.

24   Briefly describe what we mean by secure supplies and UPS with regard to IT equipment.

25   Emergency supplies are often provided by storage batteries that are secondary cells. What is the advantage of a secondary cell in these circumstances?

26   Use bullet points to list very briefly the requirements of testing and commissioning an installation following the repair of a fault.

27   Use bullet points to list very briefly the six reasons for carrying out functional testing following the repair of a fault.

28   The IET Regulation 612.13 advises us to check the effectiveness of four assemblies to show that they are properly mounted, adjusted and installed. Name them.

29   State the reasons why it is important to make good any damage to the fabric of a building as a result of your electrical activities. State how you would make good damage to a brick wall and a concrete floor where a 100×100 mm trunking passes through.

30   Very briefly state the responsibilities of an electrical company with regard to the disposal of waste material.

31   State what we mean by 'ordinary waste' and 'hazardous waste' and give examples of each.

32   Very briefly describe the system of 'Waste Transfer Notes'.

33   Very briefly describe four points you would discuss with a client when handing over a repaired faulty system, circuit or piece of equipment.

Electrical installations: Inspection, testing and commissioning

Unit 304 of the City and Guilds 2365-03 Syllabus

Learning outcomes – when you have completed this chapter you should be able to:

- implement safe systems of work before inspecting and testing an electrical installation;
- understand the inspection process and schedules of inspection;
- understand the testing process and recognize satisfactory results;
- explain the purpose of certification documentation;
- state the requirements for commissioning electrical systems.

This chapter has free associated content, including animations and instructional videos, to support your learning.

When you see the logo, visit the website below to access this material: www.routledge.com/cw/linsley

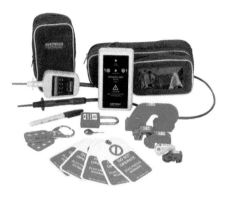

Figure 5.1 Lock-out kit.

Safe working procedures – Inspection and Testing

Whether you are carrying out the inspection and test procedure (i) as part of a new installation, (ii) upon completion of an extension to an existing installation, (iii) because you are trying to discover the cause of a fault on an installation, or (iv) because you are carrying out an Electrical Installation Condition report of a building, you must always be aware of your safety, the safety of others using the building and the possible damage which your testing might cause to other systems in the building.

For your own safety:

- Always use 'approved' test instruments and probes.
- Ensure that the test instrument carries a valid calibration certificate; otherwise the results may be invalid.
- Secure all isolation devices in the 'off' position.
- Put up warning notices so that other workers will know what is happening.
- Notify everyone in the building that testing is about to start and for approximately how long it will continue.
- Obtain a 'permit-to-work' if this is relevant.
- Obtain approval to have systems shut down which might be damaged by your testing activities. For example, computer systems may 'crash' when supplies are switched off. Ventilation and fume-extraction systems will stop working when you disconnect the supplies.

For the safety of other people:

- Fix warning notices around your work area.
- Use cones and highly visible warning tape to screen off your work area.
- Make an effort to let everyone in the building know that testing is about to begin. You may be able to do this while you carry out the initial inspection of the installation.
- Obtain verbal or written authorization to shut down information technology, emergency operation or standby circuits.

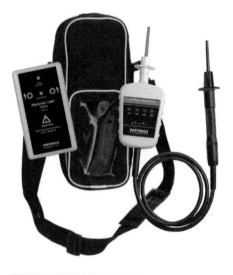

Figure 5.2 Safe isolation kit.

To safeguard other systems:

- Computer systems can be severely damaged by a loss of supply or the injection of a high test voltage from, for example, an insulation resistance test. Computer systems would normally be disconnected during the test period but this will generally require some organization before the testing begins. Commercial organizations may be unable to continue to work without their computer systems and, in these circumstances, it may be necessary to test outside the normal working day.
- Any resistance measurements made on electronic equipment or electronic circuits must be achieved with a battery-operated ohmmeter in order to avoid damaging the electronic circuits.
- Farm animals are creatures of habit and may become very grumpy to find you testing their milking parlour equipment at milking time.

- Hospitals and factories may have emergency standby generators which re-energize essential circuits in the event of a mains failure. Your isolation of the circuit for testing may cause the emergency systems to operate. Discuss any special systems with the person authorizing the work before testing begins.

Why inspect and test an electrical installation?

The reasons for carrying out the inspection and testing of an electrical installation are:

1 To ensure the safety of people and livestock.
2 To ensure the protection of property from fire and heat.
3 To confirm that the property has not been damaged or deteriorated so as to reduce its safety.
4 To identify any defects in the installation and the subsequent need for improvement.

All electrical equipment deteriorates with age as well as with wear and tear from being used. It also deteriorates as a result of excessive loading and environmental influences leading to, for example, corrosion. The electrical installation must therefore be inspected and tested periodically during its lifetime to confirm that it remains safe to use at least until the next inspection and test is carried out.

Organizations insuring buildings often require evidence that the electrical installation is safe and that the insurer's risk is therefore low. For example, insurance companies and mortgage lenders ask for electrical inspection and test reports when:

- there has been a change of building occupancy;
- there is a change of use of the building;
- alterations have been carried out;
- there is reason to believe that damage may have been caused to the building as a result of, for example, flooding.

In the event of an injury or fire alleged to have been caused by the electrical installation itself, the production of past certificates and reports will provide documentary evidence that the installation has been installed and subsequently maintained to a satisfactory standard of safety by competent persons.

Statutory and non-statutory regulations also clearly state the requirements for inspection and testing as part of a maintenance programme to reduce danger.

Regulation 4 (2) of the Electricity at Work Regulations advises that 'as may be necessary to prevent danger, all systems shall be maintained so as to prevent such danger'.

The Guidance Notes on the Electricity at Work Regulations advise that this regulation is concerned with the need for maintenance to ensure the safety of 'the system' and not the actual 'doing of maintenance' in a safe manner. The regular inspection of equipment including the electrical installation is an essential part of any preventative maintenance programme.

There is no specific requirement to test the installation on every inspection. Where testing requires dismantling, the tester should consider whether the risks associated with dismantling and reassembly are justified. Dismantling, and particularly disconnection of cables and components, introduces the possible risk of unsatisfactory reassembly.

IET Regulations 641 advise us that every installation shall, during erection and on completion before being put into service, be inspected and tested to verify that the regulations have been met. The 18th Edition of the IET Wiring Regulations gives an extensive checklist or examples of items that require inspection during the initial verification of new work. This can be found in Appendix 6 (page 467 of the blue book) of the Regulations and is discussed further in this chapter under the subheading 'Visual inspection'. Regulation 651 also advises that where required, periodic inspection and testing shall be carried out in order to determine if the installation is in a satisfactory condition for continued service. Regulation 642.3 gives us an inspection checklist and tells us that the inspection must include at least these 20 or more items plus, if appropriate, the particular requirements for special installations such as bathrooms described in Part 7 of the regulations.

The *On-Site Guide* at Section 9.2.2 gives the same inspection checklist.

Guidance Note 3 at Section 3.9.1 gives us guidance in using the IET checklist given in Regulation 642.3. We will look at this checklist later in this chapter under the subheading 'Inspection and testing techniques'.

The Electricians Guide to the Building Regulations Part P at Section 6.2 follows the requirements given in Section 641 of the IET Regulations.

Appendix 6 of the IET Regulations gives us a Condition Report Inspection Schedule (schedule means a planned programme of work) for a domestic or similar premises with up to 100 Amp supply. This inspection schedule checklist is to be used when carrying out a periodic inspection of an existing installation for the purpose of completing an Electrical Installation Condition Report (see page 476 of the blue edition of the Regs.).

Periodic Inspection – Completion of an Electrical Installation Condition Report. Regulation 653 and Appendix 6

The 18th Edition of the IET Regulations has introduced changes to the Periodic Inspection Report. This has now become the Electrical Installation Condition Report and incorporates changes to the classification coding system and a new inspection schedule which you will find in Appendix 6 of the IET Regulations. The inspection schedule for a domestic installation includes over 60 items which must be inspected and the electrician carrying out the inspection must comment on the condition of each item using the coding system described below.

The objective in producing this new Electrical Installation Condition Report inspection schedule is to provide the electrician carrying out the inspection with guidelines, so that the inspection is done in a structured and consistent way and for the client to better understand the result of the inspection. The 18th edition of the Regulations includes a requirement to carry out an inspection of accessible roof spaces where electrical equipment is present (page 473 of the blue edition).

The classification codes to be used on the Condition Report inspection schedule are C1, C2, C3 and FI:

Let us assume that while carrying out a visual inspection, the electrician observes a badly damaged socket outlet. On close inspection he realizes that it is possible to touch a live part through the cracked plate. This fault will attract a C1 code because there is immediate danger.

Alternatively, if on close inspection of another damaged socket outlet the electrician realizes it is possible to touch an earth connection through the cracked plate, it will attract a C2 code; there is potential danger here because the earth wire may become live under fault conditions.

✓ meaning an acceptable condition. The item inspected has been classified as acceptable.
C1 meaning an unacceptable condition. The item inspected has been classified as unacceptable. Immediate danger is present and the safety of those using the installation is at risk. For example, live parts are directly accessible. *Immediate remedial action is required.*
C2 meaning an unacceptable condition. The item inspected has been classified as unacceptable. Potential danger is present and the safety of those using the installation may be at risk. For example, the absence of the main protective bonding conductors. *Urgent remedial action is required.*
C3 meaning improvement is recommended. The item inspected is not dangerous for continued use but the inspector recommends that improvements be made. For example, the installation does not have RCDs installed for additional protection. *Improvements are recommended.*
FI meaning Further Investigation is required. This investigation has uncovered a deficiency. The item inspected has been classified as unacceptable. The deficiency cannot be identified at this time and further investigation is required. *Urgent remedial action is required.*

The presence of a C1, C2 or FI code will result in the overall condition of the electrical installation being reported as unsatisfactory – it has failed to meet the standard. The inspector should report their findings to the duty holder or person responsible for the installation immediately, both verbally and in writing that a risk of injury exists. If possible, immediately dangerous situations, given a C1 code, should be made safe, isolated or rectified on discovery.

A code C3 in itself would not warrant an unsatisfactory result. It is simply stating that while the installation is not compliant with the latest edition of the IET Regulations, it is compliant with a previous edition, and the item is not unsafe and does not necessarily require upgrading, and so is acceptable, for now.

An FI code means the inspection has revealed apparent deficiencies which could not be fully identified because of the limitations of the inspection. This will mean the condition of the installation is unsatisfactory.

The Electrical Installation Condition Report should only be used for an existing building. The report should include:

- schedules of both the inspection and test results in section A;
- the reasons for producing the report such as a change of occupancy or landlord's maintenance should be identified in section B;
- the extent and limitations of the report should be stated in section D;
- the inspector producing the final report should give a summary of the condition of the installation in terms of the safety of the installation in section

E, stating how any C1, C2, C3 and FI codes given for items in the schedule may effect the overall condition of the installation being classified as unsatisfactory;

* the recommended date of the next Electrical Installation Condition Report should be given in section F.

Further information on the new coding system can be downloaded free from the Electricity Safety Council's Best Practice Guide 4.

Sampling when carrying out inspections

The inspection required of an electrical installation must be a 'detailed examination' of the installation without dismantling or with only partial dismantling as is required (IET Regulation 651.2). A thorough visual inspection should be made of all electrical equipment which is not concealed and should include the accessible internal condition of a sample of the electrical equipment. It is not practicable to inspect every joint and termination in an electrical installation. Nevertheless, a detailed examination of the installation, without dismantling, should be made of all switches, switchgear, distribution boards, luminaire points and socket outlets to ensure that all terminal connections of the conductors are properly installed and secure. Any signs of overheating of conductors, terminations or equipment should be thoroughly investigated and included in the report (Guidance Note 3, Sections 3.8.3 and 3.9.1).

Considerable care and engineering judgement must be exercised when deciding upon the extent of the sample to receive the 'covers removed' visual inspection.

The factors to be considered when determining the size of the sample are:

* the availability of previous certificates;
* the age and condition of the installation;
* any evidence of ongoing maintenance;
* the time elapsed since the previous inspection;
* the size of the installation.

If no previous certificates are available and the installation is looking tired or abused it would be necessary to inspect and test a larger percentage of the installation and in some cases 100% of the installation.

The degree or extent of the sampling must be agreed with the person ordering the work before work commences. It may be impractical to inspect and test the whole of a large installation at one time. If this is the case and sampling is agreed with the person ordering the work, it is important to examine different parts of the installation in subsequent inspections. It would not be acceptable for the same parts of the installation to be repeatedly inspected to the exclusion of other parts.

Human senses and the inspection process

The senses of sight, hearing, smell and touch are powerful human forces which we can use when carrying out the 'covers removed' electrical inspection or when we are attempting to trace a fault on a circuit. Connections can become hot under fault conditions which may be as a result of loose connections or a circuit overload. The heat generated by the electrical fault might heat up the insulation of the conductors and the terminal block holding the connections. In these circumstances, the bakelite insulation used in the construction of terminal blocks

Figure 5.3 Clearly an electrical fault is hiding behind this switch plate.

and switchgear becomes hot and gives off a pungent smell, like rotting fish. So, we can use our powerful human sense of smell to guide us to the source of possible faults. Conductor insulation blackens when it becomes very hot and becomes brittle and stiff. Once more, we can use the human senses of sight and touch to detect these faults when carrying out an internal inspection of socket outlets, switches and luminaries.

Electrical testing

The electrical contractor is charged with a responsibility to carry out a number of tests on an electrical installation and electrical equipment. The individual tests are dealt with in Part 6 of the IET Regulations and described later in this chapter.

The reasons for testing the installation are:

- to ensure that the installation complies with the IET Regulations;
- to ensure that the installation meets the specification;
- to ensure that the installation is safe to use.

Those who are to carry out the electrical tests must first consider the following safety factors:

- an assessment of safe working practice must be made before testing begins;
- all safety precautions must be put in place before testing begins;
- everyone must be notified that the test process is about to take place, for example, the client and other workers who may be affected by the tests;
- 'Permits-to-work' must be obtained where relevant;
- all sources of information relevant to the tests have been obtained;
- the relevant circuits and equipment have been identified;
- safe isolation procedures have been carried out – care must be exercised here, in occupied premises, not to switch off computer systems without first obtaining permission;
- those who are to carry out the tests are skilled (electrically) or instructed (electrically) to do so.

Chapter 64 of the Regulations describe all the tests required and we will look at these later in this chapter.

The electrical contractor is required by the IET Regulations to test all new installations and major extensions during erection and upon completion before being put into service. The contractor may also be called upon to test installations and equipment in order to identify and remove faults. These requirements imply the use of appropriate test instruments, and in order to take accurate readings consideration should be given to the following points:

- Is the instrument suitable for this test?
- Has the correct scale been selected?
- Is the test instrument correctly connected to the circuit?

Many commercial instruments are capable of making more than one test or have a range of scales to choose from. A range selector switch is usually used to choose the appropriate scale. A scale range should be chosen which suits the range of the current, voltage or resistance being measured. For example, when taking a reading in the 8 or 9 V range, the obvious scale choice would be one giving 10 V full-scale deflection. To make this reading on an instrument with 100 V full-scale deflection would lead to errors, because the deflection is too small.

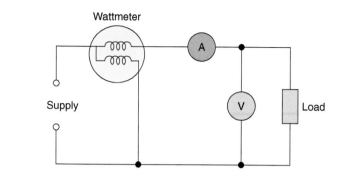

Figure 5.4 Wattmeter, ammeter and voltmeter correctly connected to a load.

Ammeters must be connected in series with the load, and voltmeters in parallel across the load, as shown in Fig. 5.4. The power in a resistive load may be calculated from the readings of voltage and current since $P = VI$. This will give accurate calculations on both a.c. and d.c. supplies, but when measuring the power of an a.c. circuit which contains inductance or capacitance, a wattmeter must be used because the voltage and current will be out of phase.

Measurement of power in a three-phase circuit

One-wattmeter method

When three-phase loads are balanced, for example, in motor circuits, one wattmeter may be connected into any phase, as shown in Fig. 5.5. This wattmeter will indicate the power in that phase and, since the load is balanced, the total power in the three-phase circuit will be given by:

$$\text{Total power} = 3 \times \text{Wattmeter reading}$$

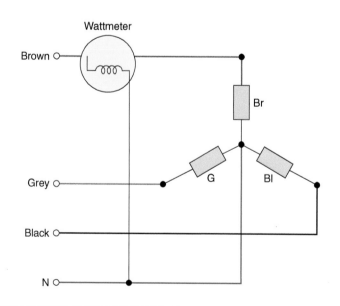

Figure 5.5 One-wattmeter measurement of power.

Two-wattmeter method

This is the most commonly used method for measuring power in a three-phase, three-wire system since it can be used for both balanced and unbalanced loads connected in either star or delta. The current coils are connected to any two of the lines, and the voltage coils are connected to the other line, the one without a current coil connection, as shown in Fig. 5.6. Then,

$$\text{Total power} = W_1 + W_2$$

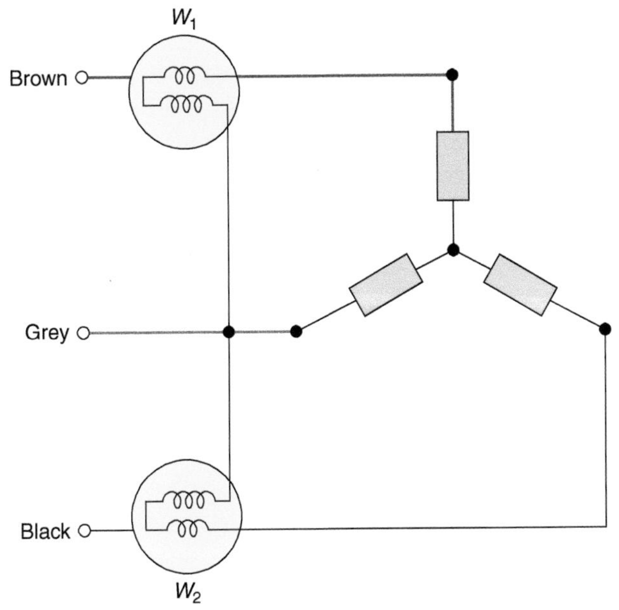

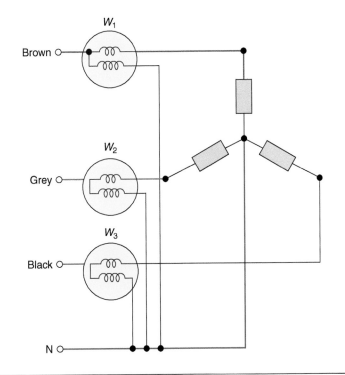

Figure 5.6 Two-wattmeter measurement of power.

Figure 5.7 Three-wattmeter measurement of power.

This equation is true for any three-phase load, balanced or unbalanced, star or delta connection, provided there is no fourth wire in the system.

Three-wattmeter method

If the installation is four-wire, and the load on each phase is unbalanced, then three-wattmeter readings are necessary, connected as shown in Fig. 5.7. Each wattmeter measures the power in one phase and the total power will be given by:

$$\text{Total power} = W_1 + W_2 + W_3$$

Tong tester

The tong tester, or clip-on ammeter, offers an easy method of measuring current with out any connections being made. It works on the same principle as the transformer. The laminated core of the transformer can be opened and passed over the busbar or single-core cable. In this way a measurement of the current being carried can be made without disconnection of the supply. The construction is shown in Fig. 5.8.

Figure 5.8 Tong tester or clip-on ammeter.

Test equipment used by electricians

The Health and Safety Executive (HSE) has published a guidance note (GS 38) which advises electricians and other electrically competent people on the selection of suitable test probes, voltage-indicating devices and measuring instruments. This is because they consider suitably constructed test equipment to be as vital for personal safety as the training and practical skills of the electrician. In the past, unsatisfactory test probes and voltage indicators have frequently been the cause of accidents, and therefore all test probes must now incorporate the following features:

1 The probes must have finger barriers or be shaped so that the hand or fingers cannot make contact with the live conductors under test.
2 The probe tip must not protrude more than 2 mm, and preferably only 1 mm, be spring-loaded and screened.
3 The lead must be adequately insulated and coloured so that one lead is readily distinguished from the other.
4 The lead must be flexible and sufficiently robust.
5 The lead must be long enough to serve its purpose but not too long.
6 The lead must not have accessible exposed conductors even if it becomes detached from the probe or from the instrument.
7 Where the leads are to be used in conjunction with a voltage detector they must be protected by a current-limiting resistor.

A suitable probe and lead is shown in Fig. 5.11.

GS 38 also tells us that where the test is being made simply to establish the presence or absence of a voltage, the preferred method is to use a proprietary test lamp or voltage indicator which is suitable for the working voltage, rather than a multimeter. Accident history has shown that incorrectly set multimeters or makeshift devices for voltage detection have frequently caused accidents. Figure 5.11 shows a suitable voltage indicator. Test lamps and voltage indicators are not fail-safe, and therefore GS 38 recommends that they should be regularly proved, preferably before and after use, as described in the method statement or flowchart for a safe isolation procedure shown in Fig 4.12.

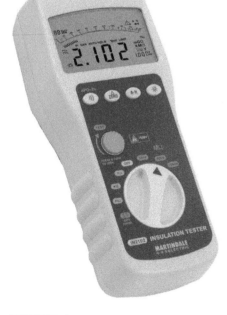

Figure 5.9 Multi voltage insulation and continuity tester.

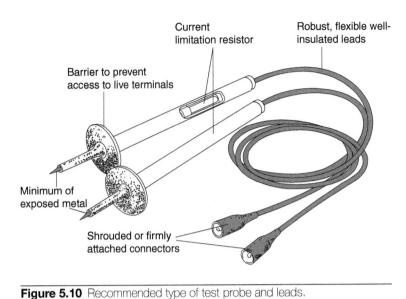

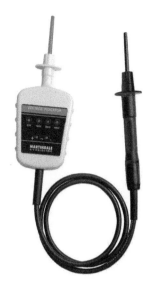

Figure 5.10 Recommended type of test probe and leads.

Figure 5.11 Typical voltage indicator.

Test procedures

1 The circuits must be isolated using a 'safe isolation procedure', such as that described below, before beginning to test.
2 All test equipment must be 'approved' and connected to the test circuits by recommended test probes as described by GS 38. The test equipment used must also be 'proved' on a known supply or by means of a proving unit such as that shown in Fig. 4.7.
3 Isolation devices must be 'secured' in the 'off' position as shown in Fig. 5.12.
4 Warning notices must be posted.
5 All relevant safety and functional tests must be completed before restoring the supply.

Live testing

The **Electricity at Work Regulations** tell us that it is 'preferable' that supplies be made dead before work commences (Regulation 4(3)). However, it does acknowledge that some work, such as fault finding and testing, may require the electrical equipment to remain energized. Therefore, if the fault finding and testing can only be successfully carried out 'live', then the person carrying out the fault diagnosis must:

• be trained so that he understands the equipment and the potential hazards of working live and can, therefore, be deemed to be 'electrically skilled' to carry out the activity;
• only use approved test equipment;
• set up barriers and warning notices so that the work activity does not create a situation dangerous to others.

Note that while live testing may be required in order to find the fault, live repair work must not be carried out. The individual circuit or item of equipment must first be isolated.

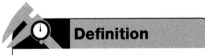

Figure 5.12 Secure isolation of a supply.

> **Definition**
>
> The *Electricity at Work Regulations* tell us that it is 'preferable' that supplies be made dead before work commences (Regulation 4(3)).

Electrical circuits-safe isolation procedure

The Electricity at Work Regulations are very specific in describing the procedure to be used for isolation of the electrical supply. Regulation 12(1) tells us that **isolation** means the disconnection and separation of the electrical equipment from every source of electrical energy in such a way that this disconnection and separation is secure. Regulation 4(3) tells us that we must also prove the conductors dead before work commences and that the test instrument used for this purpose must itself be proved immediately before, and immediately after, testing the conductors. To isolate an individual circuit or item of equipment successfully, competently and safely we must follow a procedure such as that given by the flow chart in Fig. 4.12. Start at the top and work your way down the flowchart. When you get to the heavy-outlined amber boxes, pause and ask yourself whether everything is satisfactory up to this point. If the answer is yes, move on. If no, go back as indicated by the diagram.

Inspection and testing techniques

The testing of an installation implies the use of instruments to obtain readings. However, a test is unlikely to identify a cracked socket outlet, a chipped or loose switch plate, a loose or badly fixed piece of equipment or a missing conduit-box lid or saddle, so it is also necessary to make a visual inspection of the installation.

All new installations must be inspected and tested during erection and upon completion before being put into service. All existing installations should be periodically inspected and tested to ensure that they are safe and meet the IET Regulations (IET Regulations 641–653).

The method used to test an installation may inject a current into the system. This current must not cause danger to any person or equipment in contact with the installation, even if the circuit being tested is faulty. The test results must be compared with any relevant data, including the tables in the IET Regulations, and the test procedures must be followed carefully and in the correct sequence, as indicated by IET Regulation 643. This ensures that the protective conductors are correctly connected and secure before the circuit is energized.

Visual inspection

The installation must be visually inspected before testing begins. The aim of the **initial verification** is to confirm that all equipment and accessories are undamaged and comply with the relevant British and European Standards, and also that the installation has been securely and correctly erected. IET Regulation 642.3 gives a schedule or checklist of examples of items requiring inspection during the initial verification of the installation, including:

* Connection of conductors;
* Identification of conductors;
* Routing of cables according to Reg 522;
* Selection of cables for size and volt drop;
* Connection of single pole devices in line conductor;
* Correct connection of accessories and equipment;
* Method of protection against electric shock appropriate;
* Selection and installation of earthling conductors confirmed;
* Prevention of mutual detrimental influence;

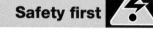

Safety first

Note that while live testing may be required in order to find the fault, live repair work must not be carried out. The individual circuit or item of equipment must first be isolated.

Definition

Isolation means the disconnection and separation of the electrical equipment from every source of electrical energy in such a way that this disconnection and separation is secure.

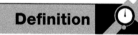

Definition

The aim of the *visual inspection* is to confirm that all equipment and accessories are undamaged and comply with the relevant British and European Standards, and also that the installation has been securely and correctly erected.

- Isolation and switching devices correctly located;
- Supplementary bonding and earthing arrangements;
- Exposed conductive parts connected to earthing arrangements;
- Presence of appropriate devices for isolation and switching;
- Selection of protective measures appropriate to external influences;
- Adequate access to switchgear and equipment;
- Appropriate labelling and warning signs;
- Erection method appropriate.

The checklist is a guide, it is not exhaustive or detailed, but should be used to identify relevant items for inspection which might then be expanded upon. For example, the first item on the list, connection of conductors might be further expanded to include:

- are conductors secure?
- are the connections correct?
- is the cable adequately supported, without strain to the conductors?
- does the outer sheath enter the accessory?
- is the insulation undamaged?
- does the insulation proceed up to but not into the terminal?

See IET Regulation 642.3 for the complete list of items requiring inspection during the initial verification procedure.

It is thought that the new inspection schedule introduced by the 18th Edition for all new work will help electricians carry out the work in an organized and efficient manner. The schedule of inspections must be appended to the electrical installation certificate and given to the customer or client upon completion.

Following the inspection, those tests which are relevant to the installation must then be carried out in the sequence given in IET Regulation 643.1 for reasons of safety and accuracy. These tests are as follows:

Before the supply is connected:

1 Test for continuity of protective conductors, including main protective and supplementary bonding.
2 Test the continuity of all ring final circuit conductors.
3 Test for insulation resistance.
4 Test for polarity using the continuity method.
5 Test the earth electrode resistance if appropriate.

With the supply connected:

6 Recheck (or verify) polarity using a voltmeter or approved test lamp.
7 Test the earth fault loop impedance.
8 Carry out additional protection testing (e.g. operation of residual current devices, RCDs).
9 Check phase sequence of three-phase supplies.
10 Functional testing.

If any test fails to comply with the IET Regulations, then *all* the preceding tests must be repeated after the fault has been rectified. This is because the earlier test results may have been influenced by the fault (IET Regulation 643.1).

There is an increased use of electronic devices in electrical installation work; for example, in dimmer switches and ignitor circuits of discharge lamps. These devices should be temporarily disconnected so that they are not damaged by the test voltage of, for example, the insulation resistance test (IET Regulation 643.3).

Key fact

Testing
- All new installations must be inspected and tested during erection and upon completion.
- All existing installations must be inspected and tested periodically.

IET Regulations 610–634.

Safety first

Live working
- **NEVER** work **LIVE**
- Some 'live testing' is allowed by 'competent persons'.
- Otherwise, isolate and secure the isolation.
- Prove the supply dead before starting work.

Definition

The *test instruments and test leads* used by the electrician for testing an electrical installation must meet all the requirements of the relevant regulations.

Definition

Calibration certificates usually last for a year. Test instruments must, therefore, be tested and recalibrated each year by an approved supplier.

Approved test instruments

The **test instruments and test leads** used by the electrician for testing an electrical installation must meet all the requirements of the relevant regulations. The HSE has published guidance note GS 38 for test equipment used by electricians. The IET Regulations (BS 7671) also specify the test voltage or current required to carry out particular tests satisfactorily. All test equipment must be chosen to comply with the relevant parts of BS EN 61557. All testing must, therefore, be carried out using an 'approved' test instrument if the test results are to be valid. The test instrument must also carry a calibration certificate, otherwise the recorded results may be void. **Calibration certificates** usually last for a year. Test instruments must, therefore, be tested and recalibrated each year by an approved supplier. This will maintain the accuracy of the instrument to an acceptable level, usually within 2% of the true value.

Modern digital test instruments are reasonably robust, but to maintain them in good working order they must be treated with care. An approved test instrument costs as much as a good-quality camera; it should, therefore, receive the same care and consideration.

Let us now look at the requirements of four often used test instruments.

Continuity tester

To measure accurately the resistance of the conductors in an electrical installation, we must use an instrument which is capable of producing an open-circuit voltage of between 4 and 24 V a.c. or d.c., and delivering a short-circuit current of not less than 200 mA (IET Regulation 643.2). The functions of continuity testing and insulation resistance testing are usually combined in one test instrument.

Insulation resistance tester

The test instrument must be capable of detecting insulation leakage between live conductors and between live conductors and earth. To do this and comply with IET Regulation 643.3 the test instrument must be capable of producing a test voltage of 250, 500 or 1000 V and delivering an output current of not less than 1 mA at its normal voltage.

Earth fault loop impedance tester

The test instrument must be capable of delivering fault currents as high as 25 A for up to 40 ms using the supply voltage. During the test, the instrument does an Ohm's law calculation and displays the test result as a resistance reading.

RCD tester

Where circuits are protected by an RCD we must carry out a test to ensure that the device will operate very quickly under fault conditions and within the time limits set by the IET Regulations. The instrument must, therefore, simulate a fault and measure the time taken for the RCD to operate. The instrument is, therefore, calibrated to give a reading measured in milliseconds to an in-service accuracy of 10%.

If you purchase good-quality 'approved' test instruments and leads from specialist manufacturers they will meet all the regulations and standards and therefore give valid test results. However, to carry out all the tests required by the IET Regulations will require a number of test instruments and this will represent a major capital investment in the region of £1000.

Let us now consider the individual tests.

Figure 5.13 Multi voltage insulation and continuity tester.

1 Testing for continuity of protective conductors, including main and supplementary protective bonding (IET Regulation 643.2)

The object of the test is to ensure that the circuit protective conductor (CPC) is correctly connected, is electrically sound and has a total resistance which is low enough to permit the overcurrent protective device to operate within the disconnection time requirements of IET Regulation 411.4.201 should an earth fault occur. Every protective conductor must be separately tested from the main earthing terminal of the installation to verify that it is electrically sound and correctly connected, including the main protective and supplementary protective bonding conductors as shown in Fig. 5.14 and Fig. 4.33. The IET Regulations describe the need to consider additional protection by supplementary protective bonding in situations where there is a high risk of electric shock such as kitchens and bathrooms (IET Regulation 415.2).

A d.c. test using an ohmmeter continuity tester is suitable where the protective conductors are of copper or aluminium up to $35\,mm^2$. The test is made with the supply disconnected, measuring from the main earthing terminal of the installation to the far end of each CPC, as shown in Fig. 5.15. The resistance of

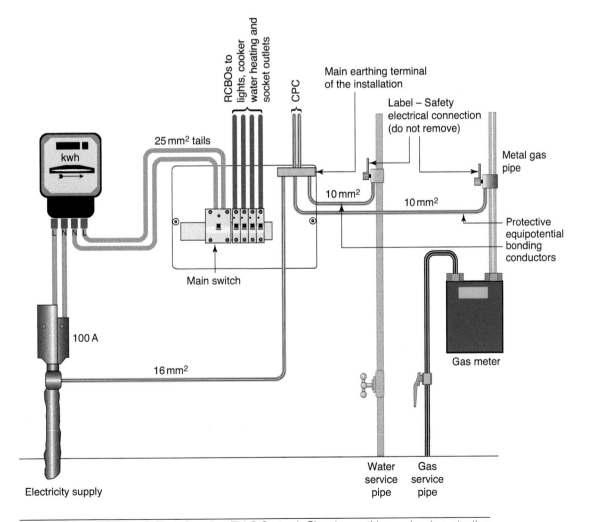

Figure 5.14 Cable Sheath Earth Supplies (TN-S System): Showing earthing and main protective equipotential bonding arrangements.

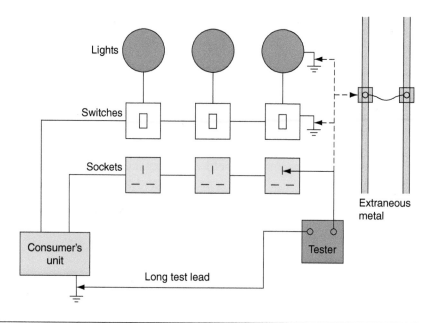

Figure 5.15 Testing continuity of protective conductors.

the long test lead is subtracted from these readings to give the resistance value of the CPC. The result is recorded on an installation schedule such as that given in Appendix 6 of the IET Regulations.

A satisfactory test result for the bonding conductors will be in the order of 0.05 Ω or less (IET Guidance Note 3).

Where steel conduit or trunking forms the protective conductor, the standard test described above may be used, but additionally the enclosure must be visually checked along its length to verify the integrity of all the joints.

If the inspecting engineer has grounds to question the soundness and quality of these joints then the phase earth loop impedance test described later in this chapter should be carried out.

If, after carrying out this further test, the inspecting engineer still questions the quality and soundness of the protective conductor formed by the metallic conduit or trunking, then a further test can be done using an a.c. voltage not greater than 50V at the frequency of the installation and a current approaching 1.5 times the design current of the circuit, but not greater than 25A.

This test can be done using a low-voltage transformer and suitably connected ammeters and voltmeters, but a number of commercial instruments are available, such as the Clare tester, which give a direct reading in ohms.

Because fault currents will flow around the earth fault loop path, the measured resistance values must be low enough to allow the overcurrent protective device to operate quickly. For a satisfactory test result, the resistance of the protective conductor should be consistent with those values calculated for a line conductor of similar length and cross-sectional area. Values of resistance per metre for copper and aluminium conductors are given in Table I1 of the *On Site Guide*. The resistances of some other metallic containers are given in Table 5.1 of this book.

Table 5.1 Resistance values of some metallic containers

Metallic sheath	Size (mm)	Resistance at 20°C (mΩ/m)
Conduit	20	1.25
	25	1.14
	32	0.85
Trunking	50 × 50	0.949
	75 × 75	0.526
	100 × 100	0.337

Example

The CPC for a ring final circuit is formed by a 1.5 mm² copper conductor of 50 m approximate length. Determine a satisfactory continuity test value for the CPC using the value given in Table I1 of the *On Site Guide*.

Table I1 gives resistance/metre for a 1.5 mm² copper conductor
$$= 12.10\,m\Omega/m$$

Therefore, the resistance of $50m = 50 \times 12.10 \times 10^{-3}$
$$= 0.605\,\Omega$$

The protective conductor resistance values calculated by this method can only be an approximation since the length of the CPC can only be estimated. Therefore, in this case, a satisfactory test result would be obtained if the resistance of the protective conductor was about 0.6 Ω. A more precise result is indicated by the earth fault loop impedance test which is carried out later in the sequence of tests.

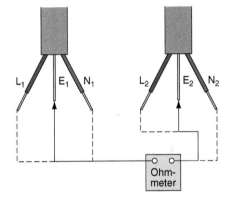

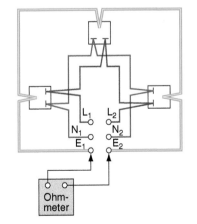

2 Testing for continuity of ring final circuit conductors (IET Regulation 643.2)

The object of the test is to ensure that all ring circuit cables are continuous around the ring; that is, that there are no breaks and no interconnections in the ring, and that all connections are electrically and mechanically sound. This test also verifies the polarity of each socket outlet.

The test is made with the supply disconnected, using an ohmmeter as follows.

Disconnect and separate the conductors of both legs of the ring at the main fuse. There are three steps to this test.

Step 1

Measure the resistance of the line conductors (L_1 and L_2), the neutral conductors (N_1 and N_2) and the protective conductors (E_1 and E_2) at the mains position, as shown in Fig. 5.16. End-to-end live and neutral conductor readings should

Figure 5.16 Step 1 test: measuring the resistance of phase, neutral and protective conductors.

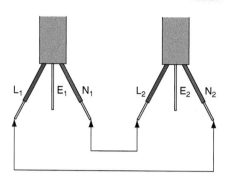

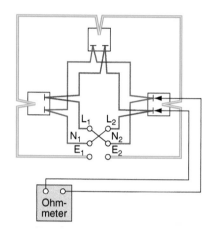

Figure 5.17 Step 2 test: connection of mains conductors and test circuit conditions.

be approximately the same (i.e. within $0.05\,\Omega$) if the ring is continuous. The protective conductor reading will be 1.67 times as great as these readings if 2.5/1.5 mm cable is used. Record the results on a table such as that shown in Table 5.2.

Step 2

The live and neutral conductors should now be temporarily joined together, as shown in Fig. 5.17. An ohmmeter reading should then be taken between live and neutral at *every* socket outlet on the ring circuit. The readings obtained should be substantially the same, provided that there are no breaks or multiple loops in the ring. Each reading should have a value of approximately half the live and neutral ohmmeter readings measured in Step 1 of this test. Sockets connected as a spur will have a slightly higher value of resistance because they are fed by only one cable, while each socket on the ring is fed by two cables. Record the results on a table such as that shown in Table 5.2.

Step 3

Where the CPC is wired as a ring, for example, where twin and earth cables or plastic conduit is used to wire the ring, temporarily join the live and CPCs together, as shown in Fig. 5.18. An ohmmeter reading should then be taken between live and earth at *every* socket outlet on the ring. The readings obtained should be substantially the same provided that there are no breaks or multiple loops in the ring. This value is equal to $R_1 + R_2$ for the circuit. Record the results on an installation schedule such as that given in Appendix 6 of the IET Regulations or a table such as that shown in Table 5.2. The Step 3 value of $R_1 + R_2$ should be equal to $(r_1 + r_2)/4$, where r_1 and r_2 are the ohmmeter readings from Step 1 of this test (see Table 5.2).

3 Testing insulation resistance (IET Regulation 643.3)

The object of the test is to verify that the quality of the insulation is satisfactory and has not deteriorated or short-circuited. The test should be made at the consumer's unit with the mains switch off, all fuses in place and all switches closed. Neon lamps, capacitors and electronic circuits should be disconnected, since they will respectively glow, charge up and be damaged by the test.

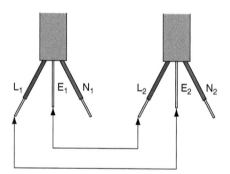

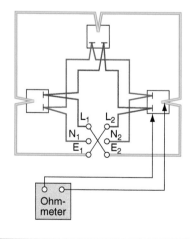

Figure 5.18 Step 3 test: connection of mains conductors and test circuit conditions.

Table 5.2 Table which may be used to record the readings taken when carrying out the continuity of ring final circuit conductors tests according to IET Regulation 643.2

Test	Ohmmeter connected to	Ohmmeter readings	This gives a value for
Step 1	L_1 and L_2		r_1
	N_1 and N_2		
	E_1 and E_2		r_2
Step 2	Live and neutral at each socket		
Step 3	Live and earth at each socket		$R_1 + R_2$
As a check $(R_1 + R_2)$ value should equal $(r_1 + r_2)/4$.			

There are two tests to be carried out using an insulation resistance tester which must have a test voltage of 500V d.c. for 230V and 400V installations. These are line and neutral conductors to earth and between line conductors. The procedures are as follows.

Line and neutral conductors to earth:

1 Remove all lamps.
2 Close all switches and circuit-breakers.
3 Disconnect appliances.
4 Test separately between the line conductor and earth *and* between the neutral conductor and earth, for *every* distribution circuit at the consumer's unit, as shown in Fig. 5.19(a). Record the results on a schedule of test results such as that given in Appendix 6 of the IET Regulations.

Between line conductors:

1 Remove all lamps.
2 Close all switches and circuit breakers.
3 Disconnect appliances.
4 Test between line and neutral conductors of *every* distribution circuit at the consumer's unit as shown in Fig. 5.19(b) and record the result.

The insulation resistance readings for each test must be not less than 1.0 MΩ for a satisfactory result (IET Regulation 643.3.2).

Where the circuit includes electronic equipment which might be damaged by the insulation resistance test, a measurement between all live conductors (i.e. live and neutral conductors connected together) and the earthing arrangements may be made. The insulation resistance of these tests should be not less than 1.0 MΩ (IET Regulation 643.3.2).

Although an insulation resistance reading of 1.0 MΩ complies with the regulations, the IET guidance notes tell us that much higher values than this

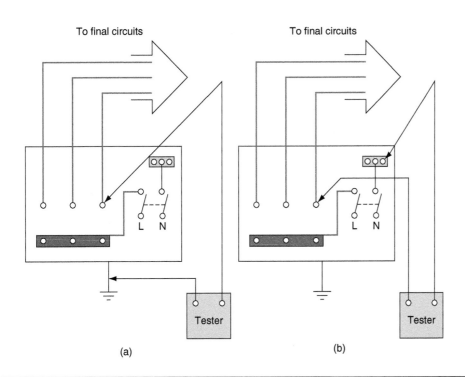

(a) (b)

Figure 5.19 Insulation resistance test.

can be expected and that a reading of less than 2 MΩ might indicate a latent, but not yet visible, fault in the installation. In these cases each circuit should be separately tested to obtain a reading greater than 2 MΩ.

4 Testing polarity (IET Regulation 643.6)

The object of this test is to verify that all fuses, circuit-breakers and switches are connected in the line or live conductor only, that all socket outlets are correctly wired and that Edison screw-type lamp holders have the centre contact connected to the live conductor. It is important to make a polarity test on the installation since a visual inspection will only indicate conductor identification.

The test is done with the supply disconnected using an ohmmeter or continuity tester as follows:

1 Switch off the supply at the main switch.
2 Remove all lamps and appliances.
3 Fix a temporary link between the line and earth connections on the consumer's side of the main switch.
4 Test between the 'common' terminal and earth at each switch position.
5 Test between the centre pin of any Edison screw lamp holders and any convenient earth connection.
6 Test between the live pin (i.e. the pin to the right of earth) and earth at each socket outlet, as shown in Fig. 5.20.

For a satisfactory test result the ohmmeter or continuity meter should read very close to zero for each test.

Remove the test link and record the results on a schedule of test results such as that given in Appendix 6 of the IET Regulations.

5 Testing earth electrode resistance (IET Regulation 643.7.2)

When an earth electrode has been sunk into the general mass of earth, it is necessary to verify the resistance of the electrode. The general mass of earth can be considered as a large conductor which is at zero potential. Connection to this

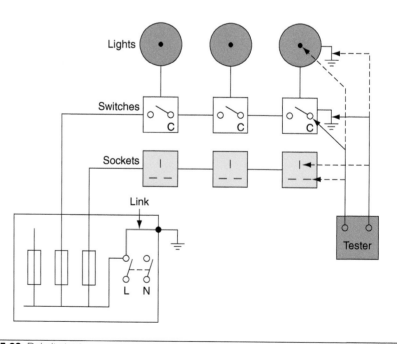

Figure 5.20 Polarity test.

mass through earth electrodes provides a reference point from which all other voltage levels can be measured. This is a technique which has been used for a long time in power distribution systems.

The resistance to earth of an electrode will depend upon its shape, size and the resistance of the soil. Earth rods form the most efficient electrodes. A rod of about 1 m will have an earth electrode resistance of between 10 and 200 Ω. Even in bad earthing conditions, a rod of about 2 m will normally have an earth electrode resistance which is less than 500 Ω in the United Kingdom. In countries which experience long, dry periods of weather the earth electrode resistance may be thousands of ohms.

In the past, electrical engineers used the metal pipes of water mains as an earth electrode, but the recent increase in the use of PVC pipe for water mains now prevents the use of water pipes as the means of earthing in the United Kingdom, although this practice is still permitted in some countries. IET Regulation 542.2.2 recognizes the use of the following types of earth electrodes:

- earth rods or pipes;
- earth tapes or wires;
- earth plates;
- earth electrodes embedded in foundations;
- welded metallic reinforcement of concrete structures;
- other suitable underground metalwork;
- lead sheaths or other metallic coverings of cables.

The earth electrode is sunk into the ground, but the point of connection should remain accessible (IET Regulation 542.4.2). The connection of the earthing conductor to the earth electrode must be securely made with a copper conductor complying with Table 54.1 and IET Regulation 542.3.2, as shown in Fig. 5.21.

The installation site must be chosen so that the resistance of the earth electrode does not increase above the required value due to climatic conditions such as the soil drying out or freezing, or from the effects of corrosion (IET Regulations 542.2.1 and 4).

Under fault conditions the voltage appearing at the earth electrode will radiate away from the electrode like the ripples radiating away from a pebble thrown into a pond. The voltage will fall to a safe level in the first 2 or 3 m away from the point of the earth electrode.

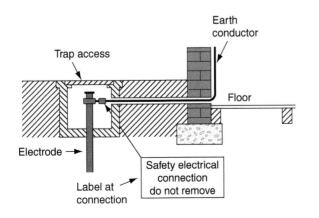

Figure 5.21 Termination of an earth electrode.

The basic method of measuring earth electrode resistance is to pass a current into the soil through the electrode and to measure the voltage required to produce this current.

IET Regulation 643.7.2 demands that where earth electrodes are used they should be tested.

If the electrode under test forms part of the earth return for a TT installation in conjunction with an RCD, Guidance Note 3 of the IET Regulations describes the following method:

1 Disconnect the installation protective bonding conductors from the earth electrode to ensure that the test current passes only through the earth electrode.
2 Switch off the consumer's unit to isolate the installation.
3 Using a line earth loop impedance tester, test between the incoming line conductor and the earth electrode.
4 Reconnect the protective bonding conductors when the test is completed.

Record the result on a schedule of test results such as that given in Appendix 6 of the IET Regulations.

The IET Guidance Note 3 tells us that an acceptable value for the measurement of the earth electrode resistance would be less than 200 Ω.

Providing the first five tests were satisfactory, the supply may now be switched on and the final tests completed with the supply connected.

6 Testing polarity – supply connected

Using an approved voltage indicator such as that shown at Fig. 5.11 or test lamp and probes which comply with the HSE Guidance Note GS 38, again carry out a polarity test to verify that all fuses, circuit-breakers and switches are connected in the live conductor. Test from the common terminal of switches to earth, the live pin of each socket outlet to earth and the centre pin of any Edison screw lamp holders to earth. In each case the voltmeter or test lamp should indicate the supply voltage for a satisfactory result.

7 Testing earth fault loop impedance – supply connected (IET Regulation 643.7.3)

The object of this test is to verify that the impedance of the whole earth fault current loop line to earth is low enough to allow the overcurrent protective device to operate within the disconnection time requirements of IET Regulations 411.3.2.2, 411.4.201 and 411.4.202, should a fault occur.

The whole earth fault current loop examined by this test comprises all the installation protective conductors, the main earthing terminal of the installation and protective bonding conductors, the earthed neutral point and the secondary winding of the supply transformer and the line conductor from the transformer to the point of the fault in the installation.

The test will, in most cases, be done with a purpose-made line earth loop impedance tester which circulates a current in excess of 10 A around the loop for a very short time, so reducing the danger of a faulty circuit. The test is made with the supply switched on, and carried out from the furthest point of *every* final circuit, including lighting, socket outlets and any fixed appliances. Record the results on a schedule of test results.

Purpose-built testers give a readout in ohms and a satisfactory result is obtained when the loop impedance does not exceed the appropriate values given in

Tables 41.2 and 41.3 of the IET Regulations. Table 41.3 gives a value of 2.73 ohm maximum for a circuit protected by a 16A MCB and 1.37 ohm maximum for a circuit protected by a 32 A MCB.

8 Additional protection: Testing of RCD – supply connected (IET Regulation 643.8)

The object of the test is to verify the effectiveness of the RCD, that it is operating with the correct sensitivity and proving the integrity of the electrical and mechanical elements. The test must simulate an appropriate fault condition and be independent of any test facility incorporated in the device.

When carrying out the test, all loads normally supplied through the device are disconnected.

The testing of a ring circuit protected by a general-purpose RCD to BS EN 61008 in a split-board consumer unit is carried out as follows:

1 Using the standard lead supplied with the test instrument, disconnect all other loads and plug in the test lead to the socket at the centre of the ring (i.e. the socket at the furthest point from the source of supply).
2 Set the test instrument to the tripping current of the device and at a phase angle of 0°.
3 Press the test button – the RCD should trip and disconnect the supply within 200 ms.
4 Change the phase angle from 0° to 180° and press the test button once again. The RCD should again trip within 200 ms. Record the highest value of these two results on a schedule of test results such as that given in Appendix 6 of the IET Regulations.
5 Now set the test instrument to 50% of the rated tripping current of the RCD and press the test button. The RCD should *not trip* within two seconds. This test is testing the RCD for inconvenience or nuisance tripping.
6 Finally, the effective operation of the test button incorporated within the RCD should be tested to prove the integrity of the mechanical elements in the tripping device. This test should be repeated every three months.

If the RCD fails any of the above tests it should be changed for a new one.

Where the RCD has a rated tripping current not exceeding 30 mA and has been installed to reduce the risk associated with 'basic' and/or 'fault' protection, as indicated in IET Regulation 411.1, a residual current of 150 mA should cause the circuit-breaker to open within 40 ms.

Wherever RCDs are installed, a label shall be fixed near to each RCD stating 'The device must be tested 6 monthly' (IET Regulation 514.12.2). Note In the 17th Edition the test period was three monthly.

9. Check for phase sequence (IET Regulation 643.9)

Phase sequence is the order in which each phase of a three-phase supply reaches its maximum value. The normal phase sequence for a three-phase supply is brown–black–grey, which means that first brown, then black and finally the grey phase reaches its maximum value. A test must be made to verify this phase sequence.

A phase sequence tester can be an indicator which is, in effect, a miniature induction motor, with three clearly colour-coded connection leads. A rotating disc with a pointed arrow shows the normal rotation for phase sequence

Figure 5.22 Phase rotation indicator instrument.

brown–black–grey. If the sequence is reversed, the disc rotates in the opposite direction to the arrow.

Alternatively, an indicator lamp showing the phase rotation by L1–L2–L3 may be used, as shown in Fig. 5.22.

Instruments are also available which contain both of the above indications.

The phase sequence is first of all checked at the point of entry to the building of the three-phase supply and the phase rotation noted.

The phase sequence is then checked at any other three-phase distribution boards within the building. The objective is to obtain confirmation that the phase rotation of every three-phase distribution board is the same as that at the origin of the installation.

Phase rotation and therefore direction of rotation of any three-phase motors affected by this action can be reversed by swapping over any two phases.

10. Functional testing (IET Regulation 643.10)

The objective of the functional test is to confirm that all switchgear, controls and switches work as they were intended to work and are properly installed, mounted and adjusted. The little test button on any RCDs must also be activated to verify that it will trip under fault conditions and that the test label is visible and adjacent.

Certification for Initial Verification (IET Regulation 644)

Following the completion of all new electrical work or additional work to an existing installation, the installation must be inspected and tested and an installation certificate issued and signed by a person who is skilled (electrically) or instructed (electrically). That person must have a sound knowledge of the type of work undertaken, be fully versed in the inspection and testing procedures contained in the IET Regulations (BS 7671) and employ adequate testing equipment.

Figure 5.23 Testing and commissioning an electrical installation.

The electrical installation certificate of test results and a copy of the completed inspection schedule shall be issued to those ordering the work in the format given in Appendix 6 of the IET Regulations.

There are three model forms recognized by BS 7671 for certifying and reporting on electrical installations: the Electrical Installation Certificate, the Minor Works Certificate and the Electrical Installation Condition report. Appendix 6 of The IET Regulations contains model forms and I will summarize their use below.

1. Electrical Installation Certificates

The IET Regulations require an Electrical Installation Certificate, including a record of the items that have been inspected and a copy of the results of testing, to be issued upon completion of a new building or additions and alterations to an existing building in order to verify the inspection and test results (Regulation 644.1).

Examples of the correct use of an Electrical Installation Certificate

- a new electrical installation;
- installation of a new final circuit or distribution circuit;
- replacement of a consumer's unit or distribution fuseboard;
- additions or alterations to an existing installation which is beyond the scope of the Minor Works Certificate described below.

Examples of the incorrect use of an Electrical Installation Certificate

- reporting upon the condition of an existing installation;
- certifying electrical installation work carried out by others.

2. Minor Electrical Installation Works Certificate

The IET Regulations require a Minor Works Certificate to be issued instead of an Electrical Installation Certificate to verify the inspection and test results of a small or minor electrical installation. A minor or small electrical installation is described as one which does not include the provision of a new circuit (Regulation 644.4.201).

Examples of the correct use of the Minor Works Certificate
- small modifications to an existing installation such as the addition of one or two lighting points, or switches or socket outlets;
- upgrading the existing main protective bonding;
- replacement of a section of damaged cable on a like-for-like basis (that is, size and type).

Examples of the incorrect use of the Minor Works Certificate
- certification of a new circuit;
- replacement of a protective device for one of a different type or rating;
- replacement of a consumer's unit.

---

3. Electrical Installation Condition Report

The IET Regulations require an Electrical Installation Condition Report, including a record of the items that have been inspected and a copy of the results of testing, to be issued following a periodic inspection and testing of an **existing installation**. The 18th Edition of the IET Wiring Regulations now also requires an inspection within any accessible roof space where electrical equipment is present in that roof space.

Examples of the correct use of the Condition Report
- reporting on the condition of an existing installation or part of an existing installation.

Examples of the incorrect use of a Condition Report
- certification of any new work;
- alterations to an existing installation;
- installation or replacement of a consumer's unit.

The documentation handed over to the client upon completion must include details of the extent of the installation covered by the report, a record of the inspection schedule and a record of the test results plus a recommendation for the interval until the next periodic inspection (IET Regulation 653.2 and 4). Inspection schedules are discussed in more detail at the beginning of this chapter.

In both cases the certificate must include the test values obtained from a 'calibrated' instrument which verify that the installation complies with the IET Regulations at the time of testing.

Suggested frequencies of periodic inspection intervals are given below:
- domestic installations – 10 years;
- commercial installations – 5 years;

- industrial installations – 3 years;
- agricultural installations – 3 years;
- caravan site installations – 1 year;
- caravans – 3 years but every year if used regularly;
- temporary installations on construction sites – 3 months.

Commissioning electrical systems

The commissioning of the electrical and mechanical systems within a building is a part of the 'handing-over' process of the new building by the architect and main contractor to the client or customer in readiness for its occupation and intended use. To 'commission' means to give authority to someone to check that everything is in working order. If it is out of commission, it is not in working order.

Following the completion, inspection and testing of the new electrical installation, the functional operation of all the electrical systems must be tested before they are handed over to the customer. It is during the commissioning period that any design or equipment failures become apparent, and this testing is one of the few quality controls possible on a building services installation.

This is the role of the commissioning engineer, who must ensure that all the systems are in working order and that they work as they were designed to work. This engineer must also instruct the client's representative, or the staff who will use the equipment, in the correct operation of the systems as part of the handover arrangements.

The commissioning engineer must test the operation of all the electrical systems, including the motor controls, the fan and air-conditioning systems, the fire alarm and emergency lighting systems. However, before testing the emergency systems, it is important to first notify everyone in the building of the intention to test the alarms so that they may be ignored during the period of testing.

Commissioning has become one of the most important functions within the building project's completion sequence. The commissioning engineer will therefore have access to all relevant contract documents, including the building specifications and the electrical installation certificates as required by the IET Regulations (BS 7671), and have a knowledge of the requirements of the Electricity at Work Regulations and the Health and Safety at Work Act.

The building will only be handed over to the client if the commissioning engineer is satisfied that all the building services meet the design specification in the contract documents and all the safety requirements.

Portable electric appliance testing

A quarter of all serious electrical accidents involve portable electrical appliances; that is, equipment which has a cable lead and plug and which is normally moved around or can easily be moved from place to place. This includes, for example, floor cleaners, kettles, heaters, portable power tools, fans, televisions, desk lamps, photocopiers, fax machines and desktop computers. There is a requirement under the Health and Safety at Work Act for employers to take adequate steps to protect users of portable appliances from the hazards of electric shock and fire. The responsibility for safety applies equally to small as well as to large companies. The Electricity at Work Regulations 1989 also place a duty of care upon employers to ensure that the risks associated with the use of electrical equipment are controlled.

Against this background the HSE has produced guidance notes HS(G) 107 *Maintaining Portable and Transportable Electrical Equipment* and leaflets *Maintaining Portable Electrical Equipment in Offices* and *Maintaining Portable Electrical Equipment in Hotels and Tourist Accommodation.* In these publications the HSE recommends that a three-level system of inspection can give cost-effective maintenance of portable appliances. These are:

*   user checking;
*   visual inspection by an appointed person;
*   combined inspection and testing by a competent person or contractor.

A user visually checking the equipment is probably the most important maintenance procedure. About 95% of faults or damage can be identified by just looking. The user should check for obvious damage using common sense. The use of potentially dangerous equipment can then be avoided. Possible dangers to look for are as follows:

*   Damage to the power cable or lead which exposes the colours of the internal conductors, which are brown, blue and green with a yellow stripe.
*   Damage to the plug itself. The plug pushes into the wall socket, usually a square pin 13 A socket in the United Kingdom, to make an electrical connection. With the plug removed from the socket the equipment is usually electrically 'dead'. If the bakelite plastic casing of the plug is cracked, broken or burned, or the contact pins are bent, do not use it.
*   Non-standard joints in the power cable, such as taped joints.
*   Poor cable retention. The outer sheath of the power cable must be secured and enter the plug at one end and the equipment at the other. The coloured internal conductors must not be visible at either end.
*   Damage to the casing of the equipment such as cracks, pieces missing, loose or missing screws or signs of melted plastic, burning, scorching or discoloration.
*   Equipment which has previously been used in unsuitable conditions such as a wet or dusty environment.

If any of the above dangers are present, the equipment should not be used until the person appointed by the company to make a 'visual inspection' has had an opportunity to do so.

A **visual inspection** will be carried out by an appointed person within a company, such person having been trained to carry out this task. In addition to the user checks described above, an inspection could include the removal of the plug top cover to check that:

*   a fuse of the correct rating is being used and also that a proper cartridge fuse is being used and not a piece of wire, a nail or silver paper;
*   the cord grip is holding the sheath of the cable and not the coloured conductors;
*   the wires (conductors) are connected to the correct terminals of the plug top, as shown in Fig. 5.24;
*   the coloured insulation of each conductor wire goes right up to the terminal so that no bare wire is visible;
*   the terminal fixing screws hold the conductor wires securely and the screws are tight;

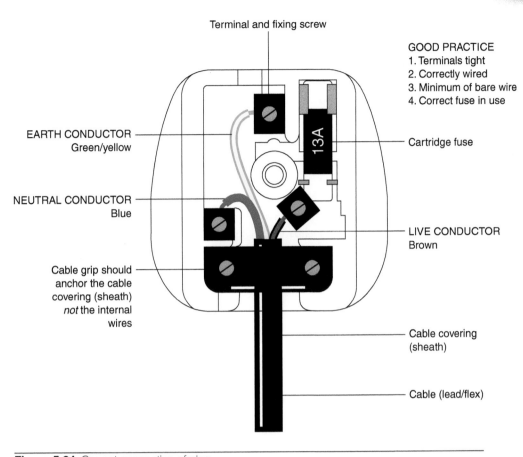

Figure 5.24 Correct connection of plug.

- all the conductor wires are secured within the terminal;
- there are no internal signs of damage such as overheating, excessive 'blowing' of the cartridge fuse or the intrusion of foreign bodies such as dust, dirt or liquids.

The above inspection cannot apply to 'moulded plugs', which are moulded on to the flexible cable by the manufacturer in order to prevent some of the bad practice described above. In the case of a moulded plug top, only the fuse can be checked. The visual inspection checks described above should also be applied to extension leads and their plugs. The HSE recommends that a simple procedure be written to give guidance to the 'appointed person' carrying out the visual inspection.

Combined inspection and testing is also necessary on some equipment because some faults cannot be seen by just looking – for example, the continuity and effectiveness of earth paths. For some portable appliances rated as Class 1 equipment the earth is essential to the safe use of the equipment and, therefore, all earthed equipment and most extension leads should be periodically tested and inspected for these faults. All portable appliance test instruments (PAT Testers) will carry out two important tests, earth bonding and insulation resistance.

Earth bonding tests apply a substantial test current, typically about 25 A, down the earth pin of the plug top to an earth probe, which should be connected to any exposed metalwork on the portable appliance being tested. The PAT Tester will then calculate the resistance of the earth bond and either give an actual reading or indicate pass or fail. A satisfactory result for this test would typically be a reading of less than 0.1 Ω. The earth bond test is, of course, not required for double-insulated portable appliances because there will be no earthed metalwork.

Insulation resistance tests apply a substantial test voltage, typically 500 V, between the live and neutral bonded together and the earth. The PAT Tester then calculates the insulation resistance and either gives an actual reading or indicates pass or fail. A satisfactory result for this test would typically be a reading greater than 2 MΩ.

Some PAT Testers offer other tests in addition to the two described above. These are described below.

A *flash test* tests the insulation resistance at a higher voltage than the 500 V test described above. The flash test uses 1.5 kV for Class 1 portable appliances, that is, earthed appliances, and 3 kV for Class 2 appliances, which are double insulated. The test establishes that the insulation will remain satisfactory under more stringent conditions but must be used with caution, since it may overstress the insulation and will damage electronic equipment. A satisfactory result for this test would typically be less than 3 mA.

A *fuse test* tests that a fuse is in place and that the portable appliance is switched on prior to carrying out other tests. A visual inspection will be required to establish that the *size* of the fuse is appropriate for that particular portable appliance.

An *earth leakage test* measures the leakage current to earth through the insulation. It is a useful test to ensure that the portable appliance is not deteriorating and liable to become unsafe. It also ensures that the tested appliances are not responsible for nuisance 'tripping' of RCDs. A satisfactory reading is typically less than 3 mA.

An *operation test* proves that the preceding tests were valid (i.e. that the unit was switched on for the tests), that the appliances will work when connected to the appropriate voltage supply and will not draw a dangerously high current from that supply. A satisfactory result for this test would typically be less than 3.2 kW for 230 V equipment and less than 1.8 kW for 110 V equipment.

All PAT Testers are supplied with an operating manual, giving step-by-step instructions for their use and pass and fail scale readings. The HSE-suggested intervals for the three levels of checking and inspection of portable appliances in offices and other low-risk environments are given in Table 5.3.

Who does what?

When actual checking, inspecting and testing of portable appliances takes place will depend upon the company's safety policy and risk assessments. In low-risk environments such as offices and schools, the three-level system of checking, inspection and testing recommended by the HSE should be carried out. Everyone can use common sense and carry out the user checks described earlier. Visual inspections must be carried out by a 'competent person' but that person does not need to be an electrician or electronics service engineer. Any sensible member of staff who has received training can carry out this duty. They will need to know what to look for and what to do, but, more importantly, they will need to be able to avoid danger to themselves and to others. The HSE recommends that the appointed person follows a simple written procedure for each visual inspection. A simple tick sheet would meet this requirement. For example:

1 Is the correct fuse fitted?   Yes/No
2 Is the cord grip holding the cable sheath?   Yes/No

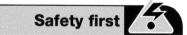

Safety first

Power tools
- Look at the power tools you use at work.
- Do they have a PAT Test label?
- Is it 'in date'?

Table 5.3 HSE-suggested intervals for checking, inspecting and testing of portable appliances in offices and other low-risk environments

Equipment/ environment	User checks	Formal visual inspection	Combined visual inspection and electrical testing
Battery-operated: (less than 20 V)	No	No	No
Extra low voltage: (less than 50 V a.c.) e.g. telephone equipment, low-voltage desk lights	No	No	No
Information technology: e.g. desktop computers, VDU screens	No	Yes, 2–4 years	No if double insulated – otherwise up to 5 years
Photocopiers, fax machines: *not* hand-held, rarely moved	No	Yes, 2–4 years	No if double insulated – otherwise up to 5 years
Double-insulated equipment: *not* hand-held, moved occasionally, e.g. fans, table lamps, slide projectors	No	Yes, 2–4 years	No
Double insulated equipment: *hand-held*, e.g. power tools	Yes	Yes, 6 months to 1 year	No
Earthed equipment (Class 1): e.g. electric kettles, some floor cleaners, power tools	Yes	Yes, 6 months to 1 year	Yes, 1–2 years
Cables (leads) and plugs connected to the above	Yes	Yes, 6 months to 4 years depending on the type of equipment it is connected to	Yes, 1–5 years depending on the type of equipment it is connected to
Extension leads (mains voltage)	Yes	As above	As above

The tick sheet should incorporate all the appropriate visual checks and inspections described earlier.

Testing and inspection require a much greater knowledge than is required for simple checks and visual inspections. This more complex task need not necessarily be carried out by a qualified electrician or electronics service engineer. However, the person carrying out the test must be trained to use the equipment and to interpret the results. Also, greater knowledge will be required for the inspection of the range of portable appliances which might be tested.

Keeping records

Records of the inspecting and testing of portable appliances are not required by law but within the Electricity at Work Regulations 1989 it is generally accepted that some form of recording of results is required to implement a quality control system. The control system should:

- ensure that someone is nominated to have responsibility for portable appliance inspection and testing;
- maintain a log or register of all portable appliance test results to ensure that equipment is inspected and tested when it is due;
- label tested equipment with the due date for its next inspection and test, as shown in Fig. 5.25. If it is out of date, don't use the equipment.

Any piece of equipment which fails a PAT Test should be disabled and taken out of service (usually by cutting off the plug), labelled as faulty and sent for repair.

The register of PAT Test results will help managers to review their maintenance procedures and the frequency of future visual inspections and testing. Combined inspection and testing should be carried out where there is a reason to suspect that the equipment may be faulty, damaged or contaminated but this cannot be verified by visual inspection alone. Inspection and testing should also be carried out after any repair or modification to establish the integrity of the equipment or at the start of a maintenance system, to establish the initial condition of the portable equipment being used by the company.

Figure 5.25 Typical PAT Test labels.

Check your understanding

When you have completed these questions, check out the answers at the back of the book.

Note: more than one multiple-choice answer may be correct.

1   A tong test instrument can also correctly be called:
    a.   a continuity tester
    b.   a clip-on ammeter
    c.   an insulation resistance tester
    d.   a voltage indicator.

2   All electrical test probes and leads must comply with the standards set by the:
    a.   BS EN 60898
    b.   BS 7671
    c.   HSE Guidance Note GS 38
    d.   IET Regulations Part 2.

3   When making a test to determine the presence or absence of a voltage, the HSE recommends that for our own safety we should use:
    a.   any old tester bought at a car-boot sale
    b.   a multimeter set to the correct voltage
    c.   a proprietary test lamp
    d.   a voltage indicator.

4   For electrical test results to be valid the test instruments used:
    a.   must be new
    b.   must be of an approved type
    c.   must have a calibration certificate
    d.   must have a digital readout.

5   The test required by the IET Regulations to ascertain that the CPC is correctly connected is called:
    a.   a basic protection
    b.   continuity of ring final circuit conductors
    c.   continuity of protective conductors
    d.   earth electrode resistance.

6   One objective of the polarity test is to verify that:
    a.   lamp holders are correctly earthed
    b.   final circuits are correctly fused
    c.   the CPC is continuous throughout the installation
    d.   the protective devices are connected in the live conductor.

7   When testing a 230 V installation an insulation resistance tester must supply
    a voltage of:
    a.   less than 50 V
    b.   500 V
    c.   less than 500 V
    d.   greater than twice the supply voltage but less than 1000 V.

8   The value of a satisfactory insulation resistance test on each final circuit of a
    230 V installation must be:
    a.   less than 1 Ω
    b.   less than 0.5 MΩ
    c.   not less than 0.5 MΩ
    d.   not less than 1 MΩ.

9   Instrument calibration certificates are usually valid for a period of:
    a.   3 months
    b.   1 year
    c.   3 years
    d.   5 years.

10 The maximum inspection and re-test period for a domestic electrical
    installation is:
    a.   3 months
    b.   3 years
    c.   5 years
    d.   10 years.

11 A visual inspection of a new installation must be carried out:
    a.   during the erection period
    b.   during testing upon completion
    c.   after testing upon completion
    d.   before testing upon completion.

12 'To ensure that all the systems within a building work as they were intended
    to work' is one definition of the purpose of:
    a.   testing electrical equipment
    b.   inspecting electrical systems
    c.   commissioning electrical systems
    d.   isolating electrical systems.

13 Use bullet points to state three reasons for testing a new electrical
    installation.

14 State five of the most important safety factors to be considered before
    electrical testing begins.

15 State the seven requirements of GS 38 when selecting probes, voltage
    indicators and measuring instruments.

16 Use bullet points to describe a safe isolation procedure of a final circuit fed
    from an MCB in a distribution board.

17 IET Regulation 611.3 gives a checklist of about 20 items to be considered in the initial visual inspection of an electrical installation. Make a list of ten of the most important items to be considered in the visual inspection process of a domestic property (perhaps by joining together similar items).

18 State three reasons why electricians must only use 'approved' test instruments.

19 State the first five tests to be carried out on a new electrical installation following a satisfactory 'inspection'. For each test:
   a. state the object (reason for) the test
   b. state a satisfactory test result.

20 State the certification process for a:
   a. new electrical installation and
   b. an electrical installation that is being re-tested in order to provide an Electrical Installation Condition Report
   c. what will be indicated on the test certificates
   d. who will receive the test certificates and
   e. who will issue the certificates
   f. finally, who will carry out the actual testing (A.................person).

21 State three safe working procedures relevant to your own safety when carrying out electrical testing.

22 State three safe working procedures relevant to the safety of other people when carrying out electrical testing.

23 State three safe working procedures relevant to the safety of other electrical systems when carrying out electrical testing.

24 State the two important tests that a PAT tester carries out on a portable appliance.

25 Use bullet points to state the reasons for commissioning a new building upon its completion.

Electrical systems design

Unit 305 of the City and Guilds 2365-03 syllabus

Learning outcomes – when you have completed this chapter you should:

- describe the importance of sustainable design;
- identify the information required for electrical systems design;
- calculate the requirements for electrical systems in a domestic property;
- calculate maximum demand, diversity and current-carrying capacity of conductors;
- describe the factors affecting the selection of electrical wiring systems;
- verify the coordination of protective devices;
- understand the procedures for connecting complex electrical systems;
- explain how to carry out a programme of work for the installation of electrical systems;
- specify methods of illustrating work programmes;
- describe the commissioning procedure, snagging and sign-off.

Advanced Electrical Installation Work. 978-0-367-35976-8
© 2019 Trevor Linsley. Published by Taylor & Francis. All rights reserved.

This chapter has free associated content, including animations and instructional videos, to support your learning.

When you see the logo, visit the website below to access this material: www.routledge.com/cw/linsley

Importance of sustainable design

The stages of human civilisation along the centuries, from early Stone Age up to today, can roughly be traced through the progress made in materials science. The Stone Age gave way to the Bronze Age which gave way to the Iron Age. In our recent history we had the industrial revolution which started in 1760 and was driven by invention. In particular, iron and steel production, the manufacture and utilisation of steam engines, spinning mills and the construction of railways and canals. It is less than 200 years (1831) since Michael Faraday gave us electromagnetic induction, the basis of our present electricity generation systems and Alexander Graham Bell made the first telephone call (1877). Today we live in the Hydrocarbon Age, which is fuelled by coal, oil and gas. These fuels supply our energy needs and make possible the materials which define our civilisation, steel, concrete and plastic. However, we now know that if we are to avoid damaging our planet with plastic waste and global warming, we must reduce our use of hydrocarbons and move to an economy based upon sustainable materials. Sustainability, means meeting the needs of the present without compromising the ability of future generations to meet their own. Sustainable development is a way for people to use resources without those resourses running out.

In 2018 steel production accounted for about 3% of the world's greenhouse gas emissions and concrete production about 5%. In 2016 the Transport Sector, that is planes trains shipping lorries busses and cars, overtook Energy as the biggest source of greenhouse gas emissions. Petrochemicals also make plastic and fertilisers.

So, how do these problems impact upon the construction industry?

In April 2006, Part L of the Building Regulations (England and Wales) was revised in order to raise energy performance standards and to reduce CO2 emissions from buildings. Part L, Conservation of Fuel and Power, now requires all new and existing buildings to be given an energy rating and for all new buildings to meet a minimum level of energy efficiency. Under this provision the electrical contractor must make 'reasonable provision' to provide lighting systems with energy-efficient lamps and sufficient controls so that electrical energy can be used efficiently. The current provision requires one energy-efficient luminaire for every $25\,m^2$ of floor area or one energy-efficient luminaire for every four fixed luminaires.

External lighting fixed to the building, including lighting in porches, but not garages or carports, should provide reasonable provision for energy-efficient lamps such as LEDs fluorescent tubes or CFLs. These lamps should automatically extinguish in daylight and, when not required at night, be controlled by passive infrared (PIR) detectors.

The traditional carbon filament lamp, called a GLS (general lighting service) lamp, is hopelessly bad in energy-efficiency terms, producing only 14 lumens of light output for every electrical watt input. Fluorescent tubes and CFLs produce more than 40 lumens of light output for every electrical watt input and LEDs more than 80 lumens per watt.

In addition, the electrical installer must have an appreciation of how the building regulations in general might affect the electrical installation in particular. For example:

- Part A Structure – the basic requirement for those installing electrical installations in a building is not to cut, drill, chase, penetrate or in any way interfere with the structure so as to cause significant reduction in its load-bearing capability. Approved document A provides practical guidance with pictures. This document can be found in the Electricians Guide to the Building Regulations published by the IET.
- Part B Fire Safety – the 'standard house' with no floor area exceeding 200 m^2 must be fitted with smoke alarms to each level. A floor area greater than 200 m^2 is considered to be a 'large house'. If the kitchen cannot be isolated from the other rooms by a door, then a compatible interlinked heat detector must also be installed in the kitchen.
- Part M Access and Facilities for the Disabled – this requires switches and socket outlets in dwellings to be installed so that all persons, including those whose reach is limited, can easily reach them. The recommendation is that they should be installed in habitable rooms at a height of between 450 and 1200 mm from the finished floor level, as shown in the Electricians Guide to the Building Regulations.

The important change in the 2006 Regulations is that **compliance is now based on the the whole building's carbon emissions,** meaning that the building designer must now consider the impact of **both the constructional elements** of the building **as well as the energy-using services within the building** such as lighting, heating, hot water and ventilation.

To achieve compliance, the building designer must show that the predicted annual carbon emissions from the building are less than, or equal to, a Target Carbon Emissions Rate for a 'standard national building' of the same floor area and shape as the one being designed.

The use of energy to provide heat for central heating and hot water in our homes is responsible for 60% of a typical family's energy bill. Heating accounts for over half of Britain's entire use of energy and carbon emissions. If Britain is to reduce its carbon footprint and achieve energy security, we must revolutionize the way we keep warm in the home.

At present 69% of our home heating comes from burning gas, 11% from oil, 3% from solid fuels such as coal and 14% from electricity which is mainly generated from these same three fossil fuels. Only 1% is currently provided by renewable sources. If Britain is to meet its clean energy targets, renewable sources will have to increase, and the revolution will have to start in the home because the country's dwellings currently provide more than half of the total demand, almost entirely for hot water and central heating.

There are about 20 million homes in the UK and a review of present buildings has found that about six million homes have inadequately lagged lofts, eight million have uninsulated cavity walls and a further seven million homes with solid walls would benefit from better insulation. If the country is to achieve its 2016 reduced carbon emissions targets, these existing homes must be heavily insulated to reduce energy demand and then supplied with renewable heat.

We cannot sustain the present level of carbon emissions without disastrous ecological consequences in the future. Low-carbon homes are sustainable homes.

HRH The Prince of Wales has entered the debate, saying,

Key fact

Heating accounts for over half of Britain's entire use of energy and carbon emissions.

Becoming more sustainable is possibly the greatest challenge humanity has faced and I am convinced that it is therefore, the most remarkable chance to secure a prosperous future for everyone. We must strive harder than ever before to convince people that by living sustainably we will improve our quality of life and our health; that by living in harmony with nature we will protect the intricate, delicate balance of the natural systems that ultimately sustain us.

(Daily Telegraph, 31 July 2010).

The Code for Sustainable Homes (see Fig. 6.1) measures the sustainability of a home against categories of sustainable design, rating the whole home as a complete package, including building materials and services within the building. The Code uses a 1 to 6-star rating to communicate the overall sustainability performance of a new home and sets minimum standards for energy and water use at each level.

Since May 2008 all new homes are required to have a Code Rating and a Code Certificate. From 2016 all new homes must be built to zero-carbon standards, which will be achieved through:

- Step by step tightening of the Building Regulations.
- New Planning Regulations will force architects and builders to build new homes with higher standards of energy efficiency; that is, better insulation in walls, floors and lofts. These are called passive measures. Individual households can now obtain free loft insulation and some homes are eligible for free insulation to walls as a part of the government initiative to reduce carbon emissions.
- Government grants will be available, called the 'Green deal', to encourage individual homeowners to make energy-saving improvements to their homes. These improvements might include energy generation using microgenerating systems such as solar PV and solar hot water as described in Chapter 2 of this book. These are called active measures.
- At a time when energy prices are rising above the rate of inflation homeowners want more energy-efficient homes. The average family is spending between £1000 and £2000 each year on fuel bills alone.

All new homes must now reach the government target level 6 of the Code shown in Fig 6.1. Zero-carbon homes are those which reach code level 6, the highest

Definition

The Code for Sustainable Homes measures the sustainability of a home against categories of sustainable design, rating the whole home as a complete package, including building materials and services within the building.

Key fact

Sustainable homes will be achieved through step by step tightening of the Building Regulations

Date	2010	2013	2016
Energy efficiency improvement of the dwelling campared to 2006 (part L Building Regulations)	2.5%	44%	Zero carbon
Equivalent Standard within the Code	Code level 3 ★★★☆☆ THE CODE FOR SUSTAINABLE HOMES	Code level 4 ★★★★☆ THE CODE FOR SUSTAINABLE HOMES	Code level 6 ★★★★★★ THE CODE FOR SUSTAINABLE HOMES

Figure 6.1 The Code for Sustainable Homes.

level attainable. These levels can only be achieved through a combination of active and passive measures and clever design on the part of the architect.

If we look at sustainability from a manufacturing point of view, sustainable manufacture is based on the principle of meeting the needs of the current generation without those resources running out and therefor compromising the ability of future generations to meet their needs.

Electrical systems design: An overview

The designer of an electrical installation must ensure that the design meets the requirements of the IET Wiring Regulations for electrical installations and any other regulations which may be relevant to a particular installation. The designer may be a professional technician or engineer whose job is to design electrical installations for a large contracting firm. In a smaller firm, the designer may also be the electrician who will carry out the installation to the customer's requirements. The designer of any electrical installation is the person who interprets the electrical requirements of the customer within the regulations, identifies the appropriate types of installation, the most suitable methods of protection and control and the size of cables to be used.

A large electrical installation may require many meetings with the customer and his professional representatives in order to identify a specification of what is required. The designer can then identify the general characteristics of the electrical installation and its compatibility with other services and equipment, as indicated in Part 3 of the regulations. The protection and safety of the installation, and of those who will use it, must be considered, with due regard to Part 4 of the regulations. An assessment of the frequency and quality of the maintenance to be expected will give an indication of the type of installation which is most appropriate.

The size and quantity of all the materials, cables, control equipment and accessories can then be determined. This is called a 'bill of quantities'.

It is common practice to ask a number of electrical contractors to **tender or submit a price** for work specified by the bill of quantities. **A tender is a formal offer to supply goods or carry out work at a stated price**. The contractor must cost all the materials, assess the labour cost required to install the materials and add on profit and overhead costs in order to arrive at a final estimate for the work. This is the 'tender' price. The contractor tendering the lowest cost is usually, but not always, awarded the contract.

To complete the contract in the specified time the electrical contractor must use the management skills required by any business to ensure that men and materials are on-site as and when they are required. If alterations or modifications are made to the electrical installation as the work proceeds which are outside the original specification, then a **variation order** must be issued so that the electrical contractor can be paid for the additional work.

The specification for the chosen wiring system will be largely determined by the building construction and the activities to be carried out in the completed building.

An industrial building, for example, will require an electrical installation which incorporates flexibility and mechanical protection. This can be achieved by a conduit, tray or trunking installation.

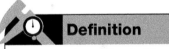

Definition

The *designer* of any electrical installation is the person who interprets the electrical requirements of the customer within the regulations.

Definition

The size and quantity of all the materials, cables, control equipment and accessories can then be determined. This is called a '*bill of quantities*'.

Definition

By definition a *competent person* is one who has the ability to perform a particular task properly and safely.

Key fact

Definitions

A person can be described as

- ordinary
- competent
- instructed or
- skilled

depending upon that person's skill or ability.

In a block of purpose-built flats, all the electrical connections must be accessible from one flat without intruding upon the surrounding flats. A loop-in conduit system, in which the only connections are at the light switch and outlet positions, would meet this requirement.

For a domestic electrical installation an appropriate lighting scheme and multiple socket outlets for the connection of domestic appliances, all at a reasonable cost, are important factors which can usually be met by a PVC-insulated and sheathed wiring system.

The final choice of a wiring system must rest with those designing the installation and those ordering the work, but whatever system is employed, good workmanship by competent persons is essential for compliance with the regulations (HSE Regulation 16).

By definition a **competent person** is one who has the ability to perform a particular task properly and safely.

Generally speaking, an electrician will have the necessary skills to perform a wide range of electrical activities competently.

HSE Regulation 16 states that persons 'must be competent to prevent danger … so that the person themselves or others are not placed at risk due to a lack of skill when dealing with electrical equipment'.

The 18th Edition of the IET Wiring Regulations also makes the following definitions relating to people:

A skilled person (electrically) is one who has relevant education training and practical skills and who is able to perceive risks and to avoid the hazards which electricity can create.

An instructed person (electrically) is a person adequately advised or supervised by electrically skilled persons to enable that person to perceive risks and to avoid the hazards which electricity can create.

An ordinary person is a person who is neither a skilled person nor an instructed person.

The 18th Edition of the IET Wiring Regulations at 134.1.1 makes the following comment: 'good workmanship by skilled (electrically) or instructed (electrically) persons and proper materials shall be used in the erection of the electrical installation.'

Try this

Definitions

People may be described as an ordinary person, a skilled person, or an instructed person. Place people you know into each of these categories – for example, yourself, your parents, your supervisor, etc.

The necessary skills can be acquired by an electrical trainee who has the correct attitude and dedication to his or her craft. NVQ Level 3 is 'skilled craft level' or the level required to be considered 'competent' or 'skilled'.

Energy efficiency: a new part 8 of the regulations

The 18th edition of the Regulations gives us **a new Appendix 17, Energy Efficiency**. It is intended that this new Appendix will become a new Part

8 in future Amendments to the regulations. The new Appendix provides **recommendations** for the design and erection of electrical installations. All new electrical installation design, and modifications to existing installations must continue to meet the required levels of safety and capacity which we will calculate in this section, but must **NOW ALSO** optimise electrical efficiency, providing the lowest levels of electricity consumption.

The new Appendix considers design and maintenance from the context of improving the efficiency of the electrical installation. There is a change of emphasis in this new section, to incorporate energy efficiency into the electrical installation design as a prerequisite, and not just as an aspiration. The standard makes it clear that any measures taken to make the electrical installation more efficient must not compromise the safety of any occupants, property or livestock, and that much of this appendix will not apply to domestic and similar installations.

Electrical systems design: the process

Part 1 of Chapter 13, Section 132 of the IET Regulations details the fundamental principles for the design of an electrical installation. Let us now look at electrical design in some detail as required by the City and Guilds syllabus and Section 132 of the IET Regulations.

Section 132 of the IET regulations gives us 16 regulations which make up the design process. The headings are as follows:

Regulations	Topic	Page number in this book
132.1	General	305
132.2	Characteristics of the supply	305
132.3	Nature of the demand	320
132.4	Supplies for safety systems	324
132.5	Environmental conditions	325
132.6	Conductor size calculations	326
132.7	Type of wiring and method of installation	330
132.8	Protective equipment	354
132.9	Emergency control	361
132.10	Disconnecting devices	361
132.11	Prevention of mutual detrimental influences	361
132.12	Accessibility of electrical equipment	361
132.13	Documentation for the electrical installation	361
132.14	Protective devices and switches	361
132.15	Isolation and switching	362
132.16	Additions and alterations to an installation	364

Let us now consider some of the design process regulations.

132.1 General

In the introduction, the IET Regulation tells us that the electrical installation must be designed to provide for:

- the protection of people, livestock and property; and
- to meet the intended use required of the installation.

132.2 The characteristics of the supply

One of the first considerations for the designer is to determine the nature or characteristics of the supply and the earthing arrangements. Inspection of the existing supply or by requesting details from the supply distributor will be necessary to determine:

- the supply voltage;
- the type of supply and earthing arrangements, that is, TN-S, TN-C-S or TT system;
- the value of the earth fault loop impedance, Z_e for the supply;
- the location of the supply.

For new electrical installations or for significant alterations to existing ones, the Electricity Safety Quality and Continuity Regulations require the electricity distributor to install the cut-out and meter in a 'safe location' where they can be mechanically protected and safely maintained. In compliance with this 'safe location' requirement the electricity distributor and installer may in some areas be required to take account of the risk of flooding. To comply, the distributors' equipment and the installations consumer unit or fuse boards must be installed above the flood level. Upstairs power and lighting circuits and downstairs lighting should be installed above the flood level. Upstairs and downstairs circuits should have separate protective devices. Consumer units or fuse boards must be generally accessible for use by responsible persons in the household. They should not be installed where young people might interfere with them (IET Guide to the Building Regs 2.1 and 3.1).

Figure 6.2 The electricity grid power supply mostly travels along conductors supported by pylons.

The 18th Edition of the IET Wiring Regulations has introduced Regulation 421.1.201 requiring switchgear assemblies, including consumer unit enclosures, to be:

- manufactured from non-combustible or not readily combustible material, or be
- enclosed in a cabinet or enclosure that is constructed of non-combustible or not readily combustible material.
- The 18th Edition of the IET Regulations at 532.6 and 421.1.7 has introduced a new topic, Arc Fault Detection Devices (AFDD). These are compulsory in the EU Countries but **only recommended** in this country by the IET and the British standards. They are recognized as giving additional protection against fires caused by arc faults. Arc faults can be formed by cable insulation defects, damage to cables by impact or penetration by nails and screws, and loose terminal connections. AFDDs detect faults which MCBs and RCDs cannot detect. They are installed at the origin of the final circuit to be protected.

These new Regulations have been introduced to protect against fire that can result from the overheating of connections within consumer units. Overheating may occur because connections have not been properly made or that they have become loose.

Upon request, the supply distributor will give the following information for a typical domestic supply:

- single-phase a.c. supply;
- maximum value of the distributor's protective device is 100 A;
- the normal voltage will be 230 V to earth;
- frequency 50 cycles per second;
- the maximum or worst case external earth fault loop impedance (Z_e) will be:
  - (i) 0.8 ohms for cable sheath (TN-S) supplies;
  - (ii) 0.35 ohms for PME (TN-C-S) supplies;
  - (iii) for no earth provided (TT) supplies. The value given by the distributor for the earth electrode resistance at the supply transformer is usually 21 ohm. The resistance of the consumer's installation earth electrode should be as low as possible but a value exceeding 200 ohm might not be stable (IET OSG 1.1);
- the maximum or worst case prospective fault current at the origin of the installation will normally be 16 kA. If a protective device is to operate safely, its rated short-circuit capacity must be not less than the prospective fault current at the point where it is installed. It is recommended that the electrical installation design should be based on the maximum fault current value provided by the distributor and not on measured values. Except for London and some other major city centres, the maximum fault current for 230 V single-phase supplies up to 100 A will not exceed 16 kA (IET OSG 7.2.1i and 10.3.7).

Figure 6.3 Legrand 'rapid' earth bonding clamp.

Electrical supplies and earthing arrangements

We know from earlier chapters in this book that using electricity is one of the causes of accidents in the workplace. Using electricity is a hazard because it has the potential to cause harm. Therefore, the provision of protective devices in an electrical installation is fundamental to the whole concept of the safe use of electricity in buildings. The electrical installation as a whole must be protected against overload or short-circuit, and the people using the building must be

Figure 6.4 Adjustable earth bonding straps.

protected against the risk of shock, fire or other risks arising from their own misuse of the installation or from a fault. The installation and maintenance of adequate and appropriate protective measures is a vital part of the safe use of electrical energy. I want to look at protection against an electric shock by both basic and fault protection, at protection by protective bonding and automatic disconnection of the supply.

Let us first define some of the words we will be using. Chapter 54 of the IET Regulations describes the earthing arrangements for an electrical installation. It gives the following definitions:

Earth – the conductive mass of the earth whose electrical potential is taken as zero.

Earthing – the act of connecting the exposed conductive parts of an installation to the main protective earthing terminal of the installation.

Bonding conductor – a protective conductor providing equipotential bonding.

Bonding – the linking together of the exposed or extraneous metal parts of an electrical installation.

Circuit protective conductor (CPC) – a protective conductor connecting exposed conductive parts of equipment to the main earthing terminal. This is the green and yellow insulated conductor in twin and earth cable.

Exposed conductive parts – this is the metalwork of an electrical appliance or the trunking and conduit of an electrical system which can be touched because they are not normally live, but which may become live under fault conditions.

Extraneous conductive parts – this is the structural steelwork of a building and other service pipes such as gas, water, radiators and sinks. They do not form a part of the electrical installation but may introduce a potential, generally earth potential, to the electrical installation.

Shock protection – protection from electric shock is provided by basic protection and fault protection.

Basic protection – this is provided by the insulation of live parts in accordance with Section 416 of the IET Regulations.

Fault protection – this is provided by protective bonding and automatic disconnection of the supply (by a fuse or MCB) in accordance with IET Regulations 411.3–6.

Protective bonding – this is protective bonding for the purpose of safety and is shown in Figs 6.6 to 6.8.

Basic protection and fault protection

The human body's movements are controlled by the nervous system. Very tiny electrical signals travel between the central nervous system and the muscles, stimulating operation of the muscles, which enable us to walk, talk and run; and remember that the heart is also a muscle.

If the body becomes part of a more powerful external circuit, such as the electrical mains, and current flows through it, the body's normal electrical operations are disrupted. The shock current causes unnatural operation of the muscles and the result may be that the person is unable to release the live conductor causing the shock, or the person may be thrown across the room.

The current which flows through the body is determined by the resistance of the human body and the surface resistance of the skin on the hands and feet.

This leads to the consideration of exceptional precautions where people with wet skin or wet surfaces are involved, and the need for special consideration in bathroom installations.

Two types of contact will result in a person receiving an electric shock. Direct contact with live parts involves touching a terminal or line conductor that is actually live. The regulations call this **basic protection** (131.2.1). Indirect contact results from contact with an exposed conductive part such as the metal structure of a piece of equipment that has become live as a result of a fault. The regulations call this **fault protection** (131.2.2).

The touch voltage curve in Fig. 6.5 shows that a person in contact with 230V must be released from this danger in 40ms if harmful effects are to be avoided. Similarly, a person in contact with 400V must be released in 15ms to avoid being harmed.

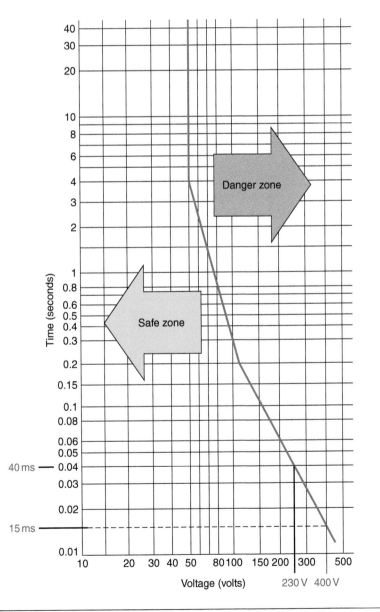

Figure 6.5 Touch voltage curve.

In installations operating at normal mains voltage, the primary method of protection against direct contact (basic protection) is by insulation. All live parts are enclosed in insulating material such as rubber or plastic, which prevent contact with those parts. The insulating material must, of course, be suitable for the circumstances in which they will be used and the stresses to which they will be subjected.

Other methods of basic protection include the provision of barriers or enclosures which can only be opened by the use of a tool, or when the supply is first disconnected. Protection can also be provided by fixed obstacles such as a guardrail around an open switchboard or by placing live parts out of reach as with overhead lines.

Fault protection

Protection against indirect contact, called fault protection (IET Regulation 131.2.2), is achieved by connecting exposed conductive parts of equipment to the main earthing terminal of the installation, as shown in Figs 6.6 to 6.8.

In Chapter 13 of the IET Regulations we are told that where the metalwork of electrical equipment may become charged with electricity in such a manner as to cause danger, that metalwork will be connected with earth so as to discharge the electrical energy without danger.

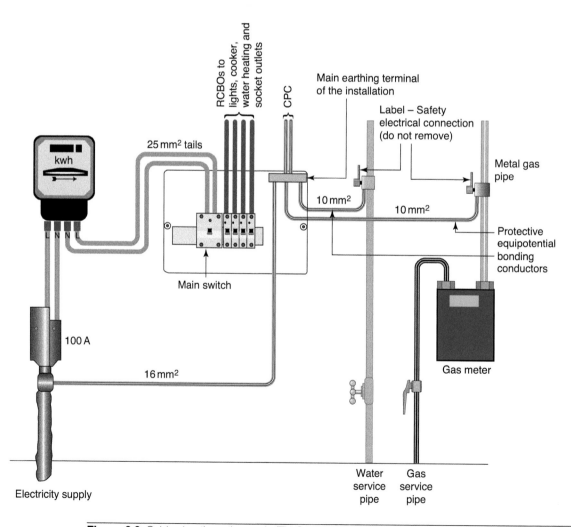

Figure 6.6 Cable sheath earth supply (TN-S system) showing earthing and bonding arrangements.

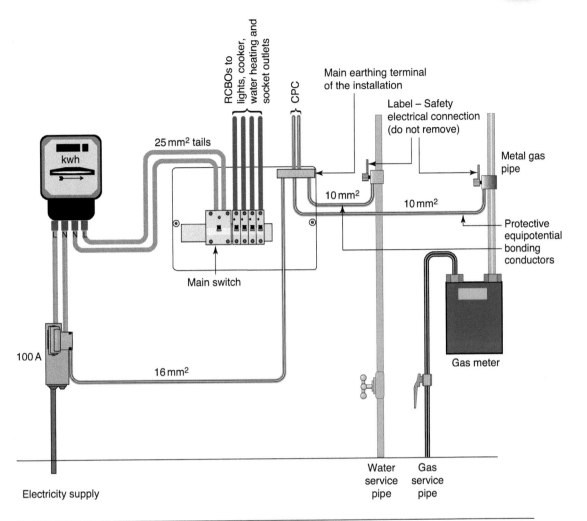

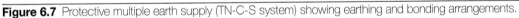

Figure 6.7 Protective multiple earth supply (TN-C-S system) showing earthing and bonding arrangements.

In connection with fault protection, protective equipotential bonding is one of the important principles for safety.

There are five methods of protection against contact with metalwork which has become unintentionally live; that is, indirect contact with exposed conductive parts recognized by the IET Regulations. These are:

1 Protective bonding coupled with automatic disconnection of the supply.
2 The use of Class II (double insulated) equipment.
3 The provision of a non-conducting location.
4 The use of earth-free protective bonding.
5 Electrical separation.

Methods 3 and 4 are limited to special situations under the effective supervision of trained personnel.

Method 5, electrical separation, is little used but does find an application in the domestic electric shaver supply unit which incorporates an isolating transformer.

Method 2, the use of Class II insulated equipment, is limited to single pieces of equipment such as tools used on construction sites, because it relies upon effective supervision to ensure that no metallic equipment or extraneous earthed metalwork enters the area of the installation.

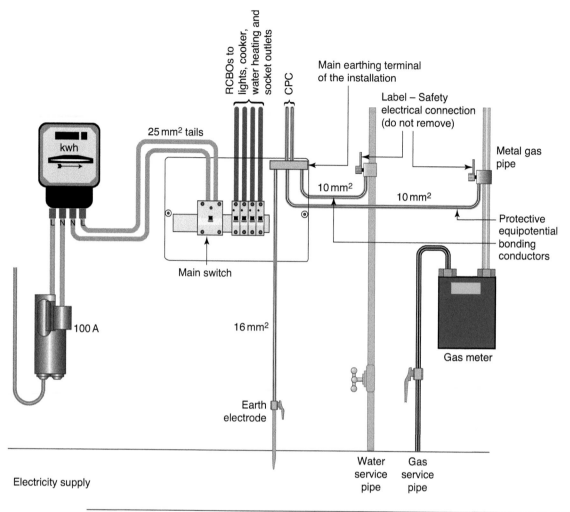

Figure 6.8 No earth provided supply (TT systems) showing earthing and bonding arrangements.

The method most universally used in the United Kingdom is, therefore, Method 1 – protective equipotential bonding coupled with automatic disconnection of the supply.

This method relies upon all exposed metalwork being electrically connected together to an effective earth connection. Not only must all the metalwork associated with the electrical installation be so connected, that is, conduits, trunking, metal switches and the metalwork of electrical appliances, but also IET Regulation 411.3.1.2 tells us to connect the extraneous metalwork of water service pipes, gas and other service pipes and ducting, central heating and air-conditioning systems, exposed metallic structural parts of the building and lightning protective systems to the main protective earthing terminal. In this way the possibility of a voltage appearing between two exposed metal parts is removed. Protective bonding is shown in Figs 6.6 to 6.8.

Figures 6.6 to 6.8 show protective equipotential bonding conductors connected to gas and water supplies. This is necessary if the incoming supplies are metallic. However, the 18th Edition of the IET Regulations at 411.3.1.2 tells us that metal pipes entering a building having an insulated section at their entry to the building need not be connected to the equipotential bonding of the installation.

However, the premises gas, water and electricity supplies must be bonded on the consumers hard metal pipe work, at the point of entry to the building

Key fact

Bonding

When carrying out earthing and bonding activities:

- use a suitable bonding clamp
- connect to a cleaned pipe
- make sure all connections are tight
- fix a label 'Safety Electrical Connection' close to the connection

IET Regulation 514.13.

as shown in Figs. 6.6 to 6.8 of this book, and Figs. 2.1 (i) to 2.1 (iii) of the On Site Guide (IET Regulation 544.1.2).

You will also see in Figs. 6.6 to 6.8 that a consumer unit is being used for the control and distribution of electrical energy. However, recent fire statistics have shown that a large number of domestic fires have involved plastic consumer units as the source of the fire. Consumer units are often located at the entrance or exit door of the home or under the stairs, raising the possibility that a fire starting as a result of faulty wiring could block the emergency exit route. Regulation 421.1.201 of the 18th edition of the Regulations now requires that consumer units be manufactured from non-combustible material, for example, metal, or be enclosed in a non-combustible enclosure.

The second element of this protection method is the provision of a means of automatic disconnection of the supply in the event of a fault occurring that causes the exposed metalwork to become live. IET Regulation 411.3.2 tells us that for final circuits not exceeding 32 A, the maximum disconnection time shall not exceed 0.4 seconds.

The achievement of these disconnection times is dependent upon the type of protective device used, fuse or circuit-breaker, the resistance of the circuit conductors to the fault and the provision of adequate protective bonding. The resistance, or we call it the impedance, of the earth fault loop must be less than the values given in Tables 41.2 to 41.4 of the IET Regulations. (Table 6.1 shows the maximum value of the earth fault loop impedance for circuits protected by miniature circuit-breakers – MCBs to BS EN 60898.) Chapter 54 of the IET Regulations gives details of the earthing arrangements to be incorporated into the supply system to meet these regulations and these are described below.

Supply system earthing arrangements

The British government agreed on 1 January 1995 that the electricity supplies in the United Kingdom would be harmonized with those of the rest of Europe. Thus the voltages used previously in low-voltage supply systems of 415 and 240 V have become 400 V for three-phase supplies and 230 V for single-phase supplies. The Electricity Supply Regulations 1988 have also been amended to permit a range of variation from the newly declared nominal voltage. From January 1995 the permitted tolerance is the nominal voltage +10% or −6%. Previously it was ±6%. This gives a voltage range of 216–253 V for a nominal voltage of 230 V and 376–440 V for a nominal voltage of 400 V (IET Regulations Appendix 2).

It is further proposed that the tolerance levels will be adjusted to ±10% of the declared nominal voltage. All EU countries will have to adjust their voltages to comply with a nominal voltage of 230 V single phase and 400 V three phase.

The supply to a domestic, commercial or small industrial consumer's installation is usually protected at the incoming service cable position with a 100 A BS 88-2 high-breaking capacity (HBC) fuse. Other items of equipment at this position are the energy meter and the consumer's distribution unit, providing the protection for the final circuits and the earthing arrangements for the installation.

An efficient and effective earthing system is essential to allow protective devices to operate. The limiting values of earth fault loop impedance are given in Tables 41.2 to 41.4 of the IET Regulations. Table 6.1 in this chapter gives the maximum Z_s values for a Type B MCB. Chapter 54 and the wiring systems of Part 2 of the regulations give details of the earthing arrangements to be incorporated into

Table 6.1 Maximum earth fault loop impedance Z_s (Ω) when the overcurrent protective device is an MCB Type B to BS EN 60898

	MCB rating (A)					
	6	10	16	20	25	32
For 0.4 s disconnection Z_S (Ω)	7.28	4.37	2.73	2.19	1.75	1.37

the supply system to meet the requirements of the regulations. Five systems are described in the definitions but only the TN-S, TN-C-S and TT systems are suitable for public supplies.

A system consists of an electrical installation connected to a supply. Systems are classified by a capital letter designation.

The supply earthing

The supply earthing arrangements are indicated by the first letter, where T means one or more points of the supply are directly connected to earth and I means the supply is not earthed or one point is earthed through a fault-limiting impedance.

The installation earthing

The installation earthing arrangements are indicated by the second letter, where T means the exposed conductive parts are connected directly to earth and N means the exposed conductive parts are connected directly to the earthed point of the source of the electrical supply.

The earthed supply conductor

The earthed supply conductor arrangements are indicated by the third letter, where S means a separate neutral and protective conductor and C means that the neutral and protective conductors are combined in a single conductor.

- a PE (protective conductor) is a conductor used for some measure of protection against electric shock, as shown in Figs. 6.9 to 6.11, and
- a PEN conductor is a conductor combining the functions of both protective conductor and neutral conductor, as shown in Fig. 6.10.

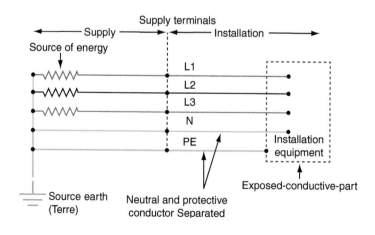

Figure 6.9 TN-S system.

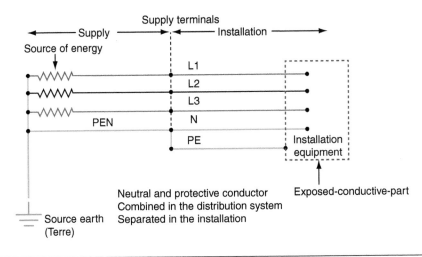

Figure 6.10 TN-C-S system (PME).

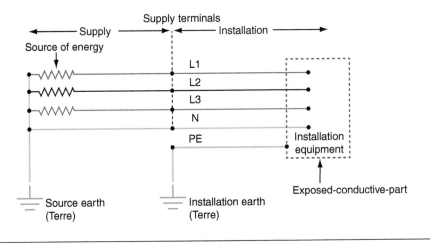

Figure 6.11 TT system.

Cable sheath earth supply (TN-S system)

This is one of the most common types of supply system to be found in the United Kingdom where the electricity companies' supply is provided by underground cables. The neutral and protective conductors are separate throughout the system. The protective earth conductor (PE) is the metal sheath and armour of the underground cable, and this is connected to the consumer's main earthing terminal. All exposed conductive parts of the installation, gas pipes, water pipes and any lightning protective system are connected to the protective conductor via the main earthing terminal of the installation. The arrangement is shown in Figs. 6.6 and 6.9.

(PME) Protective multiple earthing supply (TN-C-S system)

This type of underground supply is becoming increasingly popular to supply new installations in the United Kingdom. It is more commonly referred to as protective multiple earthing (PME). The supply cable uses a combined protective earth and neutral (PEN) conductor. At the supply intake point a consumer's main earthing terminal is formed by connecting the earthing terminal to the neutral conductor. All exposed conductive parts of the installation, gas pipes, water pipes and any lightning protective system are then connected to the main earthing terminals. Thus phase to earth faults are effectively converted into phase to neutral faults. The arrangement is shown in Figs. 6.7 and 6.10.

No earth provided supply (TT system)

This is the type of supply more often found when the installation is fed from overhead cables. The supply authorities do not provide an earth terminal and the installation's CPCs must be connected to earth via an earth electrode provided by the consumer. Regulation 542.2.3 lists the type of earth rod, earth plates or earth tapes recognized by BS 7671. An effective earth connection is sometimes difficult to obtain and in most cases a residual current device (RCD) is provided when this type of supply is used. The arrangement is shown in Figs. 6.8 and 6.11.

Figures 6.6 to 6.8 show the layout of a typical domestic service position for these three supply systems. They show circuits protected by RCBOs. The use of RCBOs will minimize inconvenience because, in the event of a fault occurring, only the faulty circuit will disconnect. However, alternative consumer unit arrangements using MCBs and RCDs are shown in the *On Site Guide* at Section 3.6.3. The TN-C and IT systems of supply do not comply with the supply regulations and therefore cannot be used for public supplies. Their use is restricted to private generating plants and for this reason I shall not include them here, but they can be seen in Part 2 of the IET Regulations.

 Visit the companion website for more on this topic.

Earth fault loop impedance Z_S

In order that an overcurrent protective device can operate successfully, it must meet the required disconnection times of IET Regulation 411.3.2.2; that is, final circuits not exceeding 32 A shall have a disconnection time not exceeding 0.4 s. To achieve this, the earth fault loop impedance value measured in ohms must be less than those values given in Appendix B of the *On Site Guide* and Tables 41.2 and 41.3 of the IET Regulations. The value of the earth fault loop impedance may be verified by means of an earth fault loop impedance test, as described in Chapter 5 of this book. The formula is:

$$Z_S = Z_E + (R_1 + R_2)\,(\Omega)$$

Here Z_E is the impedance of the supply side of the earth fault loop. The actual value will depend upon many factors: the type of supply, the ground conditions, the distance from the transformer, etc. The value can be obtained from the area electricity companies, but typical values are $0.35\,\Omega$ for TN-C-S (protective multiple earthing, PME) supplies and $0.8\,\Omega$ for TN-S (cable sheath earth) supplies, as described a little earlier in this chapter. Also in the above formula, R_1 is the resistance of the line conductor and R_2 is the resistance of the earth conductor. The complete earth fault loop path is shown in Fig. 6.12.

Values of $R_1 + R_2$ have been calculated for copper and aluminium conductors and are given in Table I1 of the *On Site Guide,* as shown in Table 6.2 of this chapter.

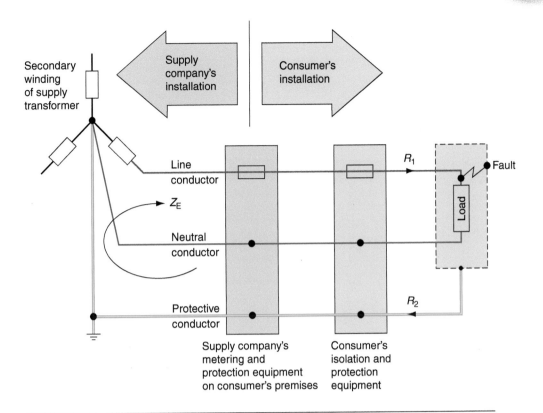

Figure 6.12 Earth fault loop path for a TN-S system.

Example

A 20 A radial socket outlet circuit is wired in 2.5 mm^2 PVC cable incorporating a 1.5 mm^2 CPC. The cable length is 30 m installed in an ambient temperature of 20°C and the consumer's protection is by 20 A MCB Type B to BS EN 60898. The earth fault loop impedance of the supply is 0.5 Ω. Calculate the total earth fault loop impedance Z_S, and establish that the value is less than the maximum value permissible for this type of circuit.

We have:

$$Z_S = Z_E + (R_1 + R_2)\ (\Omega)$$
$$Z_E = 0.5\,\Omega\,(\text{value given in the question})$$

From the value given in Table I1 of the *On Site Guide* and reproduced in Table 6.2, a 2.5 mm phase conductor with a 1.5 mm protective conductor has an $(R_1 + R_2)$ value of 19.51 $\times$ 10$^{-3}\,\Omega$/m.

$$\text{For 30 m cable } (R_1 + R_2) = 19.51 \times 10^{-3}\,\Omega/\text{m} \times 30\,\text{m} = 0.585\,\Omega$$

However, under fault conditions, the temperature and therefore the cable resistance will increase. To take account of this, we must multiply the value of cable resistance by the factor given in Table I3 of the *On Site Guide*. In this case the factor is 1.20 and therefore the cable resistance under fault conditions will be:

$$0.585\,\Omega \times 1.20 = 0.702\,\Omega$$

The total earth fault loop impedance is therefore:

$$Z_S = 0.5\,\Omega + 0.702\,\Omega = 1.202\,\Omega$$

Table 6.2 Value of resistance/metre for copper conductors and of $R_1 + R_2$ per metre at 20°C in milliohms/metres

Cross-sectional area (mm²)		Resistance/metre or $(R_1 + R_2)$/metre (mΩ/m)
Phase conductor	Protective conductor	Copper
1	–	18.10
1	1	36.20
1.5	–	12.10
1.5	1	30.20
1.5	1.5	24.20
2.5	–	7.41
2.5	1	25.51
2.5	1.5	19.51
2.5	2.5	14.82
4	–	4.61
4	1.5	16.71

Table 6.3 Maximum earth fault loop impedance Z_s (Ω) when the overcurrent protective device is an MCB Type B to BS EN 60898

	MCB rating (A)					
	6	10	16	20	25	32
For 0.4 s disconnection Z_S (Ω)	7.28	4.37	2.73	2.19	1.75	1.37

The maximum permitted value given in Table 41.3 of the IET Regulations for a 20 A MCB protecting a socket outlet is 2.19 Ω, as shown in Table 6.3 above. The circuit earth fault loop impedance is less than this value and therefore the protective device will operate within the required disconnection time of 0.4 s.

The value of the earth fault loop impedance can be measured using the test described in IET Regulation 619.2 and compared with the tables of 41.2 and 41.3 of the IET Regulations and Appendix B of the *On Site Guide*. The test is also described in Chapter 5 of this book at test number seven.

Protective conductor size

The CPC forms an integral part of the total earth fault loop impedance, so it is necessary to check that the cross-section of this conductor is adequate. If the cross-section of the CPC complies with Table 54.7 of the IET Regulations, there is no need to carry out further checks. Where line and protective conductors are made from the same material, Table 54.7 tells us that:

- for line conductors equal to or less than 16 mm², the protective conductor should equal the line conductor;
- for line conductors greater than 16 mm² but less than 35 mm², the protective conductor should have a cross-sectional area of 16 mm²;
- for line conductors greater than 35 mm², the protective conductor should be half the size of the line conductor.

However, where the conductor cross-section does not comply with this table, then the formula given in IET Regulation 543.1.3 must be used:

$$S = \frac{\sqrt{I^2 t}}{k} \ (\text{mm}^2)$$

where

S = cross-sectional area in mm^2

I = value of maximum fault current in amperes

t = operating time of the protective device

k = a factor for the particular protective conductor (see Tables 54.2 to 54.4 of the IET Regulations).

Example 1

A 230 V ring main circuit of socket outlets is wired in 2.5 mm single PVC copper cables in a plastic conduit with a separate 1.5 mm CPC. An earth fault loop impedance test identifies Z_s as 1.15 Ω. Verify that the 1.5 mm CPC meets the requirements of IET Regulation 543.1.3 when the protective device is a 30 A semi-enclosed fuse.

$$I = \text{Maximum fault current} = \frac{V}{Z_s}(A)$$

$$\therefore \frac{230}{1.15} = 200\,A$$

t = maximum operating time of the protective device for a circuit not exceeding 32 A is 0.4 s from IET Regulation 411.3.2.2. From Fig. 3A2(a) in Appendix 3 of the IET Regulations you can see that the time taken to clear a fault of 200 A is about 0.4 s

k = 115 (from Table 54.3)

$$S = \frac{\sqrt{I^2 t}}{k}(\text{mm}^2)$$

$$S = \frac{\sqrt{(200\,A)^2 \times 0.4\,s}}{115} = 1.10\,\text{mm}^2$$

A 1.5 mm^2 CPC is acceptable since this is the nearest standard-size conductor above the minimum cross-sectional area of 1.10 mm^2 found by calculation.

Example 2

A TN supply feeds a domestic immersion heater wired in 2.5 mm^2 PVC insulated copper cable and incorporates a 1.5 mm^2 CPC. The circuit is correctly protected with a 16 A semi-enclosed fuse to BS 3036. Establish by calculation that the CPC is of an adequate size to meet the requirements of IET Regulation 543.1.3. The characteristics of the protective device are given in IET Regulation Fig. 3A2(a) of Appendix 3.

For final circuits less than 32 A the maximum operating time of the protective device is 0.4 s. From Fig. 3A2(a) of Appendix 3 it can be seen that a current

(Continued)

Example 2 (Continued)

of about 90 A will trip the 15 A fuse in 0.4 s. The small insert table on the top right of Fig. 3A2(a) of Appendix 3 of the IET Regulations gives the value of the prospective fault current required to operate the device within the various disconnection times given.

So, in this case the table states that 90 A will trip a 16 A semi-enclosed fuse in 0.4 s

$$\therefore I = 90\,A$$
$$t = 0.4\,s$$
$$k = 115 \text{ (from Table 54.3)}$$

$$S = \frac{\sqrt{I^2 t}}{k}\,(\text{mm}^2) \text{ (from IEE Regulation 543.1.3)}$$

$$S = \frac{\sqrt{(90\,A)^2 \times 0.4\,s}}{115} = 0.49\,\text{mm}^2$$

The CPC of the cable is greater than 0.49 mm² and is therefore suitable. If the protective conductor is a separate conductor, that is, it does not form part of a cable as in this example and is not enclosed in a wiring system as in Example 1, the cross-section of the protective conductor must be not less than 2.5 mm² where mechanical protection is provided or 4.0 mm² where mechanical protection is **not** provided in order to comply with IET Regulation 544.2.3.

Table 6.4 Assumed current demand for electrical equipment and circuits

Current using equipment	Assumed current demand
Lighting points	Minimum of 100 watts/lamp holder
Discharge lighting such as fluorescent tubes	Lamp watts × 1.8 to take account of the control gear
Electric clock, shaver unit, bell transformer	May be neglected in this assessment
2A socket outlet	0.5A
Standard household –13A ring circuit	30A or 32A – max floor area 100 m² wired in 2.5 mm cable
Standard household – 13A radial circuit	30A or 32A – max floor area 75 m² wired in 4.0 mm cable
Standard household – 13A radial circuit	20A – max floor area 50 m² wired in 2.5 mm cable
Cooking appliances	The first 10A of the rated current plus 30% of the remainder of the rated current plus 5A if control unit incorporates a socket outlet
All other stationary equipment such as shower or immersion heater	British Standard rated current

Finally, it should perhaps be said that a foolproof method of giving protection to people or animals who simultaneously touch both live and neutral has yet to be devised. The ultimate safety of an installation depends upon the skill and experience of the electrical contractor and the good sense of the user.

132.3 The nature of demand

The designer must identify the number of circuits required and the expected load on these circuits.

The total current demand of any final circuit is estimated by adding together the current demands of all points of utilization such as socket and lighting points and equipment outlets.

Final circuit current demand

In this chapter we will only look at straightforward household installations but the same principles apply to shops, hotels and guest houses. All of these premises are dealt with in Appendix A of the *On Site Guide*. Let us begin by looking at the current demand to be assumed for points of utilization given in Table A1 of the *On Site Guide* and shown in Table 6.4.

Example 1

Calculate the current demand of a 7.36 kW electric shower connected to the 230 V mains supply.

The power $P = 7.36\text{kW}$ or   7360 watts
The voltage $V = 230$ volts
The power factor $= \cos \phi = 1$ for a resistive load

Now power $= V\,I \cos \phi$ so transposing for current $I = \dfrac{\text{power}}{V \cos \phi}$ amps

So the current demand $I = \dfrac{7360}{230 \times 1} = 32$ amps

---

Example 2

Calculate the current demand of an electric cooker comprising

A hob with 4 2.5 kW rings $= 10\text{kW}$
A main oven rated at 2kW and $= 02\text{kW}$
A grill/top oven rated at 2kW $= 02\text{kW}$ total 14kW

The cooker control unit incorporates a 13A socket outlet

So the total capacity of the cooker is 14kW at 230 volts

As in example 1 the current $I = \dfrac{\text{power}}{V \cos \phi}$

And $\cos \phi = 1$ for a resistive load

$$\text{Therefore } I = \frac{14000}{230 \times 1} = 60.87 \text{ amps}$$

(Continued)

Example 2 (Continued)

From Table 6.4 above we take the first 10A of the rated current, plus 30% of the remainder plus 5A if the control unit incorporates a socket outlet.

$$\text{Therefore } I = 10A + \left[\frac{30}{100} \times (60.87 - 10)\right] + 5A$$

$$I = 10A + \left[\frac{30}{100} \times 50.87\right] + 5A$$

$$I = 10 + 15.26 + 5$$

$$I = 30.26 \text{ amps}$$

Therefore a 32 amp MCB would adequately protect this circuit.

Example 3

A lighting circuit consists of 10 points. Calculate the current demand.

From Table 6.4 we must assume a minimum of 100 watts per point.

$$\text{As in the previous examples } I = \frac{\text{power}}{V \cos\phi} \text{ amps}$$

$$\text{Therefore } I = \frac{10 \times 100\,W}{230\,V \times 1} = 4.35 \text{ amps}$$

Therefore a 5 amp fuse or a 6 amp MCB would protect this circuit if ordinary GLS lamps were being used. However, if extra-low voltage or discharge lighting is connected to this circuit and protected by a type 'B' MCB we must take account of the inrush current which occurs at switch-on and sometimes causes unwanted or nuisance tripping of the MCB. To avoid this, the circuit will be adequately protected by a 10 amp MCB.

Example 4

The kitchen work surface of a domestic kitchen is to be illuminated by 10, 30 watt fluorescent tubes fixed to the underside of the cupboards above the work surface. Calculate the current demand of this circuit.

From Table 6.4 the demand is taken as the lamp watts multiplied by 1.8

$$\text{As in the previous examples } \quad I = \frac{\text{power}}{V \cos\phi} \text{ amps}$$

$$\text{Therefore } I = \frac{10 \times 30\,W \times 1.8}{230 \times 1} = 2.34 \text{ amps}$$

Therefore a 5 amp fuse or a 6 amp MCB would protect this circuit.

Example 5

Calculate the current demand of a 3kW immersion heater installed in a 30-litre water storage vessel. We know from Table 6.4 that we should take the British Standard rated current for a water heater such as this, or we can calculate the current rating as before.

$$\text{Current } I = \frac{\text{power}}{V \cos\phi} \qquad \text{therefore } I = \frac{3000\,W}{230\,V \times 1} = 13.04 \text{ amps}$$

A 15 amp fuse or 16 amp MCB will protect this circuit. We also know from Appendix H5 of the *On Site Guide* that water heaters fitted to vessels in excess of 15 litres must be supplied on their own separate circuit.

Diversity between final circuits

The current demand of a circuit is the current taken by that circuit over a period of time, say, 30 minutes. Some loads make a constant demand all the time they are switched on. A 100 watt light bulb will make a constant current demand of (100 watts ÷ 230 volts) 0.43 amperes whenever it is switched on. However, an automatic washing machine is made up of a variable speed motor, a pump and a water heater all controlled by a programmer. The washer load will not be constant for the whole time it is switched on, but will vary depending upon the wash cycle.

Diversity makes an allowance on the basis that not all of the load or connected items will be in use at the same time.

The design would be wasteful if it did not take advantage of the diversity between different loads. Let us now look at diversity as it applies to a straightforward household installation but the same principles will apply to shops, offices, business premises and hotels. Appendix A of the *On Site Guide* gives the diversity for all these different types of premises.

The allowances for diversity shown in Table A2 of the *On Site Guide* are for very specific situations and can only provide guidance. The diversity allowances for lighting circuits should be applied to 'items of equipment' connected to the consumer unit at a rate of 66%. However, standard power circuit arrangements for households, as described in Appendix H of the *On Site Guide*, can be applied to the rated current of the overcurrent protective device for the circuit. The diversity allowance for standard socket circuits is 100% of the largest circuit plus 40% of all other power circuits. It is important to ensure that the consumer's unit is of sufficient rating to take the total load connected without the application of any diversity.

Example 6

Calculate the current demand, including diversity, for a six-way consumer unit comprising the following circuits in a domestic dwelling.

Circuit 1. A lighting circuit comprising 10 points as described in Example 3, previously having a maximum demand of 4.35 A and protected by a 6 amp MCB (Note 66% diversity for lighting).

(Continued)

Example 6 (Continued)

Circuit 2. A lighting circuit comprising 10, 30 watt fluorescents as described in Example 4, previously having a maximum demand of 2.3 A and protected by a 6 amp MCB (Note again 66% diversity).

Circuit 3. A thermostatically controlled 3kW immersion heater as described in Example 5, previously having a maximum demand of 13.04A and protected by a 16 amp MCB (Note no diversity allowed).

Circuit 4. A ring circuit of 13A socket outlets installed in accordance with Section H of the *On Site Guide* and protected by a 32 amp MCB (Note we are calling this the largest circuit, so, no diversity allowed).

Circuit 5. A radial circuit of 13A socket outlets installed in accordance with section H of the *On Site Guide* and protected by a 32 amp MCB (Note 40% diversity allowed).

Circuit 6. A radial circuit of 13A socket outlets installed in accordance with section H of the *On Site Guide* and protected by a 20 amp MCB (Note 40% diversity allowed).

Table A2 of the *On Site Guide* and the IET Design Guide give us guidance when calculating maximum demand and diversity as stated in the notes above. So, let us now make the calculation for our own example 6.

From the question

Circuit 1 has a maximum demand of	4.35 amps
Circuit 2 has a maximum demand of	2.34 amps
Circuit 3 has a maximum demand of	13.04 amps
Circuit 4 has a maximum demand of	32.00 amps
Circuit 5 has a maximum demand of	32.00 amps
Circuit 6 has a maximum demand of	20.00 amps

We can see from the above that the largest socket circuit in the consumers unit is 32 amps, so let us nominate circuit 4 as the largest circuit and apply 40% diversity to all other socket circuits.

Circuit 1 $= 4.35A \times \dfrac{66}{100} =$ 2.87A

Circuit 2 $= 2.34A \times \dfrac{66}{100} =$ 1.54A

Circuit 3 $= 13.04A$ (no diversity) 13.04A

Circuit 4 $= 32.00A$ (no diversity) 32A

Circuit 5 $= 32.00A \times \dfrac{40}{100} =$ 12.80A

Circuit 6 $= 20.00A \times \dfrac{40}{100} =$ 8.00A

Adding the values 25.21A + 45.04A

Total installed demand with diversity $= 70.25$ amps

Example 6 (Continued)

To summerise: The total design current is 103.73 amps but, after applying diversity, because not all circuits are being used all of the time, the total demand falls to 70.25amps. Therefor a smaller cable may be used to supply this reduced load.

132.4 Electrical supplies for safety systems

Where a supply for safety services or standby electrical systems is required, the designer must determine the characteristics of the supply and the services to be supplied by the safety source.

Approved document B, fire safety, states that:

* there must be routes for people to escape to a place of safety;
* these routes must be protected from the effects of fire;
* the routes must be adequately illuminated;
* the exits are suitably signed;
* there is sufficient means of giving early warning of fire to persons in the building.
* The 18th edition of the IET Wiring Regulations has introduced new Regulations requiring robust support of wiring systems so that they will not become compromised or collapse in the event of a fire.

Note 3 to Regulation 521.10.202 tells us that it is the cable systems fixed with plastic clips or inside plastic trunking which now require our consideration. These systems can fail when subject to either direct flame or the hot products of combustion leading to wiring systems hanging down and causing an entanglement risk as a result of the fire.

This makes it impossible for us to use non metallic cable clips, cable ties or plastic trunking as the only means of support for PVC wiring system. The regulation tells us that where non- metallic cable systems are used, a suitable means of fire resistant support and retention must be used to prevent cables falling down in the event of a fire. Note 4 to this Regulation advises that suitably spaced steel or copper clips, saddles or ties are examples which will meet this requirement.

All of the above must be applied to the extent that is dependent upon the use of the building, its size and its height.

Appendix C of the Electricians Guide to the Building Regulations gives guidance on the provision required as follows:

* automatic fire detection and alarms complying with BS 5839 should be installed in institutional and other residential occupancies;
* it is essential that the fire detection systems are properly designed, installed and maintained;
* where services pass through walls, called fire separating elements, they must be sealed to prevent the passage of smoke and fire.

In dwellings not protected by automatic fire detection and alarm systems, including domestic homes, they are required to be fitted with a suitable number of smoke alarms. The Building Regulations require all new and refurbished

Figure 6.13 Electrical equipment for fire safety.

dwellings to be fitted with mains-operated smoke alarms. The requirements for a single family dwelling of not more than two storeys are:

- smoke alarms should normally be positioned in the circulation spaces between sleeping spaces and where fires are most likely to start such as kitchens and living rooms;
- in a house or bungalow there should be at least one smoke alarm on every storey;
- where more than one smoke alarm is installed they should be linked so that the detection of smoke by one unit operates the alarm in all units;
- alarms should normally be ceiling-mounted and at least 300 mm from walls and light fittings;
- where the kitchen area is not separated by a door there should be a heat detector in the kitchen;
- the power supply for a smoke alarm should be derived from the dwelling's mains electricity supply;
- the cable for the power supply to each self-contained unit and the interconnecting cable need have no fire-retardant properties and need no segregation from other circuits (BS 52661–1: 2011);
- smoke alarms that include a standby power supply can operate during a mains failure and therefore may be connected to a regularly used local lighting circuit. This has the advantage that the circuit is unlikely to be disconnected for prolonged periods.

132.5 Environmental conditions

The electrical design must take into account the environmental conditions to which the installation will be subjected. Electrical equipment in surroundings susceptible to risk of fire or explosion shall be constructed or protected so as to prevent danger.

Appendix 5 of the IET Regulations gives us a concise list of external influences. Each condition is designated with a code; A is environment, B is utilization and C is construction of buildings. For example, a code AA4 signifies:

Figure 6.14 All electrical equipment must be suitable for the installed conditions.

A   = environment
AA = environment ambient temperature
AA4 = environment ambient temperature in the range –5°c to + 40°c

Section 522 of the IET Regulations details the installations requirements for:

- ambient temperature;
- external heat sources;
- the presence of water or/and humidity;
- the presence of corrosive or polluting substances;
- mechanical damage and stresses;
- vibration;
- presence of flora, fauna and mould growth;
- solar radiation.

Some installations require special consideration because of the inherent dangers listed above. Installations requiring special consideration are flameproof installations, construction sites, agricultural and horticultural buildings. All of these installations are described in detail in Part 7 of the IET Regulations and some are described in Chapter 4 of this book.

Appendix C of the *On Site Guide* gives guidance on the selection and types of cable for particular influences.

This point in the design process might also be a good time to consider:

- the environmental impact of the design;
- the importance of sustainable design as described at the beginning of this chapter;
- the possible use of environmental technology systems and renewable energy systems, as described in Chapter 2 of this book.

132.6 Conductor size calculations

The size of a cable to be used for an installation depends upon:

- the current rating of the cable under defined installation conditions, and
- the maximum permitted drop in voltage as defined by IET Regulation 525.

The factors which influence the current rating are:

1 *Design current*: cable must carry the full load current.
2 *Type of cable*: PVC, MICC, copper conductors or aluminium conductors.
3 *Installed conditions*: clipped to a surface or installed with other cables in a trunking.
4 *Surrounding temperature*: cable resistance increases as temperature increases and insulation may melt if the temperature is too high.
5 *Type of protection*: for how long will the cable have to carry a fault current?

IET Regulation 525 states that the drop in voltage from the supply terminals to the fixed current-using equipment must not exceed 3% for lighting circuits and 5% for other uses of the mains voltage. That is a maximum of 6.9 V for lighting and 11.5 V for other uses on a 230 V installation. The volt drop for a particular cable may be found from:

$$VD = \text{Factor} \times \text{Design current} \times \text{Length of run}$$

The factor is given in the tables of Appendix 4 of the IET Regulations and Appendix F of the *On Site Guide*. They are also given in Table 6.7 in this book.

The cable rating, denoted I_t, may be determined as follows:

$$I_t = \frac{\text{Current rating of protective device}}{\text{Any applicable correction factors}}$$

The cable rating must be chosen to comply with IET Regulation 433.1. The correction factors which may need applying are given below as:

Ca the ambient or surrounding temperature correction factor, which is given in Tables 4B1 and 4B2 of Appendix 4 of the IET Regulations. They are also shown in Table 6.5 of this book.

Cg the grouping correction factor given in Tables 4C1 to 4C5 of the IET Regulations and Table 6C of the *On Site Guide*.

Cf the 0.725 correction factor to be applied when semi-enclosed fuses protect the circuit as described in item 5.1.1 of the preface to Appendix 4 of the IET Regulations.

Ci the correction factor to be used when cables are enclosed in thermal insulation. IET Regulation 523.6.6 gives us three possible correction values:

- Where one side of the cable is in contact with thermal insulation we must read the current rating from the column in the table which relates to reference method A (see Table 6.6).
- Where the cable is *totally* surrounded over a length greater than 0.5 m we must apply a factor of 0.5.
- Where the cable is *totally* surrounded over a short length, the appropriate factor given in Table 52.2 of the IET Regulations or Table F2 of the *On Site Guide* should be applied.

Note: **A cable should preferably not be installed in thermal insulations.**

Visit the companion website for more on this topic.

Table 6.5 Ambient air temperature correction factors

Type of insulation	Conductor Operating temperature	Ambient temperature (°C)			
		25	30	35	40
Thermoplastic (general purpose PVC)	70°C	1.03	1.0	0.94	0.87

Table 6.6 Current-carrying capacity of cables

Conductor cross-sectional area	Reference Method A (enclosed in conduit in an insulated wall, etc.)		Reference Method B (enclosed in conduit on a wall or ceiling, or in trunking)		Reference Method C (clipped direct)		Reference Method E (on a perforated cable tray) or in free air	
	Two cables single-phase a.c. or d.c.	Three or four cables three-phase a.c.	One two-core cable, single-phase a.c. or d.c.	One three-core cable or one four-core cable, three-phase a.c.	One two-core cable, single-phase a.c. or d.c.	One three-core cable or one four-core cable, three-phase a.c.	One two-core cable, single-phase a.c. or d.c.	One three-core cable or one four-core cable, three-phase a.c.
1	2	3	4	5	6	7	8	9
mm²	A	A	A	A	A	A	A	A
1	11	10	13	11.5	15	13.5	17	14.5
1.5	14	13	16.5	15	19.5	17.5	22	18.5
2.5	18.5	17.5	23	20	27	24	30	25
4	25	23	30	27	37	32	40	34
6	32	29	38	34	46	41	51	43
10	43	39	52	46	63	57	70	60
16	57	52	69	62	85	76	94	80
25	75	68	90	80	112	96	119	101
35	92	83	111	99	138	119	148	126

Having calculated the cable rating, I_t the smallest cable should be chosen from the appropriate table which will carry that current. This cable must also meet the voltage drop (IET Regulation 525) and this should be calculated as described earlier. When the calculated value is less than 3% for lighting and 5% for other uses of the mains voltage the cable may be considered suitable. If the calculated value is greater than this value, the next larger cable size must be tested until **a cable is found which meets both the current rating and voltage drop criteria**.

Example

A house extension has a total load of 6 kW installed some 18 m away from the mains consumer unit for lighting. A PVC insulated and sheathed twin and earth cable will provide a sub-main to this load and be clipped to the side of the ceiling joists over much of its length in a roof space which is anticipated to reach 35°C in the summer and where insulation is installed up to the top of the joists. Calculate the minimum cable size if the circuit is to be protected by a type B MCB to BS EN 60898. Assume a TN-S supply; that is, a supply having a separate neutral and protective conductor throughout.

Let us solve this question using only the tables given in the *On Site Guide*. The tables in the regulations will give the same values, but this will simplify the problem because we can refer to Tables 6.5, 6.6 and 6.7 in this book which give the relevant *On Site Guide* tables.

$$\text{Design current, } I_b = \frac{\text{Power}}{\text{Volts}} = \frac{6000\,\text{W}}{230\,\text{V}} = 26.09\,\text{A}$$

Nominal current setting of the protection for this load $I_n = 32$ A.

The cable rating I_t is given by:

$$I_t = \frac{\text{Current rating of protective device } (I_n)}{\text{The product of the correction factors}}$$

The correction factors to be included in this calculation are:

Ca ambient temperature; as shown in Table 6.5 the correction factor for 35°C is 0.94.

Cg grouping factors need not be applied.

Cf, since protection is by MCB no factor need be applied.

Ci thermal insulation demands that we assume installed Method A (see Table 6.6).

The design current is 26.09 A and we will therefore choose a 32A MCB for the nominal current setting of the protective device, I_n.

$$\text{Cable rating, } I_t = \frac{32}{0.94} = 34.04\,\text{A}$$

From column 2 in Table 6.6, a 10 mm cable, having a rating of 43A, is required to carry this current.

(Continued)

Example (Continued)

Now test for volt drop: from Table 6.7 the volt drop per ampere per metre for a 10 mm cable is 4.4 mV. So the volt drop for this cable length and load is equal to:

$$4.4 \times 10^{-3} \text{ V/Am} \times 26.09 \text{ A} \times 18 \text{m} = 2.06 \text{ V}$$

Since this is less than the maximum permissible value for a lighting circuit of 6.9 V, a 10 mm cable satisfies the current and drop in voltage requirements when the circuit is protected by an MCB. This cable is run in a loft that gets hot in summer and has thermal insulation touching one side of the cable. We must, therefore, use installed reference Method A of Table 6.6. If we were able to route the cable under the floor, clipped direct or in conduit or trunking on a wall, we may be able to use a 6 mm cable for this load. You can see how the current-carrying capacity of a cable varies with the installed method by looking at Table 6.6. Compare the values in Column 2 with those in Column 6. When the cable is clipped direct on to a wall or surface the current rating is higher because the cable is cooler. If the alternative route was longer, you would need to test for volt drop before choosing the cable. These are some of the decisions the electrical contractor must make when designing an installation which meets the requirements of the customer and the IET Regulations.

If you are unsure of the standard fuse and MCB rating of protective devices, you can refer to Fig. 3A4 of Appendix 3 of the IET Regulations.

 Visit the companion website for more on this topic.

Cable size for standard domestic circuits

Section 3.2 of Chapter 3 of the IET Electrical Installation Design Guide tells us that the basic design intent is to use standard final circuits wherever possible to avoid repeated design. Provided that earth fault loop impedances are below 0.35 ohm for TN-C-S supplies and 0.8 ohm for TN-S supplies the standard circuits can be used as the basis of all final circuits.

Appendix 4 of the IET Regulations (BS 7671) and Appendix F of the IET *On Site Guide* contain tables for determining the current-carrying capacities of conductors which we looked at in the previous section. However, for standard domestic circuits, Table 6.8 gives a guide to cable size.

In this table, I am assuming a standard 230 V domestic installation, having a sheathed earth or PME supply terminated in a 100A HBC fuse at the mains position. Final circuits are fed from a consumer unit, having Type B MCB protection and wired in PVC insulated and sheathed cables with copper conductors having a grey thermoplastic PVC outer sheath or a white thermosetting cable with LSF (low smoke and fume properties). I am also assuming that the surrounding temperature throughout the length of the circuit does not exceed 30°C and the cables are run singly and clipped to a surface.

132.7 Types of wiring and methods of installation

An electrical installation is made up of many different electrical circuits: lighting circuits, power circuits, single-phase domestic circuits and three-phase industrial or commercial circuits.

Whatever the type of circuit, the circuit conductors are contained within cables or enclosures.

Part 5 of the IET Regulations tells us that electrical equipment and materials must be chosen so that they are suitable for the installed conditions, taking into

Table 6.7 Voltage drop in cables factor

Voltage drop (per ampere per metre)			Conductor operating temperature: 70°C
Conductor cross-sectional area (mm²) 1	Two-core cable, d.c. (mV/A/m) 2	Two-core cable, single-phase a.c. (mV/A/m) 3	Three- or four-core cable, three-phase (mV/A/m) 4
1	44	44	38
1.5	29	29	25
2.5	18	18	15
4	11	11	9.5
6	7.3	7.3	6.4
10	4.4	4.4	3.8
16	2.8	2.8	2.4
25	1.75	1.75	1.50
35	1.25	1.25	1.10

Table 6.8 Cable size for standard domestic circuits

Type of final circuit	Cable size (twin and earth)	MCB rating, Type B (A)	Maximum floor area covered by circuit (m²)	Maximum length of cable run (m)
Fixed lighting	1.0	6	–	40
Fixed lighting	1.5	6	–	60
Immersion heater	2.5	16	–	30
Storage radiator	2.5	16	–	30
Cooker (oven only)	2.5	16	–	30
13A socket outlets (radial circuit)	2.5	20	50	30
13A socket outlets (ring circuit)	2.5	32	100	90
13A socket outlets (radial circuit)	4.0	32	75	35
Cooker (oven and hob)	6.0	32	–	40
Shower (up to 7.5 kw)	6.0	32	–	40
Shower (up to 9.6 kw)	10.0	40	–	40

account temperature, the presence of water, corrosion, mechanical damage, vibration or exposure to solar radiation. Therefore, PVC insulated and sheathed cables are suitable for domestic installations but for a cable requiring mechanical protection and suitable for burying underground, a PVC/SWA cable would be preferable. These two types of cable are shown in Figs 6.15 and 6.16.

MI cables are waterproof, heatproof and corrosion-resistant with some mechanical protection. These qualities often make it the only cable choice for hazardous or high-temperature installations such as oil refineries, chemical works, boiler houses and petrol pump installations. An MI cable with terminating gland and seal is shown in Fig. 6.17.

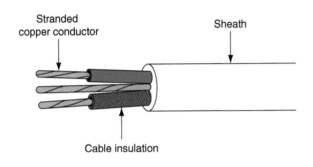

Figure 6.15 A twin and earth PVC insulated and sheathed cable.

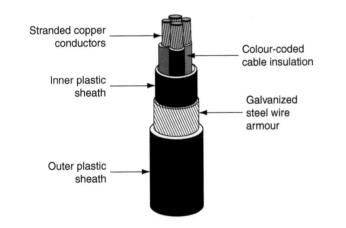

Figure 6.16 A four-core PVC/SWA cable.

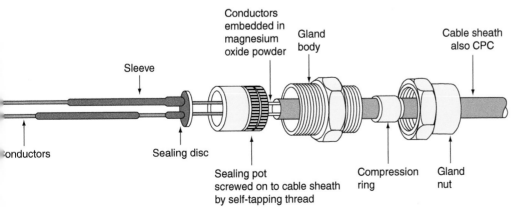

Figure 6.17 MI cable with terminating seal and gland.

 Visit the companion website for more on this topic.

Wiring systems and enclosures

The final choice of a wiring system must rest with those designing the installation and those ordering the work, but whatever system is employed, good workmanship by skilled (electrically) or instructed (electrically) persons and the use of proper materials shall be used in the erection of the electrical installation if it is to comply with the IET Regulations (IET Regulation 134.1.1). The necessary skills can be acquired by electrical trainees who have the correct attitude and dedication to their craft.

PVC insulated and sheathed cable installations

PVC insulated and sheathed wiring systems are used extensively for lighting and socket installations in domestic dwellings. Mechanical damage to the cable caused by impact, abrasion, penetration, compression or tension must be minimized during installation (IET Regulation 522.6.1). The cables are generally fixed using plastic clips incorporating a masonry nail, which means the cables can be fixed to wood, plaster or brick with almost equal ease. Cables should be run horizontally or vertically, not diagonally, down a wall. All kinks should be removed so that the cable is run straight and neatly between clips fixed at equal distances, providing adequate support for the cable so that it does not become damaged by its own weight (IET Regulation 522.8.4 and Table D1 of the *On Site Guide*). Table D1 of the *On Site Guide* is shown in Table 6.9. Where cables are bent, the radius of the bend should not cause the conductors to be damaged (IET Regulation 522.8.3 and Table D5 of the *On Site Guide*).

Terminations or joints in the cable may be made in ceiling roses, junction boxes, or behind sockets or switches, provided that they are enclosed in a non-ignitable material, are properly insulated, and are mechanically and electrically secure (IET Regulation 526). All joints must be accessible for inspection, testing and maintenance when the installation is completed (IET Regulation 526.3). However, there are some exceptions such as the maintenance-free (MF) box shown in Fig. 6.32.

Cable supports to prevent the premature collapse of cables in the event of fire was introduces by the 3rd Amendment to the IET Regulations at 521.11.201. This requires wiring systems in escape routes to be supported in a manner that they will not prematurely collapse in the event of a fire.

Wiring systems which drop and hang across escape routes due to a failure of the means of support in fire conditions have the potential to entangle persons using the escape route including fire fighters entering the building to put out the fire.

The 18th Edition of the regulations at 521.10.202 now requires the same fire proof support for all systems throughout the installation, not just escape routes.

It is the cable systems fixed with plastic clips or inside plastic trunking which will now require our consideration. These systems can fail when subject to either direct flame or the hot products of combustion leading to wiring systems hanging down and causing an entanglement risk as a result of the fire.

This makes it impossible for us to use non metallic cable clips, cable ties or plastic trunking as the only means of support for PVC wiring system. The regulation tells us that where non- metallic cable systems are used, a suitable means of fire resistant support and retention must be used to prevent cables

Table 6.9 Spacing of cable supports

Overall diameter of cable (mm)	Maximum spacings of clips								
	PVC sheathed cables				Armoured cables		Mineral insulated copper sheathed cables	
	Generally		In caravans					
	Horizontal (mm)	Vertical (mm)	Horizontal (mm)	Vertical (mm)	Horizontal (mm)	Vertical (mm)	Horizontal (mm)	Vertical (mm)
1	2	3	4	5	6	7	8	9
Not exceeding 9	250	400	250 (for all sizes)	400 (for all sizes)	–	–	600	800
Exceeding 9 and not exceeding 15	300	400			350	450	900	1200
Exceeding 15 and not exceeding 20	350	450			400	550	1500	2000
Exceeding 20 and not exceeding 40	400	550			450	600	–	–

falling down in the event of a fire. Note 4 of the Regulation advises that suitably spaced steel or copper clips, saddles or ties are examples which will met this requirement.

In the future, if an electrical contractor carrying out an Electrical Installation Condition Report observes PVC cable systems supported only by non-metallic cable supports, they are to record a code C3 (Improvement Recommended) on the report.

Where PVC insulated and sheathed cables are concealed in walls, floors or partitions, they must be provided with a box incorporating an earth terminal at each outlet position. PVC cables do not react chemically with plaster, as do some cables, and consequently PVC cables may be buried under plaster. Further protection by channel or conduit is only necessary if mechanical protection from nails or screws is required or to protect them from the plasterer's trowel. However, IET Regulation 522.6.101 now tells us that where PVC cables are to be embedded in a wall or partition at a depth of less than 50 mm they should be run along one of the permitted routes shown in Fig. 6.19. Figure 6.18 shows a typical PVC installation. To identify the most probable cable routes, IET Regulation 522.6.101 tells us that outside a zone formed by a 150 mm border all around a wall edge, cables can only be run horizontally or vertically to a point or accessory if they are contained in a substantial earthed enclosure, such as a conduit, which can withstand nail penetration, as shown in Fig. 6.19.

Where the accessory or cable is fixed to a wall which is less than 100 mm thick, protection must also be extended to the reverse side of the wall if a position can be determined.

Where none of this protection can be complied with, then the cable must be given additional protection with a 30 mA RCD (IET Regulation 522.6.201).

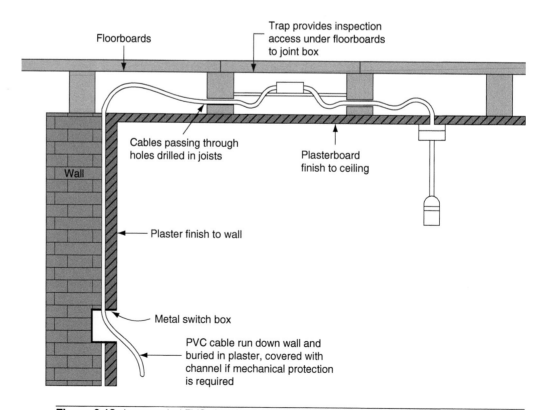

Figure 6.18 A concealed PVC sheathed wiring system.

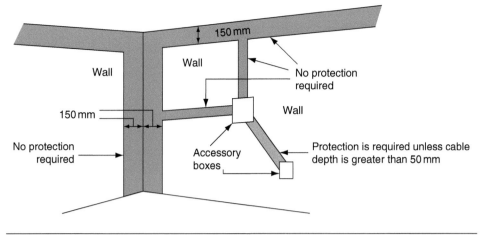

Figure 6.19 Permitted cable routes.

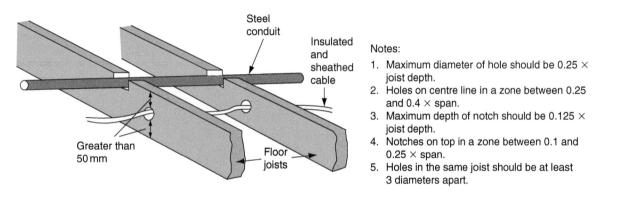

Notes:

1. Maximum diameter of hole should be 0.25 × joist depth.
2. Holes on centre line in a zone between 0.25 and 0.4 × span.
3. Maximum depth of notch should be 0.125 × joist depth.
4. Notches on top in a zone between 0.1 and 0.25 × span.
5. Holes in the same joist should be at least 3 diameters apart.

Figure 6.20 Correct installation of cables in floor joists.

Where cables pass through walls, floors and ceilings the hole should be made good with incombustible material such as mortar or plaster to prevent the spread of fire (IET Regulations 527.1.2 and 527.2.1). Cables passing through metal boxes should be bushed with a rubber grommet to prevent abrasion of the cable. Holes drilled in floor joists through which cables are run should be 50 mm below the top or 50 mm above the bottom of the joist to prevent damage to the cable by nail penetration (IET Regulation 522.6.100), as shown in Fig. 6.20. PVC cables should not be installed when the surrounding temperature is below 0°C or when the cable temperature has been below 0°C for the previous 24 hours because the insulation becomes brittle at low temperatures and may be damaged during installation.

Try this

Definitions

* In the margin write down a short definition of a 'skilled person'.

Conduit installations

A **conduit** is a tube, channel or pipe in which insulated conductors are contained. The conduit, in effect, replaces the PVC outer sheath of a cable, providing mechanical protection for the insulated conductors. A conduit

Definition

A *conduit* is a tube, channel or pipe in which insulated conductors are contained.

installation can be rewired easily or altered at any time, and this flexibility, coupled with mechanical protection, makes conduit installations popular for commercial and industrial applications. There are three types of conduit used in electrical installation work: steel, PVC and flexible.

Steel conduit

Steel conduits are made to a specification defined by BS 4568 and are either heavy gauge welded or solid drawn. Heavy gauge is made from a sheet of steel welded along the seam to form a tube and is used for most electrical installation work. Solid drawn conduit is a seamless tube which is much more expensive and only used for special gas-tight, explosion-proof or flameproof installations.

Conduit is supplied in 3.75 m lengths and typical sizes are 16, 20, 25 and 32 mm. Conduit tubing and fittings are supplied in a black enamel finish for internal use or hot galvanized finish for use on external or damp installations. A wide range of fittings is available and the conduit is fixed using saddles or pipe hooks, as shown in Fig. 6.21.

Metal conduits are threaded with stocks and dies and bent using special bending machines. The metal conduit is also utilized as the CPC and, therefore, all connections must be screwed up tightly and all burrs removed so that cables will not be damaged as they are drawn into the conduit. Metal conduits containing a.c. circuits must contain phase and neutral conductors in the same conduit to prevent eddy currents from flowing, which would result in the metal conduit becoming hot (IET Regulations 521.5.1, 522.8.1 and 522.8.11).

PVC conduit

PVC conduit used on typical electrical installations is heavy gauge standard impact tube manufactured to BS 4607. The conduit size and range of fittings are the same as those available for metal conduit. PVC conduit is most often joined by placing the end of the conduit into the appropriate fitting and fixing with a PVC solvent adhesive. PVC conduit can be bent by hand using a bending spring of the same diameter as the inside of the conduit. The spring is pushed into the conduit to the point of the intended bend and the conduit then bent over

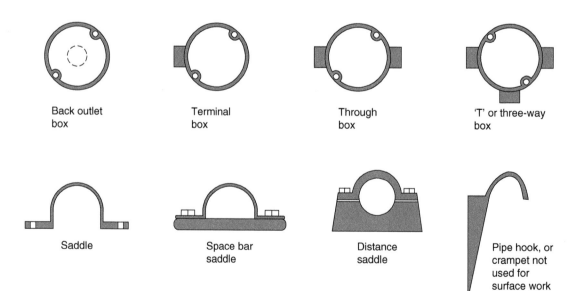

Back outlet box

Terminal box

Through box

'T' or three-way box

Saddle

Space bar saddle

Distance saddle

Pipe hook, or crampet not used for surface work

Figure 6.21 Conduit fittings and saddles.

the knee. The spring ensures that the conduit keeps its circular shape. In cold weather, a little warmth applied to the point of the intended bend often helps to achieve a more successful bend.

The advantages of a PVC conduit system are that it may be installed much more quickly than steel conduit and is non-corrosive, but it does not have the mechanical strength of steel conduit or the fire proof support of steel conduit unless it is supported by steel saddles. Since PVC conduit is an insulator it cannot be used as the CPC and a separate earth conductor must be run to every outlet. It is not suitable for installations subjected to temperatures below 25°C or above 60°C. Where luminaires are suspended from PVC conduit boxes, precautions must be taken to ensure that the lamp does not raise the box temperature or that the mass of the luminaire supported by each box does not exceed the maximum recommended by the manufacturer (IET Regulations 522.1 and 522.2). PVC conduit also expands much more than metal conduit and so long runs require an expansion coupling to allow for conduit movement and to help prevent distortion during temperature changes.

All conduit installations must be erected first before any wiring is installed (IET Regulation 522.8.2). The radius of all bends in conduit must not cause the cables to suffer damage, and therefore the minimum radius of bends given in Table D5 of the *On Site Guide* applies (IET Regulation 522.8.3). All conduits should terminate in a box or fitting and meet the boxes or fittings at right angles, as shown in Fig. 6.22. Any unused conduit-box entries should be blanked off and all boxes covered with a box lid, fitting or accessory to provide complete enclosure of the conduit system. Conduit runs should be separate from other services, unless intentionally bonded, to prevent arcing from occurring from a faulty circuit within the conduit, which might cause the pipe of another service to become punctured.

When drawing cables into conduit they must first be *run off* the cable drum. That is, the drum must be rotated as shown in Fig. 6.23 and not allowed to *spiral off*, which will cause the cable to twist.

Cables should be fed into the conduit in a manner which prevents any cable from crossing over and becoming twisted inside the conduit. The cable insulation must not be damaged on the metal edges of the draw-in box. Cables can be pulled in on a draw wire if the run is a long one. The draw wire itself may be drawn in on a fish tape, which is a thin spring steel or plastic tape.

A limit must be placed on the number of bends between boxes in a conduit run and the number of cables which may be drawn into a conduit to prevent the cables from being strained during wiring. Appendix E of the *On Site Guide* gives a guide to the cable capacities of conduits and trunking.

Flexible conduit

Flexible conduit manufactured to BS 731-1: 1993 is made of interlinked metal spirals often covered with a PVC sleeving. The tubing must not be relied upon to provide a continuous earth path and, consequently, a separate CPC must be run either inside or outside the flexible tube (IET Regulation 543.2.7).

Flexible conduit is used for the final connection to motors so that the vibrations of the motor are not transmitted throughout the electrical installation and to allow for modifications to be made to the final motor position and drive belt adjustments.

Definition

Flexible conduit manufactured to BS 731-1:1993 is made of interlinked metal spirals often covered with a PVC sleeving.

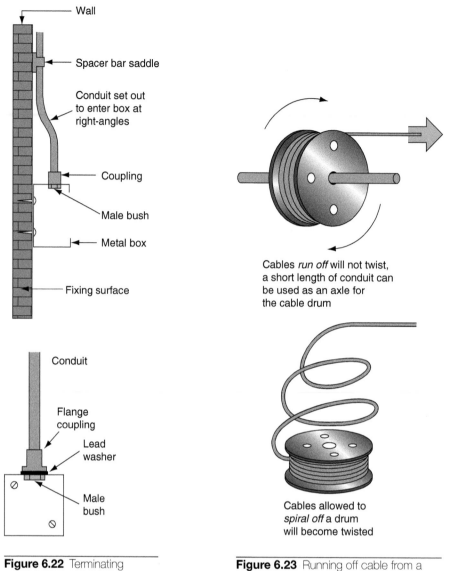

Figure 6.22 Terminating conduits.

Figure 6.23 Running off cable from a drum.

Definition

Single-PVC insulated conductors are usually drawn into the installed conduit to complete the installation.

Conduit capacities

Single-PVC insulated conductors are usually drawn into the installed conduit to complete the installation. Having decided upon the type, size and number of cables required for a final circuit, it is then necessary to select the appropriate size of conduit to accommodate those cables.

The tables in Appendix E of the *On-Site Guide* describe a 'factor system' for determining the size of conduit required to enclose a number of conductors. Similar tables are shown in Tables 6.10 and 6.11. The method is as follows:

- Identify the cable factor for the particular size of conductor; see Table 6.10.
- Multiply the cable factor by the number of conductors, to give the sum of the cable factors.
- Identify the appropriate part of the conduit factor table given by the length of run and number of bends; see Table 6.11.
- The correct size of conduit to accommodate the cables is that conduit which has a factor equal to or greater than the sum of the cable factors.

Table 6.10 Conduit cable factors

Cable factors for conduit in long straight runs over 3 m, or runs of any length incorporating bends	
Conductor CSA (mm²)	Cable factor
1	16
1.5	22
2.5	30
4	43
6	58
10	105
16	145

Example 1

Six 2.5 mm² PVC insulated cables are to be run in a conduit containing two bends between boxes 10 m apart. Determine the minimum size of conduit to contain these cables.

From Table 6.10:

$$\text{The factor for one 2.5mm}^2 \text{ cable} = 30$$
$$\text{The sum of the cable factors} = 6 \times 30$$
$$= 180$$

From Table 6.11, a 25 mm conduit, 10 m long and containing two bends, has a factor of 260. A 20 mm conduit containing two bends only has a factor of 141 which is less than 180, the sum of the cable factors and, therefore, 25 mm conduit is the minimum size to contain these cables.

Example 2

Ten 1.0 mm² PVC insulated cables are to be drawn into a plastic conduit which is 6 m long between boxes and contains one bend. A 4.0 mm PVC insulated CPC is also included. Determine the minimum size of conduit to contain these conductors.

From Table 6.10:

$$\text{The factor for one 1.0mm cable} = 16$$
$$\text{The factor for one 4.0mm cable} = 43$$
$$\text{The sum of the cable factors} = (10 \times 16) + (1 \times 43)$$
$$= 203$$

From Table 6.11, a 20 mm conduit, 6 m long and containing one bend, has a factor of 233. A 16 mm conduit containing one bend only has a factor of 143 which is less than 203, the sum of the cable factors and, therefore, 20 mm conduit is the minimum size to contain these cables.

Table 6.11 Conduit factors

Length of run (m)	Conduit diameter (mm)											
	16	20	25	32	16	20	25	32	16	20	25	32
	Straight				One bend				Two bends			
3.5	179	290	521	911	162	263	475	837	136	222	404	720
4	177	286	514	900	158	256	463	818	130	213	388	692
4.5	174	282	507	889	154	250	452	800	125	204	373	667
5	171	278	500	878	150	244	442	783	120	196	358	643
6	167	270	487	857	143	233	422	750	111	182	333	600
7	162	263	475	837	136	222	404	720	103	169	311	563
8	158	256	463	818	130	213	388	692	97	159	292	529
9	154	250	452	800	125	204	373	667	91	149	275	500
10	150	244	442	783	120	196	358	643	86	141	260	474

Trunking installations

A **trunking** is an enclosure provided for the protection of cables which is normally square or rectangular in cross-section, having one removable side. Trunking may be thought of as a more accessible conduit system and for industrial and commercial installations it is replacing the larger conduit sizes. A trunking system can have great flexibility when used in conjunction with conduit; the trunking forms the background or framework for the installation, with conduits running from the trunking to the point controlling the current-using apparatus. When an alteration or extension is required it is easy to drill a hole in the side of the trunking and run a conduit to the new point. The new wiring can then be drawn through the new conduit and the existing trunking to the supply point.

Trunking is supplied in 3 m lengths and various cross-sections measured in millimetres from 50 × 50 up to 300 × 150. Most trunking is available in either steel or plastic.

Metallic trunking

Metallic trunking is formed from mild steel sheet, coated with grey or silver enamel paint for internal use or a hot-dipped galvanized coating where damp conditions might be encountered and made to a specification defined by BS EN 500 85. A wide range of accessories is available, such as 45° bends, 90° bends, three- and four-way junctions, for speedy on-site assembly. Alternatively, bends may be fabricated in lengths of trunking, as shown in Fig. 6.24. This may be necessary or more convenient if a bend or set is non-standard, but it does take more time to fabricate bends than merely to bolt on standard accessories. Insulated non- sheathed cables are permitted in a trunking system which provides at least the degree of protection IPXXD (which means total protection) or IP4X which means protection from a solid object greater than 1.0 mm such as a thin wire or strip. For site fabricated joints such as that shown in Fig. 6.24, the installer must confirm that the completed item meets at least IPXXD (IET Regulation 521.10).

When fabricating bends the trunking should be supported with wooden blocks for sawing and filing, in order to prevent the sheet-steel from vibrating or

Definition

A *trunking* is an enclosure provided for the protection of cables which is normally square or rectangular in cross-section, having one removable side. Trunking may be thought of as a more accessible conduit system.

Definition

Metallic trunking is formed from mild steel sheet, coated with grey or silver enamel paint for internal use or a hot-dipped galvanized coating where damp conditions might be encountered.

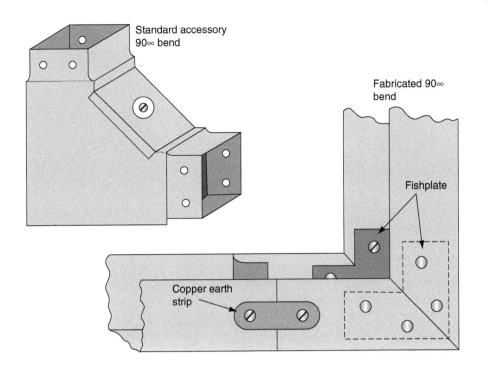

Figure 6.24 Alternative trunking bends.

becoming deformed. Fish plates must be made and riveted or bolted to the trunking to form a solid and secure bend. When manufactured bends are used, the continuity of the earth path must be ensured across the joint by making all fixing screw connections very tight, or fitting a separate copper strap between the trunking and the standard bend. The trunking installation must be treated as an exposed conductive part and be properly earthed in accordance with Regulation 411.3.1.1. Care must be taken to provide reliable earth continuity and an adequate earth fault current path by making all joints electrically and mechanically secure. If an earth continuity test on the trunking is found to be unsatisfactory, an insulated CPC must be installed inside the trunking. The size of the protective conductor will be determined by the largest cable contained in the trunking, as described by Table 54.7 of the IET Regulations. If the circuit conductors are less than 16 mm², then a 16 mm² CPC will be required.

Non-metallic trunking

Trunking and trunking accessories are also available in high-impact PVC. The accessories are usually secured to the lengths of trunking with a PVC solvent adhesive. PVC trunking, like PVC conduit, is easy to install and is non-corrosive. A separate CPC will need to be installed and non-metallic trunking may require more frequent fixings because it is less rigid than metallic trunking. All trunking fixings should use round-headed screws to prevent damage to cables since the thin sheet construction makes it impossible to countersink screw heads.

The new 18th Edition fire proof support for non-metallic cable systems (Regulations 521.10.202) makes it impossible for us to use non metallic cable clips, cable ties or **plastic trunking** as the only means of support for a PVC wiring system. **The regulation tells us that where non- metallic cable systems are used, a suitable means of fire resistant support and retention must be used** to prevent cables falling down in the event of a fire. Note 4 to this Regulation advises that suitably spaced steel or copper clips, saddles or ties are examples which will meet this requirement. Electrical wholesalers are now able

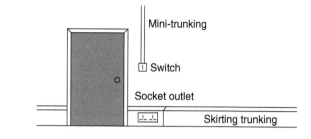

Figure 6.25 Typical installation of skirting trunking and mini-trunking.

to supply metal insert fixings for all standard plastic trunking systems which meet the regulation requirements.

Mini-trunking

Mini-trunking is very small PVC trunking, ideal for surface wiring in domestic and commercial installations such as offices. The trunking has a cross-section of 16 × 16mm, 25 × 16mm, 38 × 16mm or 38 × 25mm and is ideal for switch drops or for housing auxiliary circuits such as telephone or audio equipment wiring. The modern square look in switches and sockets is complemented by the mini-trunking which is very easy to install (see Fig. 6.25).

Skirting or dado trunking

Skirting trunking is a trunking manufactured from PVC or steel in the shape of a skirting board and is frequently used in commercial buildings such as hospitals, laboratories and offices. The trunking is fitted around the walls of a room at either the skirting board level or at the working surface level where it is called dado trunking and contains the wiring for socket outlets, computer and telephone points which are mounted on the lid, as shown in Fig. 6.25.

Where any trunking passes through walls, partitions, ceilings or floors, short lengths of lid should be fitted so that the remainder of the lid may be removed later without difficulty. Any damage to the structure of the buildings must be made good with mortar, plaster or concrete in order to prevent the spread of fire. Fire barriers must be fitted inside the trunking every 5m, or at every floor level or room-dividing wall if this is a shorter distance, as shown in Fig. 6.26(a).

Where trunking is installed vertically, the installed conductors must be supported so that the maximum unsupported length of non-sheathed cable does not exceed 5m. Figure 6.26(b) shows cables woven through insulated pin supports, which is one method of supporting vertical cables.

PVC insulated cables are usually drawn into an erected conduit installation or laid into an erected trunking installation. Table E4 of the *On Site Guide* only gives factors for conduits up to 32mm in diameter, which would indicate that conduits larger than this are not in frequent or common use. Where a cable enclosure greater than 32mm is required because of the number or size of the conductors, it is generally more economical and convenient to use trunking.

Trunking capacities

The **ratio** of the space occupied by all the cables in a conduit or trunking to the whole space enclosed by the conduit or trunking is known as the **space factor**. Where sizes and types of cable and trunking are not covered by the tables in the

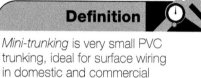

Definition

Mini-trunking is very small PVC trunking, ideal for surface wiring in domestic and commercial installations such as offices.

Definition

Skirting trunking is a trunking manufactured from PVC or steel in the shape of a skirting board and is frequently used in commercial buildings such as hospitals, laboratories and offices.

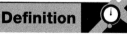

Definition

The *ratio* of the space occupied by all the cables in a conduit or trunking to the whole space enclosed by the conduit or trunking is known as the *space factor*.

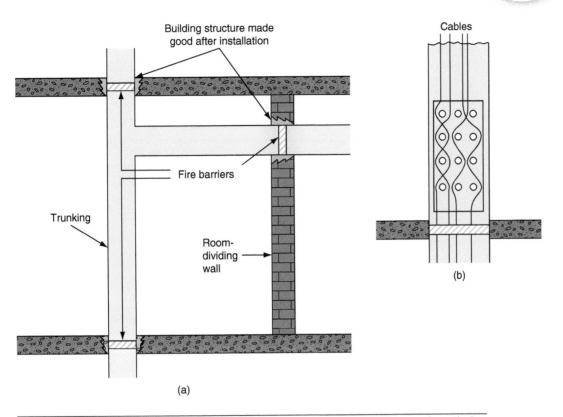

Figure 6.26 Installation of trunking: (a) fire barriers in trunking and (b) cable supports in vertical trunking.

Table 6.12 Trunking cable factors

Type of conductor	Conductor CSA (mm²)	PVC Cable factor	Thermosetting Cable factor
Solid	1.5	8.0	8.6
	2.5	11.9	11.9
Stranded	1.5	8.6	9.6
	2.5	12.6	13.9
	4	16.6	18.1
	6	21.2	22.9
	10	35.3	36.3

On Site Guide, a space factor of 45% must not be exceeded. This means that the cables must not fill more than 45% of the space enclosed by the trunking. The tables take this factor into account.

To calculate the size of trunking required to enclose a number of cables:

- Identify the cable factor for the particular size of conductor, see Table 6.12.
- Multiply the cable factor by the number of conductors to give the sum of the cable factors.
- Consider the factors for trunking shown in Table 6.13. The correct size of trunking to accommodate the cables is that trunking which has a factor equal to, or greater than, the sum of the cable factors.

Table 6.13 Trunking factors

Dimensions of trunking (mm × mm)	Factor
50 × 38	767
50 × 50	1037
75 × 25	738
75 × 38	1146
75 × 50	1555
75 × 75	2371
100 × 25	993
100 × 38	1542
100 × 50	2091
100 × 75	3189
100 × 100	4252
150 × 38	2999
150 × 50	3091
150 × 75	4743
150 × 100	6394
150 × 150	9697

Example 3

Calculate the minimum size of trunking required to accommodate the following single-core PVC cables:

20 × 1.5 mm solid conductors
20 × 2.5 mm solid conductors
21 × 4.0 mm stranded conductors
16 × 6.0 mm stranded conductors

From Table 6.12, the cable factors are:

for 1.5 mm solid cable – 8.0
for 2.5 mm solid cable – 11.9
for 4.0 mm stranded cable – 16.6
for 6.0 mm stranded cable – 21.2

The sum of the cable terms is:

$(20 \times 8.0) + (20 \times 11.9) + (21 \times 16.6) + (16 \times 21.2) = 1085.8$. From Table 6.13, 75 × 38 mm trunking has a factor of 1146 and, therefore, the minimum size of trunking to accommodate these cables is 75 × 38 mm, although a larger size, say, 75 × 50 mm, would be equally acceptable if this was more readily available as a standard stock item.

Segregation of circuits

Where an installation comprises a mixture of low-voltage and very low-voltage circuits such as mains lighting and power, fire alarm and telecommunication

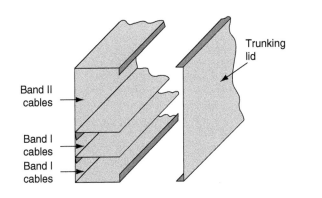

Band II cables

Band I cables

Band I cables

Trunking lid

Figure 6.27 Segregation of cables in trunking.

circuits, they must be separated or *segregated* to prevent electrical contact (IET Regulation 528.1).

For the purpose of these regulations various circuits are identified by one of two bands as follows:

- Band I: telephone, radio, bell, call and intruder alarm circuits, emergency circuits for fire alarm and emergency lighting.
- Band II: mains voltage circuits.

When Band I circuits are insulated to the same voltage as Band II circuits, they may be drawn into the same compartment.

When trunking contains rigidly fixed metal barriers along its length, the same trunking may be used to enclose cables of the separate bands without further precautions, provided that each band is separated by a barrier, as shown in Fig. 6.27.

Multi-compartment PVC trunking cannot provide band segregation since there is no metal screen between the bands. This can only be provided in PVC trunking if screened cables are drawn into the trunking.

Cable tray installations

Cable tray is a sheet-steel channel with multiple holes. The most common finish is hot-dipped galvanized but PVC-coated tray is also available. It is used extensively on large industrial and commercial installations for supporting MI and SWA cables which are laid on the cable tray and secured with cable ties through the tray holes.

Cable tray should be adequately supported during installation by brackets which are appropriate for the particular installation. The tray should be bolted to the brackets with round-headed bolts and nuts, with the round head inside the tray so that cables drawn along the tray are not damaged.

The tray is supplied in standard widths from 50 to 900 mm, and a wide range of bends, tees and reducers is available. Figure 6.28 shows a factory-made 90° bend at B. The tray can also be bent using a cable tray bending machine to create bends such as that shown at A in Fig. 6.28. The installed tray should be securely bolted with round-headed bolts where lengths or accessories are attached, so that there is a continuous earth path which may be bonded to an electrical earth. The whole tray should provide a firm support for the cables and therefore the tray fixings must be capable of supporting the weight of both the tray and cables.

Definition

Cable tray is a sheet-steel channel with multiple holes. The most common finish is hot-dipped galvanized but PVC-coated tray is also available. It is used extensively on large industrial and commercial installations for supporting MI and SWA cables which are laid on the cable tray and secured with cable ties through the tray holes.

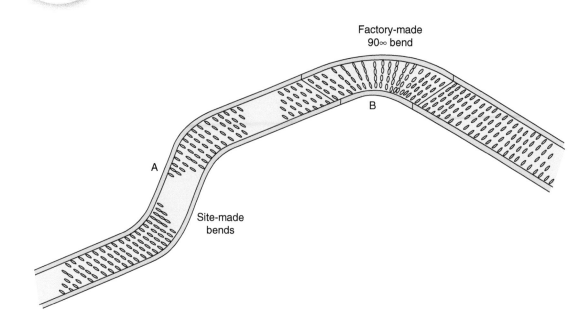

Figure 6.28 Cable tray with bends.

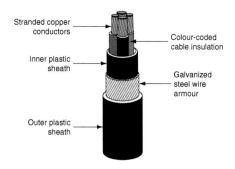

Cable basket

Cable basket installations

Cable basket is becoming very popular for commercial and industrial installations. It is made from steel wire into a basket channel with sides.

Cable basket allows maximum airflow around the cables which are laid into the basket without fixing.

Cable basket requires similar installation techniques to cable tray and should be adequately supported.

PVC/SWA cable installations

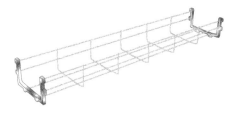
Definition

Steel wire armoured PVC insulated cables are now extensively used on industrial installations and often laid on cable tray.

Steel wire armoured PVC insulated cables are now used extensively on industrial installations and often laid on cable tray. This type of installation has the advantage of flexibility, allowing modifications to be made speedily as the need arises. The cable has a steel wire armouring giving mechanical protection and permitting it to be laid directly in the ground or in ducts, or it may be fixed directly or laid on a cable tray or basket. Figure 6.16 shows a PVC/SWA cable.

It should be remembered that when several cables are grouped together the current rating will be reduced according to the correction factors given in Appendix 4 (Table 4C1) of the IET Regulations.

The cable is easy to handle during installation, is pliable and may be bent to a radius of eight times the cable diameter. The PVC insulation would be damaged if installed in ambient temperatures over 70°C or below 0°C, but once installed the cable can operate at low temperatures.

The cable is terminated with a simple gland which compresses a compression ring on to the steel wire armouring to provide the earth continuity between the switchgear and the cable.

Stranded copper conductors

Colour-coded cable insulation

Inner plastic sheath

Galvanized steel wire armour

Outer plastic sheath

SWA cable.

MI cable installations

Mineral insulated cables are available for general wiring as:

- light-duty MI cables for voltages up to 600 V and sizes from 1.0 to 10 mm^2;
- heavy-duty MI cables for voltages up to 1000 V and sizes from 1.0 to 150 mm^2.

Figure 6.17 shows an MI cable and termination.

The cables are available with bare sheaths or with a PVC oversheath. The cable sheath provides sufficient mechanical protection for all but the most severe situations, where it may be necessary to fit a steel sheath or conduit over the cable to give extra protection, particularly near floor level in some industrial situations.

The cable may be laid directly in the ground, in ducts, on cable tray or clipped directly to a structure. It is not affected by water, oil or the cutting fluids used in engineering and can withstand very high temperatures or even fire. The cable diameter is small in relation to its current carrying capacity and it should last indefinitely if correctly installed because it is made from inorganic materials. These characteristics make the cable ideal for Band I emergency circuits, boiler houses, furnaces, petrol stations and chemical plant installations.

The cable is supplied in coils and should be run off during installation and not spiralled off, as described in Fig. 6.23 for conduit. The cable can be work hardened if over-handled or over-manipulated. This makes the copper outer sheath stiff and may result in fracture. The outer sheath of the cable must not be penetrated, otherwise moisture will enter the magnesium oxide insulation and lower its resistance. To reduce the risk of damage to the outer sheath during installation, cables should be straightened and formed by hammering with a hide hammer or a block of wood and a steel hammer. When bending MI cables the radius of the bend should not cause the cable to become damaged and clips should provide adequate support (IET Regulation 522.8.5); see Table 6.9 of this chapter.

The cable must be prepared for termination by removing the outer copper sheath to reveal the copper conductors. This can be achieved by using a rotary stripper tool or, if only a few cables are to be terminated, the outer sheath can be removed with side cutters, peeling off the cable in a similar way to peeling the skin from a piece of fruit with a knife. When enough conductor has been revealed, the outer sheath must be cut off square to facilitate the fitting of the sealing pot, and this can be done with a ringing tool. All excess magnesium oxide powder must be wiped from the conductors with a clean cloth. This is to prevent moisture from penetrating the seal by capillary action.

Cable ends must be terminated with a special seal to prevent the entry of moisture. Figure 6.17 shows a brass screw-on seal and gland assembly, which allows termination of the MI cables to standard switchgear and conduit fittings. The sealing pot is filled with a sealing compound, which is pressed in from one side only to prevent air pockets from forming, and the pot closed by crimping home the sealing disc with an MI crimping tool. Such an assembly is suitable for working temperatures up to 105°C. Other compounds or powdered glass can increase the working temperature up to 250°C.

The conductors are not identified during the manufacturing process and so it is necessary to identify them after the ends have been sealed. A simple continuity or polarity test, as described in Chapter 5, can identify the conductors which are then sleeved or identified with coloured markers.

Connection of MI cables can be made directly to motors, but to absorb the vibrations a 360° loop should be made in the cable just before the termination. If excessive vibration is expected, the MI cable should be terminated in a conduit through box and the final connection made by flexible conduit.

Copper MI cables may develop a green incrustation or patina on the surface, even when exposed to normal atmospheres. This is not harmful and should not be removed. However, if the cable is exposed to an environment which might encourage corrosion, an MI cable with an overall PVC sheath should be used.

FP 200 cable

FP 200 cable is similar in appearance to an MI cable in that it is a circular tube, or the shape of a pencil, and is available with a red or white sheath. However, it is much simpler to use and terminate than an MI cable.

The cable is available with either solid or stranded conductors that are insulated with 'insudite', a fire-resistant insulation material. The conductors are then screened by wrapping an aluminium tape around the insulated conductors; that is, between the insulated conductors and the outer sheath. This aluminium tape screen is applied metal side down and in contact with the bare CPC.

The sheath is circular and made of a robust thermoplastic low-smoke, zero-halogen material.

The cable is as easy to use as a PVC insulated and sheathed cable. No special terminations are required. The cable may be terminated through a grommet into a knock-out box or terminated through a simple compression gland.

The cable is a fire-resistant cable, intended primarily for use in fire alarms and emergency lighting installations or it may be embedded in plaster.

Terminating and connecting conductors

The entry of a cable end into an accessory, enclosure or piece of equipment is what we call a **termination**. Section 526 of the IET Regulations tells us that:

1 Every connection between conductors and equipment shall be durable, provide electrical continuity and mechanical strength and protection.
2 Every termination and joint in a live conductor shall be made within a suitable accessory, piece of equipment or enclosure that complies with the appropriate product standard.
3 Every connection shall be accessible for inspection, testing and maintenance.
4 The means of connection shall take account of the number and shape of the wires forming the conductor.
5 The connection shall take account of the cross-section of the conductor and the number of conductors to be connected.
6 The means of connection shall take account of the temperature attained in normal service.
7 There must be no mechanical strain on the conductor connections.

There is a wide range of suitable means of connecting conductors and we shall look at these in a moment. Whatever method is used to connect live conductors, the connection must be contained in an enclosed compartment such as an accessory; for example, a switch or socket box or a junction box. Alternatively, an equipment enclosure may be used; for example, a motor enclosure or an enclosure partly formed by non-combustible building material (IET Regulation 526.5). This is because faulty joints and terminations in live conductors can attain very high temperatures due to the effects of resistive heating. They might also emit arcs, sparks or hot particles with the consequent risk of fire or other harmful thermal effects to adjacent materials.

Definition

The entry of a cable end into an accessory, enclosure or piece of equipment is what we call a *termination*.

Figure 6.29 Electrical terminal control connector.

Types of terminal connection

Junction boxes

Junction boxes are probably the most popular method of making connections in domestic properties. Brass terminals are fixed inside a bakelite container. The two important factors to consider when choosing a junction box are the number of terminals required and the current rating. Socket outlet junction boxes have larger brass terminals than lighting junction boxes. See Fig. 6.30.

Strip connectors

Strip connectors or a chocolate block is a very common method of connecting conductors. The connectors are mounted in a moulded plastic block in strips of 10 or 12. The conductors are inserted into the block and secured with the grub-screw. In order that the conductors do not become damaged, the screw connection must be firm but not overtightened. The size used should relate to the current rating of the circuit. Figure 6.31 shows a strip connector.

Pillar terminal

A pillar terminal is a brass pillar with a hole through the side into which the conductor is inserted and secured with a set-screw. If the conductor is small in relation to the hole it should be doubled back. In order that the conductor does not become damaged, the screw connection should be tight but not overtightened. Figure 6.31 shows a pillar terminal.

Screwhead, nut and washer terminals

The conductor being terminated is formed into an eye, as shown in Fig. 6.31. The eye should be slightly larger than the screw shank but smaller than the outside diameter of the screwhead, nut or washer. The eye should be placed on the

Key fact

Junction boxes are probably the most popular method of making connections in domestic properties.

Key fact

Strip connectors or a chocolate block is a very common method of connecting conductors.

Figure 6.30 Standard junction boxes with screw terminals, junction box must be fixed and cables clamped.

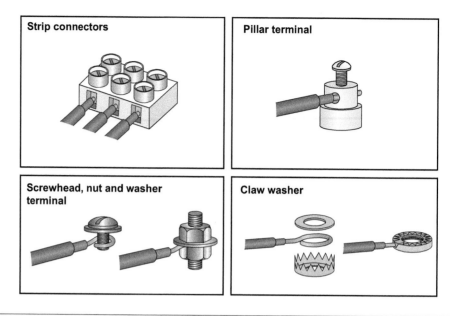

Figure 6.31 Types of terminal.

screw shank in such a way that the rotation of the screwhead or nut will tend to close the joint in the eye.

Claw washers

In order to avoid inappropriate separation or spreading of individual wires of multiwire, claw washers are used to obtain a good sound connection. The looped conductor is laid in the pressing as shown in Fig. 6.31, a plain washer is placed on top of the conductor and the metal points folded over the washer. When terminating very fine multiwire conductors, see also 526.9 of the IET Regulations which gives us the following advice:

1 To avoid separation or spreading of individual wires, suitable terminals must be used or the conductor ends treated; for example, by enclosing the individual wires of multiwire in a brass ferrule or claw washer, as shown in Fig 6.31.

2 Soldering or tinning of the whole conductor end of multiwire is not permitted if screw terminals are used.

3 Soldered or tinned conductor ends are not permissible at connection and junction points which may be subject in service to relative movement or vibration.

Crimp terminals

Crimp terminals are made of tinned sheet copper. The chosen crimp terminal is slipped over the end of the conductor and crimped with the special crimping tool. This type of connection is very effective for connecting protective bonding conductors to approved earth clamps.

Soldered joints or compression joints

Although the soldering of large underground cables is still common today, joints up to about 100A are now usually joined with a compression joint. This uses the same principle as for the crimp termination above; it is just a little larger.

If a large SWA cable must be connected and the joint placed in a position which will be inaccessible for future inspection and testing, then a compression joint encased in a resin compound filled jacket will probably provide a solution.

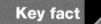
Key fact

Whatever method is used to make the connection in conductors, the connection must be both electrically and mechanically sound if we are to avoid high-resistance joints, corrosion and erosion at the point of termination.

Regulation 526.3 tells us that every connection must be accessible for inspection and testing except for the following:

1 A joint which is designed to be buried underground such as the one described above.

2 A compound filled or encapsulated joint.

3 A maintenance-free junction box marked with the symbol MF, as shown in Fig. 6.32.

The introduction of this maintenance-free junction box was a small but important change made by Amendment No 1: 2011 of the IET Regulations.

There has always been a debate as to when a junction box is accessible for inspection and testing. Is it accessible when installed under floorboards? Is a screwed down floorboard accessible? Does a fitted carpet make a difference, and who will know where it is?

A maintenance-free junction box does not have to be accessible for inspection and testing, and this new junction box will provide a solution to those difficult and unavoidable situations where there is doubt as to whether a junction box is inaccessible.

Whatever method is used to make the connection in conductors, the connection must be both electrically and mechanically sound if we are to avoid high-resistance joints, corrosion and erosion at the point of termination.

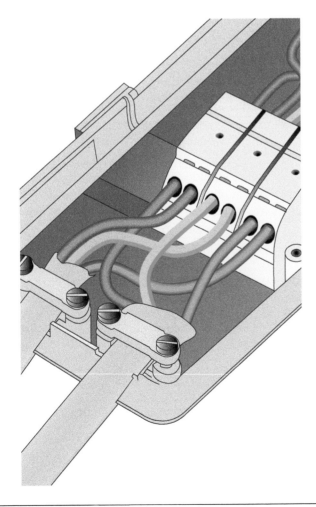

Figure 6.32 A maintenance-free junction box.

Figure 6.33 Both conductors and insulators are found in electrical cable.

Safe terminations and connections

To ensure that all electrical terminations and connections are safe, the installing electrician should give consideration to the following good practice points:

- all connections must be both electrically and mechanically secure;
- all connections must be long lasting and not fail quickly;
- the method of connection must take account of:
  (i) the size of conductor and, therefore, the current carrying capacity of that conductor;
  (ii) the material of the conductor; copper is a soft metal but aluminium is softer;
  (iii) the number of conductors being connected;
  (iv) the temperature to be attained at the point of connection in normal service;
  (v) the provision of adequate locking arrangements in situations subject to vibration.
- every connection must remain accessible for inspection and testing unless designed to be maintence free;
- every connection in a live conductor must be made within:
  (i) a suitable accessory such as a switch, socket ceiling rose or joint box;
  (ii) an equipment enclosure such as a luminaire; or
  (iii) a non-combustible enclosure designed for this purpose.
- there must be no mechanical strain put on the conductors or connections.

Section 526 of the IET Regulations deals with electrical connections.

Table 6.14 shows a comparison of the old and new cable colours.

132.8 Protective equipment

The consumer's mains equipment must provide protection against overcurrent; that is, a current exceeding the rated value (IET Regulation 430.3). Fuses provide overcurrent protection when situated in the live conductors; they must not be connected in the neutral conductor. Circuit-breakers may be used in place of fuses, in which case the circuit-breaker may also provide the means of isolation, although a further means of isolation is usually provided so that maintenance can be carried out on the circuit-breakers themselves (see IET Regulation 537).

Table 6.14 Example of conductor marking at the interface for additions and alterations to an a.c. installation identified with the old cable colours

Function	Old conductor		New conductor	
	Colour	Marking	Marking	Colour
Phase 1 of a.c.	Red	L1	L1	Brown[1]
Phase 2 of a.c.	Yellow	L2	L2	Black[1]
Phase 3 of a.c.	Blue	L3	L3	Grey[1]
Neutral of a.c.	Black	N	N	Blue
Protective conductor	Green and yellow			Green and yellow

[1]Three single-core cables with insulation of the same colour may be used if identified at the terminations.

When selecting a protective device we must give consideration to the following factors:

- the prospective fault current;
- the circuit load characteristics;
- the current-carrying capacity of the cable;
- the disconnection time requirements for the circuit.

The essential requirements for a device designed to protect against overcurrent are:

- it must operate automatically under fault conditions;
- have a current rating matched to the circuit design current;
- have a disconnection time which is within the design parameters;
- have an adequate fault-breaking capacity;
- be suitably located and identified.

We will look at these requirements below.

An overcurrent may be an overload current, or a short-circuit current. An **overload current** can be defined as a current which exceeds the rated value in an otherwise healthy circuit. Overload currents usually occur because the circuit is abused or because it has been badly designed or modified. A **short-circuit** is an overcurrent resulting from a fault of negligible impedance connected between conductors. Short-circuits usually occur as a result of an accident which could not have been predicted before the event.

An overload may result in currents of two or three times the rated current flowing in the circuit. Short-circuit currents may be hundreds of times greater than

> **Definition**
>
> An *overload current* can be defined as a current which exceeds the rated value in an otherwise healthy circuit.
>
> A *short-circuit* is an overcurrent resulting from a fault of negligible impedance connected between conductors.

Figure 6.34 MCB board.

the rated current. In both cases the basic requirements for protection are that the fault currents should be interrupted quickly and the circuit isolated safely before the fault current causes a temperature rise or mechanical effects which might damage the insulation, connections, joints and terminations of the circuit conductors or their surroundings (IET Regulation 131).

The selected protective device should have a current rating which is not less than the full load current of the circuit but which does not exceed the cable current rating. The cable is then fully protected against both overload and short-circuit faults (IET Regulation 435.1). Devices which provide overcurrent protection are:

- High breaking capacity (HBC) fuses to BS 88-2:2010. These are for industrial applications, having a maximum fault capacity of 80 kA.
- Cartridge fuses to BS 88-3:2010. These are used for a.c. circuits on industrial and domestic installations, having a fault capacity of about 30 kA.
- Cartridge fuses to BS 1362. These are used in 13 A plugs and have a maximum fault capacity of about 6 kA.
- Semi-enclosed fuses to BS 3036. These were previously called re-wirable fuses and are used mainly on domestic installations having a maximum fault capacity of about 4 kA.
- MCBs to BS EN 60898. These are miniature circuit-breakers (MCBs) which may be used as an alternative to fuses for some installations. The British Standard includes ratings up to 100 A and maximum fault capacities of 9 kA. They are graded according to their instantaneous tripping currents; that is, the current at which they will trip within 100 ms. This is less than the time taken to blink an eye.
- The 18th Edition of the IET Wiring Regulations tells us at Note 5 of Table 537.4 that circuit protective devices and RCDs are not intended for frequent load switching. However, infrequent switching of MCBs is permissible for the purpose of isolation or emergency switching.

Semi-enclosed fuses (BS 3036)

The semi-enclosed fuse consists of a fuse wire, called the fuse element, secured between two screw terminals in a fuse carrier. The fuse element is connected in series with the load and the thickness of the element is sufficient to carry the normal rated circuit current. When a fault occurs an overcurrent flows and the fuse element becomes hot and melts or 'blows'.

By definition a fuse is the weakest link in the circuit. Under fault conditions it will melt when an overcurrent flows, protecting the circuit conductors from damage.

This type of fuse is illustrated in Fig. 6.35. The fuse element should consist of a single strand of plain or tinned copper wire having a diameter appropriate to the current rating of the fuse. *This type of fuse was very popular in domestic installations, but less so these days because of the advantages of MCBs.*

Cartridge fuses (BS 88-3:2010, previously BS 1361)

The cartridge fuse breaks a faulty circuit in the same way as a semi-enclosed fuse, but its construction eliminates some of the disadvantages experienced with an open-fuse element. The fuse element is encased in a glass or ceramic tube and secured to end-caps which are firmly attached to the body of the fuse so that they do not blow off when the fuse operates. Cartridge fuse construction

Definition

By definition a *fuse* is the weakest link in the circuit. Under fault conditions it will melt when an overcurrent flows, protecting the circuit conductors from damage.

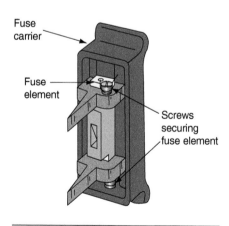

Fuse carrier

Fuse element

Screws securing fuse element

Figure 6.35 A semi-enclosed fuse.

is illustrated in Fig. 6.36. With larger size cartridge fuses, lugs or tags are sometimes brazed on the end-caps to fix the fuse cartridge mechanically to the carrier. They may also be filled with quartz sand to absorb and extinguish the energy of the arc when the cartridge is brought into operation.

Miniature circuit-breakers (BS EN 60898)

The disadvantage of all fuses is that when they have operated they must be replaced. An MCB overcomes this problem since it is an automatic switch which opens in the event of an excessive current flowing in the circuit and can be closed when the circuit returns to normal.

An MCB of the type shown in Fig. 6.37 incorporates a thermal and magnetic tripping device. The load current flows through the thermal and the electromagnetic devices in normal operation but under overcurrent conditions they activate and trip the MCB.

The circuit can be restored when the fault is removed by pressing the ON toggle. This latches the various mechanisms within the MCB and 'makes' the switch contact. The toggle switch can also be used to disconnect the circuit for maintenance or isolation or to test the MCB for satisfactory operation.

Characteristics of MCBs

MCB Type B to BS EN 60898 will trip instantly at between three and five times its rated current and is also suitable for domestic and commercial installations.

MCB Type C to BS EN 60898 will trip instantly at between five and ten times its rated current. It is more suitable for highly inductive commercial and industrial loads.

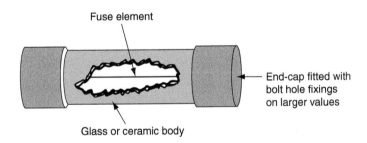

Figure 6.36 Cartridge fuse.

Figure 6.37 MCBs – B Breaker, fits Wylex standard consumer unit (courtesy of Wylex).

MCB Type D to BS EN 60898 will trip instantly at between ten and 25 times its rated current. It is suitable for welding and X-ray machines where large inrush currents may occur.

Residual current devices (RCDs)

The object of the Regulations concerning these devices (411.3.2 to 411.3.4) is to remove an earth fault current very quickly, less then 0.4 secs, for all final circuits not exceeding 32A, and to limit the voltage which might appear on any exposed metal parts under fault conditions to not more than 50 Volt. The regulations recognise RCDs as 'additional' protection in the event of a failure of Basic or Fault protection or the carelessness of the user of the installation (Reg 415.1.1). The circuit diagram and further information can be found in Chapter 3 and Fig 3.39. Wherever RCDs are installed a label shall be fixed near to each device stating '**this device must be tested 6 monthly**'. Note the test period was 3 monthly in the 17th edition of the Regulations (514.12.2).

RCBO

A residual current operated circuit-breaker with integral overcurrent protection (RCBO) provides protection against overload and/or short-circuit. RCBOs give the combined protection of an MCB and an RCD in one device.

The operating principle of all these overcurrent protective devices is discussed in Chapter 3 of this book.

AFDD (Arc Fault Detection Devices)

The 18th edition of the IET Regulations at 532.6 and 421.1.7 has introduced a new topic, Arc Fault Detection Devices (AFDD). These are compulsory in EU Countries but only recommended in this country by the IET and the British standards. They are recognized as giving additional protection against fires caused by arc faults. Arc faults can be formed by cable insulation defects, damage to cables by impact or penetration by nails and screws, and loose terminal connections. AFDDs detect faults which MCBs and RCDs cannot detect. They are installed at the origin of the final circuit to be protected.

Installing overcurrent protective devices

The general principle to be followed is that a protective device must be placed at a point where a reduction occurs in the current-carrying capacity of the circuit conductors (IET Regulations 433.2 and 434.2). A reduction may occur because of a change in the size or type of conductor or because of a change in the method of installation or a change in the environmental conditions. The only exceptions to this rule are where an overload protective device opening a circuit might cause a greater danger than the overload itself – for example, a circuit feeding an overhead electromagnet in a scrapyard.

Fault protection

The overcurrent protective device protecting circuits not exceeding 32A shall have a disconnection time not exceeding 0.4 s (IET Regulation 411.3.2.2).

The IET Regulations permit us to assume that where an overload protective device is also intended to provide short-circuit protection, and has a rated breaking capacity greater than the prospective short-circuit current at the point of its installation, the conductors on the load side of the protective device are considered to be adequately protected against short-circuit currents without further proof. This is because the cable rating and the overload rating of the device are compatible. However, if this condition is not met or if there is some doubt, it must be verified that fault currents will be interrupted quickly before they can cause a dangerously high temperature rise in the circuit conductors. IET Regulation 434.5.2 provides an equation for calculating the maximum operating time of the protective device to prevent the permitted conductor temperature rise from being exceeded as follows:

$$t = \frac{k^2 S^2}{I^2} \text{ (s)}$$

where

t = duration time in seconds

S = cross-sectional area of conductor in square millimetres

I = short-circuit r.m.s. current in amperes

k = a constant dependent upon the conductor metal and type of insulation (see Table 43 A of the IET Regulations).

Example

A 10 mm PVC sheathed mineral insulated (MI) copper cable is short-circuited when connected to a 400 V supply. The impedance of the short-circuit path is 0.1 Ω. Calculate the maximum permissible disconnection time and show that a 50 A Type B MCB to BS EN 60898 will meet this requirement.

$$I = \frac{V}{Z} \text{ (A)} \qquad I = \frac{400\,V}{0.1\Omega} = 4000\,A$$

$$\therefore \quad \text{Fault current} = 4000\,A$$

For PVC sheathed MI copper cables, Table 43.1 gives a value for k of 115. So,

$$t = \frac{k^2 S^2}{I^2} \text{ (s)}$$

$$\therefore t = \frac{115^2 \times 10^2\,mm^2}{4000\,A} = 82.66 \times 10^{-3}\,s$$

The maximum time that a 4000 A fault current can be applied to this 10 mm^2 cable without dangerously raising the conductor temperature is 82.66 ms. Therefore, the protective device must disconnect the supply to the cable in less than 82.66 ms under short-circuit conditions. Manufacturers' information and Appendix 3 of the IET Regulations give the operating times of protective devices at various short-circuit currents in the form of graphs. Let us come back to this problem in a few moments.

Time/current characteristics of protective devices

Disconnection times for various overcurrent devices are given in the form of a logarithmic graph. This means that each successive graduation of the axis represents a 10 times change over the previous graduation.

These logarithmic scales are shown in the graphs of Figs 6.38 and 6.39. From Fig. 6.38 it can be seen that the particular protective device represented by this characteristic will take 8s to disconnect a fault current of 50A and 0.08s to clear a fault current of 1000A.

Let us now go back to the problem and see if the Type B MCB will disconnect the supply in less than 82.66 ms.

Figure 6.39 shows the time/current characteristics for a Type B MCB to BS EN 60898. This graph shows that a fault current of 4000A will trip the protective device in 20ms. Since this is quicker than 82.66 ms, the 50A Type B MCB is

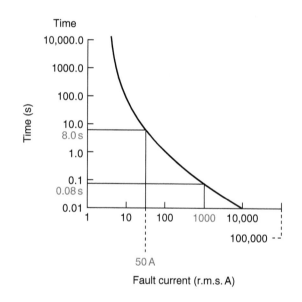

Figure 6.38 Time/current characteristic of an overcurrent protective device.

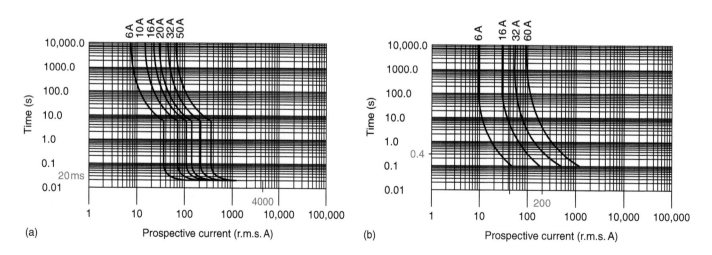

Figure 6.39 Time/current characteristics of (a) a Type B MCB to BS EN 60898 and (b) semi-enclosed fuse to BS 3036.

suitable and will clear the fault current before the temperature of the cable is raised to a dangerous level.

Appendix 3 of the IET Regulations gives the time/current characteristics and specific values of prospective short-circuit current for a number of protective devices.

These indicate the value of fault current which will cause the protective device to operate in the times indicated by IET Regulation 411.

Figures 3.1, 3.2 and 3.3 in Appendix 3 of the IET Regulations deal with fuses and Figs 3.4, 3.5 and 3.6 with MCBs.

It can be seen that the prospective fault current required to trip an MCB in the required time is a multiple of the current rating of the device. The multiple depends upon the characteristics of the particular devices. Thus:

- Type B MCB to BS EN 60898 has a multiple of 5;
- Type C MCB to BS EN 60898 has a multiple of 10;
- Type D MCB to BS EN 60898 has a multiple of 20.

Example

A 6 A Type B MCB to BS EN 60898 which is used to protect a domestic lighting circuit will trip within 0.4 s when 6 A times a multiple of 5, that is 30 A, flows under fault conditions.

Therefore if the earth fault loop impedance is low enough to allow at least 30 A to flow in the circuit under fault conditions, the protective device will operate within the time required by IET Regulation 411.

The characteristics shown in Appendix 3 of the IET Regulations give the specific values of prospective short-circuit current for all standard sizes of protective device.

Effective coordination of protective devices

IET Regulation 536 tells us that in the event of a fault occurring on an electrical installation only the protective device nearest to the fault should operate, leaving other healthy circuits unaffected. A circuit designed in this way would be considered to have effective coordination, or selectivity. Effective coordination, or selectivity, can be achieved by coordinated protection since the speed of operation of the protective device increases as the rating decreases. This can be seen in Fig. 6.39(b). A fault current of 200 A will cause a 16 A semi-enclosed fuse to operate in about 0.1 s, a 32 A semi-enclosed fuse in about 0.4 s and a 60 A semi-enclosed fuse in about 5.0 s. If a circuit is arranged as shown in Fig. 6.40 and a fault occurs on the appliance, effective coordination or selectivity will be achieved because the 16 A fuse will operate more quickly than the other protective devices if they were, for example, all semi-enclosed-type fuses with the characteristics shown in Fig. 6.39(b). In general, when over current protective devices are connected in series, only the device which the designer intended to operate, should operate. This is usually the device closest to the point at which the over current occurs.

Security of supply, and therefore effective coordination or selectivity, is an important consideration for an electrical designer and is also a requirement of IET Regulation 536.

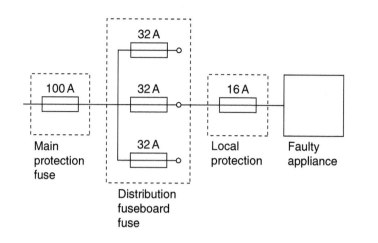

Figure 6.40 Effective coordination of the protective devices.

132.9 Emergency control

A person operating an electrically driven machine must have access to an emergency stop switch so that the machine can be stopped by that person in an emergency. See also 132.15 Isolation and switching below.

132.10 Disconnecting devices

Devices to permit switching and isolation shall be provided for operation, maintenance and repair. See also 132.15 Isolation and switching below.

132.11 Prevention of mutual detrimental influences

The electrical installation must be arranged so as to avoid damage or harm being caused to it (detrimental influences) by the presence of water, mechanical damage, electrolytic corrosion, harmonics, temperature and chemical contamination.

132.12 Accessibility of electrical equipment

Electrical equipment shall be arranged so as to afford sufficient space for the initial installation and future accessibility for operation, inspection, maintenance and repair. See Section 2.1.2.3 and 10.7 of The Electricians Guide to the Building Regulations.

132.13 Documentation for the electrical installation

Every electrical installation must be provided with inspection and testing certificates as required by Part 6 of the IET Regulations. Chapter 5 of this book deals with the inspection testing and certification of electrical installations.

132.14 Protective devices and switches

This regulation tells us that the protective devices discussed earlier in this section at 132.8 must be connected in the line conductor.

132.15 Isolation and switching

Part 4 of the IET Regulations deals with the application of protective measures for safety and Chapter 53 with the regulations for switching devices or switchgear required for protection, isolation and switching of a consumer's installation.

The consumer's main switchgear must be readily accessible to the consumer and be able to:

- isolate the complete installation from the supply;
- protect against overcurrent;
- cut off the current in the event of a serious fault occurring;
- the 18th edition of the IET Regulations has introduced Regulation 421.1.201 requiring switchgear assemblies, including consumer unit enclosures, to be:
  o manufactured from non-combustible or not readily combustible material, or
  o enclosed in a cabinet or enclosure that is constructed of non-combustible or not readily combustible material.

This new Regulation has been introduced to protect against fire that can result from the overheating of connections within consumer units. Overheating may occur because connections have not been properly made or that they have become loose. These new Regulations at 421.1.7 has introduced us to arc fault detection devices (AFDDs), which give extra protection against fires caused by arc faults. The 18th edition recommends us to use them.

The regulations identify four separate types of switching: switching for isolation, switching for mechanical maintenance, emergency switching and functional switching.

Isolation is defined as cutting off the electrical supply to a circuit or item of equipment in order to ensure the safety of those working on the equipment by making dead those parts which are live in normal service.

The purpose of isolation switching is to enable electrical work to be carried out safely on an isolated circuit or piece of equipment. Isolation is intended for use by electrically skilled or supervised persons.

An isolator is a mechanical device which is operated manually and used to open or close a circuit off load. An isolator switch must be provided close to the supply point so that all equipment can be made safe for maintenance. Isolators for motor circuits must isolate the motor and the control equipment, and isolators for discharge lighting luminaires must be an integral part of the luminaire so that it is isolated when the cover is removed or be provided with effective local isolation (IET Regulation 537.2.1.6). Devices which are suitable for isolation are isolation switches, fuse links, circuit-breakers, plugs and socket outlets. They must isolate all live supply conductors and provision must be made to secure the isolation (IET Regulation 537.2.2.4).

Isolation at the consumer's service position can be achieved by a double pole switch which opens or closes all conductors simultaneously. On three-phase supplies the switch need only break the live conductors with a solid link in the neutral, provided that the neutral link cannot be removed before opening the switch.

The **switching for mechanical maintenance** requirements is similar to those for isolation except that the control switch must be capable of switching the full load current of the circuit or piece of equipment.

Definition

Isolation is defined as cutting off the electrical supply to a circuit or item of equipment in order to ensure the safety of those working on the equipment by making dead those parts which are live in normal service.

Visit the companion website for more on this topic.

Definition

The *switching for mechanical maintenance* requirements is similar to those for isolation except that the control switch must be capable of switching the full load current of the circuit or piece of equipment.

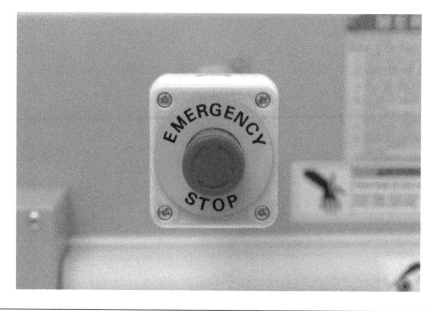

Figure 6.41 Emergency stop control.

The purpose of switching for mechanical maintenance is to enable non-electrical work to be carried out safely on the switched circuit or equipment.

Mechanical maintenance switching is intended for use by skilled but non-electrical persons. Switches for mechanical maintenance must be manually operated, not have exposed live parts when the appliance is opened, must be connected in the main electrical circuit and have a reliable on/off indication or visible contact gap (IET Regulation 537.3.2.2). Devices which are suitable for switching off for mechanical maintenance are switches, circuit-breakers, and plug and socket outlets.

Emergency switching involves the rapid disconnection of the electrical supply by a single action to remove or prevent danger.

The purpose of emergency switching is to cut off the electrical energy *rapidly* to remove an unexpected hazard.

Emergency switching is for use by anyone. The device used for emergency switching must be immediately accessible and identifiable, and be capable of cutting off the full load current (see Fig. 6.41).

Electrical machines must be provided with a means of emergency switching, and a person operating an electrically driven machine must have access to an emergency switch so that the machine can be stopped in an emergency. A remote stop/start or emergency stop arrangement can be added to a direct on-line starter by removing the link shown in Fig 3.29 (in Chapter 3 of this book) to meet the requirements for an electrically driven machine as required by IET Regulation 537.4.2.2. Devices which are suitable for emergency switching are switches, circuit-breakers and contactors. Where contactors are operated by remote control they should *open* when the coil is de-energized; that is, fail safe. Push-buttons used for emergency switching must be coloured red and latch in the stop or off position. They should be installed where danger may arise and be clearly identified as emergency switches. Plugs and socket outlets cannot be considered appropriate for emergency disconnection of supplies.

Functional switching involves the switching on or off, or varying the supply, of electrically operated equipment in normal service.

Definition

Emergency switching involves the rapid disconnection of the electrical supply by a single action to remove or prevent danger.

Definition

Functional switching involves the switching on or off, or varying the supply, of electrically operated equipment in normal service.

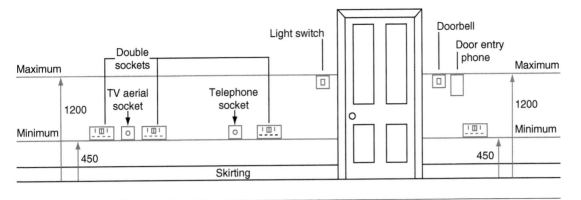

Figure 6.42 Fixing positions of switches and socket outlets.

The purpose of functional switching is to provide control of electrical circuits and equipment in normal service.

Functional switching is for the user of the electrical installation or equipment. The Building Regulations require that switches and sockets be installed so that all persons, including those whose reach is limited, can reach them. The guidance is shown in Fig. 6.42 and applies to all new dwellings but not to rewires. However, these recommendations will undoubtedly 'influence' the rewiring of existing dwellings. The device must be capable of interrupting the total steady current of the circuit or appliance. When the device controls a discharge lighting circuit it must have a current rating capable of switching an inductive load. The regulations acknowledge the growth in the number of electronic dimmer switches being used for the control and functional switching of lighting circuits. The functional switch must be capable of performing the most demanding duty it may be called upon to perform (IET Regulations 537.5.2.1 and 2).

132.16 Additions and alterations to an installation

No additions or alterations shall be made to an existing installation without first of all finding out that the condition of the existing electrical equipment will be adequate for the altered circumstances, including the earthing and bonding arrangements.

Planning the installation of electrical systems

Legal contracts

Before work commences, some form of legal contract should be agreed between the two parties; that is, those providing the work (e.g. the subcontracting electrical company) and those asking for the work to be carried out (e.g. the main building company).

A contract is a formal document which sets out the terms of agreement between the two parties. A standard form of building contract typically contains four sections:

1 *The articles of agreement* – this names the parties, the proposed building and the date of the contract period.
2 *The contractual conditions* – this states the rights and obligations of the parties concerned, for example, whether there will be interim payments for work completed, or a penalty if work is not completed on time.

3 *The appendix* – this contains details of costings, for example, the rate to be paid for extras as daywork, who will be responsible for defects, how much of the contract tender will be retained upon completion and for how long.

4 *The supplementary agreement* – this allows the electrical contractor to recoup any value-added tax paid on materials at interim periods.

In signing the contract, the electrical contractor has agreed to carry out the work to the appropriate standards in the time stated and for the agreed cost. The other party, say, the main building contractor, is agreeing to pay the price stated for that work upon completion of the installation.

If a dispute arises the contract provides written evidence of what was agreed and will form the basis for a solution.

For smaller electrical jobs, a verbal contract may be agreed, but if a dispute arises there is no written evidence of what was agreed and it then becomes a matter of one person's word against another's.

The technical information required to carry out an electrical installation is communicated to electrical personnel in lots of different ways. It comes in the form of:

* *Specifications* – these are details of the client's requirements, usually drawn up by an architect. For example, the specification may give information about the type of wiring system to be employed or detail the type of luminaires or other equipment to be used.
* *Manufacturer's data* – if certain equipment is specified, let's say a particular type of luminaire or other piece of equipment, then the manufacturer's data sheet will give specific instructions for its assembly and fixing requirements. It is always good practice to read the data sheet before fitting the equipment. A copy of the data sheet should also be placed in the job file for the client to receive when the job is completed.
* *Reports and schedules* – a report is the written detail of something that has happened or the answer to a particular question asked by another professional person or the client. It might be the details of some problem on-site.

If the report is internal to the organization, a handwritten report is acceptable, but if the final report will go outside the organization, then it must be more formal and typed.

A **schedule** gives information about a programme or timetable of work; it might be a list or a chart giving details of when certain events will take place; for example, when the electricians will start to do the 'first fix' and how many days it will take. A simple **bar chart** is an easy-to-understand schedule of work that shows how different activities interact on a project. Figure 6.43 shows a bar chart or schedule of work where activity A takes two days to complete and activity B starts at the same time as activity A but carries on for eight days, etc.

If an electrical contract is to be completed successfully, safely and efficiently it will require some planning. A schedule of work is a part of the planning process. It might be a list of what is to be done. It might indicate when certain events need to take place. In drawing up a schedule of work we:

* check the drawings and specifications and produce a list of what has to be done;
* create a logical sequence for what has to be done;
* make a list of special tools and equipment which will be required to complete the project;

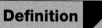

Definition

A *schedule* is a list or programme of planned work.

Definition

Bar chart – the object of any bar chart is to establish the sequence and timing of the various activities involved in the whole job.

Time	Day number													
Activity	1	2	3	4	5	6	7	8	9	10	11	12	13	14
A	▓	▓												
B	▓	▓	▓	▓	▓	▓	▓	▓						
C			▓	▓	▓	▓								
D						▓	▓	▓	▓	▓	▓	▓	▓	▓
E	▓	▓	▓	▓										
F	▓	▓	▓	▓	▓	▓								
G		▓	▓	▓	▓	▓	▓	▓						
H					▓	▓	▓	▓						
I		▓	▓											
J	▓	▓	▓											
K						▓	▓	▓	▓	▓	▓	▓		
L									▓	▓	▓	▓		
M													▓	▓

Figure 6.43 A simple bar chart or schedule of work.

- identify how many operatives or those with special skills will be required at any one time to complete the project;
- identify key dates in the installation process and perhaps put these onto a bar chart to ensure that work is completed on time.

On a large project the contracts manager, project manager and site supervisors may be involved in this process. If the project is small, it may fall to the site electrician to do it all.

A schedule of test results is a written record of the results obtained when carrying out the electrical tests required by Part 6 of the IET Regulations. The schedule brings together all the detailed information about the testing of that particular installation. In the same way, a work schedule brings together the details of the whole project.

- *User instructions* – give information about the operation of a piece of equipment. Manufacturers of equipment provide 'user instructions' and a copy should be placed in the job file for the client to receive when the project is handed over.
- *Job sheets and time sheets* – give 'on-site' information. Job sheets give information about what is to be done and are usually issued by a manager to an electrician. Time sheets are a record of where an individual worker has been spending his time, which job and for how long. This information is used to make up individual wages and to allocate company costs to a particular job. We will look at these again later under the subheading 'on-site documentation'.

Technical information is required by many of the professionals involved in any electrical activity, so who are the key people?
There is often no clear distinction between the duties of the individual employees; each does some of the others' activities.
Responsibilities vary, even by people holding the same job title, and some individuals hold more than one job title. However, let us look at some of the roles and responsibilities of those working in the electrotechnical industry.

1 Design Engineer
- meets with clients and other trade professionals to interpret their requirements
- produces the design specification which enables the cost of the project to be estimated.

2 Estimator/Cost Engineer
- measures the quantities of labour and material necessary to complete the electrical project using the plans and specifications for the project
- from these calculations and the company's fixed costs, a project cost can be agreed.

3 Contracts Manager
- may oversee a number of electrical contracts on different sites
- monitors progress in consultation with the project manager on behalf of the electrical companies
- costs out variations to the initial contract
- may have health and safety responsibilities because he or she has an overview of all company employees and contracts in progress.

4 Project Manager
- responsible for the day-to-day management of one specific contract
- has overall responsibility on that site for the whole electrical installation
- attends site meetings with other trades as the representative of the electrical contractor.

5 Service Manager
- monitors the quality of the service delivered under the terms of the contract
- checks that the contract targets are being met
- checks that the customer is satisfied with all aspects of the project
- focus is customer specific, while the project manager's focus is job specific.

6 Technician
- more office-based than site-based
- carries out surveys of electrical systems
- updates electrical drawings
- obtains quotations from suppliers
- maintains records such as ISO 9000 quality systems
- carries out test inspections and commissioning of electrical installations
- trouble-shooting.

7 Supervisor/Foreman
- will probably be a mature electrician
- responsible for small contracts
- responsible for a small part of a large contract
- leads a small team (e.g. electrician and trainee) installing electrical systems.

8 Operative or Skilled Operative
- carries out the electrical work under the direction and guidance of a supervisor
- demonstrates a high degree of skill and competence in electrical work
- has or is working towards a recognized electrical qualification and status as an electrician, approved electrician or electrical technician.

9 Mechanic/Fitter
- operative who usually has a 'core skill' or 'basic skill' and qualification in mechanical rather than electrical engineering
- in production or process work, he or she would have responsibility for the engineering and fitting aspects of the contract, while the electrician and instrumentation technician would take care of the electrical and instrumentation aspects
- all three operatives must work closely in production and process work
- 'additional skilling' or 'multi-skilling' training produces a more flexible operative for production and process plant operations.

10 Maintenance Manager/Engineer
- responsible for keeping the installed electrotechnical plant and equipment working efficiently
- takes over responsibility from the builders and contractors of maintaining all plant equipment and systems under his or her control
- may be responsible for a hospital or commercial building, university or college complex
- will set up routine and preventative maintenance programmes to reduce possible future breakdowns
- when faults or breakdowns occur he or she will be responsible for the repair using the company's maintenance staff.

11 Contractor
- takes responsibility for the whole project for the client
- may take on a subcontractor to carry out some part of the whole project on a large construction site
- the electrical contractor is usually the subcontractor.

12 Site Agent
- responsible for the smooth running of the whole project and for bringing the contract to a conclusion on schedule and within budget
- the site agent may be nominated by the architect.

13 Customer or Client
- they also are the people ordering the work to be done
- they will pay the final bill that pays everyone's wages.

On-site documentation

A lot of communications between and within larger organizations take place by completing standard forms or sending internal memos. Written messages have the advantage of being 'auditable'. An auditor can follow the paperwork trail to see, for example, who was responsible for ordering certain materials.

When completing standard forms, follow the instructions given and ensure that your writing is legible. Do not leave blank spaces on the form, always specifying 'not applicable' or 'N/A' whenever necessary. Sign or give your name and the date as asked for on the form. Finally, read through the form again to make sure you have answered all the relevant sections correctly.

Internal memos are forms of written communication used within an organization; they are not normally used for communicating with customers or suppliers. Figure 6.44 shows the layout of a typical standard memo form used by Dave Twem to notify John Gall that he has ordered the hammer drill.

FLASH-BANG ELECTRICAL

internal **MEMO**

From _Dave Twem_ To _John Gall_

Subject _Power Tool_ Date _Thurs 11 Aug. 19_

Message

Have today ordered Hammer Drill from P.S. Electrical – should be with you end of next week – Hope this is OK. Dave.

Figure 6.44 Typical standard memo form.

Letters provide a permanent record of communications between organizations and individuals. They may be handwritten for internal use but formal business letters give a better impression of the organization if they are typewritten. A letter should be written using simple, concise language, and the tone of the letter should always be polite even if it is one of complaint. Always include the date of the correspondence. The greeting on a formal letter should be 'Dear Sir/Madam' and conclude with 'Yours faithfully'. A less formal greeting would be 'Dear Mr Smith' and conclude 'Yours sincerely'. Your name and status should be typed below your signature.

Delivery notes

When materials are delivered to site, the person receiving the goods is required to sign the driver's '**delivery note**'. This record is used to confirm that goods have been delivered by the supplier, who will then send out an invoice requesting payment, usually at the end of the month.

The person receiving the goods must carefully check that all the items stated on the delivery note have been delivered in good condition. Any missing or damaged items must be clearly indicated on the delivery note before signing, because, by signing the delivery note the person is saying 'yes, these items were delivered to me as my company's representative on that date and in good condition and I am now responsible for these goods'. Suppliers will replace materials damaged in transit provided that they are notified within a set time period, usually three days. The person receiving the goods should try to quickly determine their condition. Has the packaging been damaged, does the container 'sound' like it might contain broken items? It is best to check at the time of delivery if possible, or as soon as possible after delivery and within the notifiable period. Electrical goods delivered to site should be handled carefully and stored securely until they are installed. Copies of delivery notes are sent to head office so that payment can be made for the goods received.

Time sheets

A **time sheet** is a standard form completed by each employee to inform the employer of the actual time spent working on a particular contract or site. This

Definition

A *delivery note* is used to confirm that goods have been delivered by the supplier, who will then send out an invoice requesting payment.

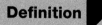

Definition

A *time sheet* is a standard form completed by each employee to inform the employer of the actual time spent working on a particular contract or site.

helps the employer to bill the hours of work to an individual job. It is usually a weekly document and includes the number of hours worked, the name of the job and any travelling expenses claimed. Office personnel require time sheets such as that shown in Fig. 6.45 so that wages can be made up.

TIME SHEET				**FLASH-BANG ELECTRICAL**		

Employee's name (Print) _____

Week ending _____

Day	Job number and/or address	Start time	Finish time	Total hours	Travel time	Expenses
Monday						
Tuesday						
Wednesday						
Thursday						
Friday						
Saturday						
Sunday						

Employee's signature _____ Date _____

Figure 6.45 Typical time sheet.

Definition

A *job sheet* or job card such as that shown in Fig. 6.46 carries information about a job which needs to be done, usually a small job.

Job sheets

A **job sheet** or job card such as that shown in Fig. 6.46 carries information about a job which needs to be done, usually a small job. It gives the name and address of the customer, contact telephone numbers, often a job reference number and a brief description of the work to be carried out. A typical job sheet work description might be:

Job 1 Upstairs lights not working.
Job 2 Funny fishy smell from kettle socket in kitchen.

An electrician might typically have a 'jobbing day' where he picks up a number of job sheets from the office and carries out the work specified.

Job 1, for example, may be the result of a blown fuse which is easily rectified, but the electrician must search a little further for the fault which caused the fuse to blow in the first place. The actual fault may, for example, be a decayed flex on a pendant drop which has become shorted out, blowing the fuse. The pendant drop would be re-flexed or replaced, along with any others in poor condition. The installation would then be tested for correct operation and the customer given an account of what has been done to correct the fault. General information and assurances about the condition of the installation as a whole might be requested and given before setting off to job 2.

The kettle socket outlet at job 2 is probably getting warm and, therefore, giving off that 'fishy' bakelite smell because loose connections are causing the bakelite

JOB SHEET **FLASH-BANG ELECTRICAL**

Job Number ----------------

Customer name --

Address of job --

Contact telephone No. --

Work to be carried out --

Any special instructions/conditions/materials used

Figure 6.46 Typical job sheet.

socket to burn locally. A visual inspection would confirm the diagnosis. A typical solution would be to replace the socket and repair any damage to the conductors inside the socket box. Check the kettle plug top for damage and loose connections. Make sure all connections are tight before reassuring the customer that all is well; then, off to the next job or back to the office.

The time spent on each job and the materials used are sometimes recorded on the job sheet, but alternatively a daywork sheet can be used. This will depend upon what is normal practice for the particular electrical company. This information can then be used to 'bill' the customer for work carried out.

Daywork sheets or variation order

Daywork is one way of recording variations to a contract; that is, work done which is outside the scope of the original contract; If daywork is to be carried out, the site supervisor must first obtain a signature from the client's representative, for example, the architect, to authorize the extra work. A careful record must then be kept on the daywork sheets of all extra time and materials used so that the client can be billed for the extra work. A typical daywork sheet or variation order is shown in Fig. 6.47.

Definition

Daywork is one way of recording variations to a contract.

Reports

On large jobs, the foreman or supervisor is often required to keep a report of the relevant events which happen on the site – for example, how many people from your company are working on-site each day, what goods were delivered, whether there were any breakages or accidents, and records of site meetings attended. Some firms have two separate documents: a site diary to record daily events and a weekly report which is a summary of the week's events extracted from the site diary. The site diary remains on-site and the weekly report is sent to head office to keep managers informed of the work's progress.

The electrical team

The electrical contractor is the subcontractor responsible for the installation of electrical equipment within the building.

Electrical installation activities include:

- installing electrical equipment and systems into new sites or locations;
- installing electrical equipment and systems into buildings that are being refurbished because of change of use;
- installing electrical equipment and systems into buildings that are being extended or updated;
- replacement, repairs and maintenance of existing electrical equipment and systems.

An electrical contracting firm is made up of a group of individuals with varying duties and responsibilities. There is often no clear distinction between the duties of the individuals, and the responsibilities carried out by an employee will vary from one employer to another. If the firm is to be successful, the individuals must work together to meet the requirements of their customers. Good customer relationships are important for the success of the firm and the continuing employment of the employee.

FLASH-BANG ELECTRICAL

VARIATION ORDER OR DAYWORK SHEET

Client name --.

Job number/ref. ---

Date	Labour	Start time	Finish time	Total hours	Office use

Materials quantity	Description	Office use

Site supervisor or F.B. Electrical Representative responsible for carrying out work ------------------

Signature of person approving work and status e.g.

Client ☐   Architect ☐   Q.S. ☐   Main contractor ☐   Clerk of works ☐

Signature --

Figure 6.47 Typical daywork sheet or variation order.

The customer or his representatives will probably see more of the electrician and the electrical trainee than the managing director of the firm and, therefore, the image presented by them is very important. They should always be polite and seen to be capable and in command of the situation. This gives a customer confidence in the firm's ability to meet his or her needs. The electrician and his trainee should be appropriately dressed for the job in hand, which probably means an overall of some kind. Footwear is also important, but sometimes a difficult consideration for a journeyman electrician. For example, if working in a factory, the safety regulations may insist that protective footwear be worn, but rubber boots with a protective insole and toe caps may be most appropriate for a building site. However, neither of these would be the most suitable footwear for an electrician fixing a new light fitting in the home of the managing director!

The electrical installation in a building is often carried out alongside other trades. It makes sound sense to help other trades where possible and to develop good working relationships with other employees.

The employer has the responsibility of finding sufficient work for his employees, paying government taxes and meeting the requirements of the Health and Safety at Work Act described in Chapter 1. The rates of pay and conditions for electricians and trainees are determined by negotiation between the Joint Industry Board and UNITE, the trade union, which will also represent their members in any disputes. Electricians are usually paid at a rate agreed for their grade as an electrician, approved electrician or technician electrician; movements through the grades are determined by a combination of academic achievement and practical experience.

The electrical team will consist of a group of professionals with varying responsibilities, as described earlier on pages 365–366.

Organizing and overseeing work programmes

Smaller electrical contracting firms will know where their employees are working and what they are doing from day to day because of the level of personal contact between the employer, employee and customer.

As a firm expands and becomes engaged on larger contracts, it becomes less likely that there is anyone in the firm with a complete knowledge of the firm's operations, and there arises an urgent need for sensible management and planning skills so that men and materials are on-site when they are required and a healthy profit margin is maintained.

When the electrical contractor is told that he has been successful in tendering for a particular contract he is committed to carrying out the necessary work within the contract period. He must therefore consider:

- by what date the job must be finished;
- when the job must be started if the completion date is not to be delayed;
- how many men will be required to complete the contract;
- when certain materials will need to be ordered;
- when the supply authorities must be notified that a supply will be required;
- if it is necessary to obtain authorization from a statutory body for any work to commence.

In thinking ahead and planning the best method of completing the contract, the individual activities or jobs must be identified and consideration given to how the various jobs are interrelated. To help in this process a number of management techniques are available. In this chapter we will consider only two: bar charts and network analysis. The very preparation of a bar chart or network analysis forces the contractor to think deeply, carefully and logically about the particular contract, and it is therefore a very useful aid to the successful completion of the work.

Bar charts

There are many different types of bar chart used by companies but the **object of any bar chart** is to establish the sequence and timing of the various activities involved in the contract as a whole. They are a visual aid in the process of

Definition

There are many different types of bar chart used by companies but the *object of any bar chart* is to establish the sequence and timing of the various activities involved in the contract as a whole.

communication. In order to be useful they must be clearly understood by the people involved in the management of a contract. The chart is constructed on a rectangular basis, as shown in Fig. 6.48.

All the individual jobs or activities which make up the contract are identified and listed separately down the vertical axis on the left-hand side, and time flows from left to right along the horizontal axis. The unit of time can be chosen to suit the length of the particular contract, but for most practical purposes either days or weeks are used.

The simple bar chart in Fig. 6.48(a) shows a particular activity A which is estimated to last two days, while activity B lasts eight days. Activity C lasts four days and should be started on day three. The remaining activities can be interpreted in the same way.

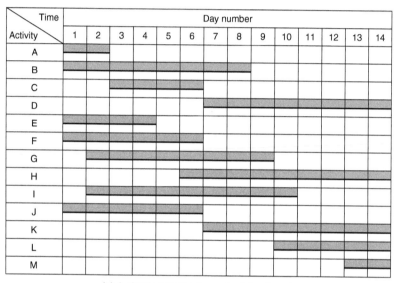

(a) A simple bar chart or schedule of work

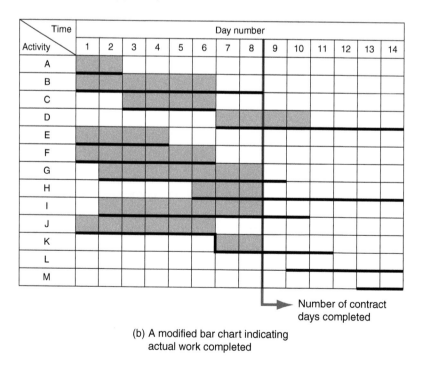

Number of contract days completed

(b) A modified bar chart indicating actual work completed

Figure 6.48 Bar charts: (a) a simple bar chart or schedule of work; (b) a modified bar chart indicating actual work completed.

With the aid of colours, codes, symbols and a little imagination, much additional information can be included on this basic chart. For example, the actual work completed can be indicated by shading above the activity line, as shown in Fig. 6.48(b) with a vertical line indicating the number of contract days completed; the activities which are on time, ahead of or behind time can easily be identified. Activity B in Fig. 6.48(b) is two days behind schedule, while activity D is two days ahead of schedule. All other activities are on time. Some activities must be completed before others can start. For example, all conduit work must be completely erected before the cables are drawn in. This is shown in Fig. 6.48(b) by activities J and K. The short vertical line between the two activities indicates that activity J must be completed before K can commence.

Useful and informative as the bar chart is, there is one aspect of the contract which it cannot display. It cannot indicate clearly the interdependence of the various activities upon each other, and it is unable to identify those activities which must strictly adhere to the time schedule if the overall contract is to be completed on time, and those activities in which some flexibility is acceptable. To overcome this limitation, in 1959 the Central Electricity Generating Board (CEGB) developed the critical path network diagram which we will now consider.

Network analysis

In large or complex contracts there are a large number of separate jobs or activities to be performed. Some can be completed at the same time, while others cannot be started until others are completed. A **network diagram** can be used to coordinate all the interrelated activities of the most complex project in such a way that all sequential relationships between the various activities, and the restraints imposed by one job on another, are allowed for. It also provides a method of calculating the time required to complete an individual activity and will identify those activities which are the key to meeting the completion date, called the critical path. Before considering the method of constructing a network diagram, let us define some of the terms and conventions we shall be using.

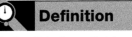

Definition

A *network diagram* can be used to coordinate all the interrelated activities of the most complex project.

Critical path

Critical path is the path taken from the start event to the end event which takes the longest time. This path denotes the time required for completion of the whole contract.

Definition

Critical path is the path taken from the start event to the end event which takes the longest time.

Float time

Float time, slack time or time in hand is the time remaining to complete the contract after completion of a particular activity.

Float time = Critical path time − Activity time

The total float time for any activity is the total leeway available for all activities in the particular path of activities in which it appears. If the float time is used up by one of the early activities in the path, there will be no float left for the remaining activities and they will become critical.

Definition

Float time, slack time or time in hand is the time remaining to complete the contract after completion of a particular activity.

Activities

Activities are represented by an arrow, the tail of which indicates the commencement and the head the completion of the activity. The length and

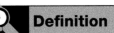

Definition

Activities are represented by an arrow, the tail of which indicates the commencement and the head the completion of the activity.

direction of the arrows have no significance: they are not vectors or phasors. Activities require time, manpower and facilities. They lead up to or emerge from events.

Dummy activities

Definition

Dummy activities are represented by an arrow with a dashed line.

Dummy activities are represented by an arrow with a dashed line. They signify a logical link only, require no time and denote no specific action or work.

Event

Definition

An *event* is a point in time, a milestone or stage in the contract when the preceding activities are finished.

An **event** is a point in time, a milestone or stage in the contract when the preceding activities are finished. Each activity begins and ends in an event. An event has no time duration and is represented by a circle which sometimes includes an identifying number or letter.

Time may be recorded to a horizontal scale or shown on the activity arrows. For example, the activity from event A to B takes nine hours in the network diagram shown in Fig. 6.49.

Example 1

Identify the three possible paths from the start event A to the finish event F for the contract shown by the network diagram in Fig. 6.49. Identify the critical path and the float time in each path.

The three possible paths are:

1 event A–B–D–F
2 event A–C–D–F
3 event A–C–E–F

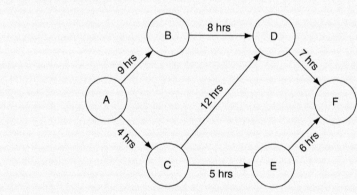

Figure 6.49 A network diagram for Example 1.

The times taken to complete these activities are:

1 path A–B–D–F = 9 + 8 + 7 = 24 hours
2 path A–C–D–F = 4 + 12 + 7 = 23 hours
3 path A–C–E–F = 4 + 5 + 6 = 15 hours

The longest time from the start event to the finish event is 24 hours, and therefore the critical path is A – B – D – F.

(Continued)

The float time is given by:

$$\text{Float time} = \text{Critical path} - \text{Activity time}$$

For path 1, A−B−D−F,

$$\text{Float time} = 24 \text{ hours} - 24 \text{ hours} = 0 \text{ hours}$$

There can be no float time in any of the activities which form a part of the critical path, since a delay on any of these activities would delay completion of the contract. On the other two paths some delay could occur without affecting the overall contract time.

For path 2, A−C−D−F,

$$\text{Float time} = 24 \text{ hours} - 23 \text{ hours} = 1 \text{ hour}$$

For path 3, A−C−E−F,

$$\text{Float time} = 24 \text{ hours} - 15 \text{ hours} = 9 \text{ hours}$$

Example 2

Identify the time taken to complete each activity in the network diagram shown in Fig. 6.50. Identify the three possible paths from the start event A to the final event G and state which path is the critical path.

The time taken to complete each activity using the horizontal scale is:

$$\text{activity A−B} = 2 \text{ days}$$
$$\text{activity A−C} = 3 \text{ days}$$
$$\text{activity A−D} = 5 \text{ days}$$
$$\text{activity B−E} = 5 \text{ days}$$
$$\text{activity C−F} = 5 \text{ days}$$
$$\text{activity E−G} = 3 \text{ days}$$
$$\text{activity D−G} = 0 \text{ days}$$
$$\text{activity F−G} = 0 \text{ days}$$

Activities D−G and F−G are dummy activities which take no time to complete but indicate a logical link only. This means that in this case, once the activities preceding events D and F have been completed, the contract will not be held up by work associated with these particular paths and they will progress naturally to the finish event.

The three possible paths are:

1 A−B−E−G
2 A−D−G
3 A−C−F−G

The times taken to complete the activities in each of the three paths are:

$$\text{path 1, A−B−E−G} = 2 + 5 + 3 = 10 \text{ days}$$
$$\text{path 2, A−D−G} = 5 + 0 = 5 \text{ days}$$
$$\text{path 3, A−C−F−G} = 3 + 5 + 0 = 8 \text{ days}$$

The critical path is path 1, A−B−E−G.

(Continued)

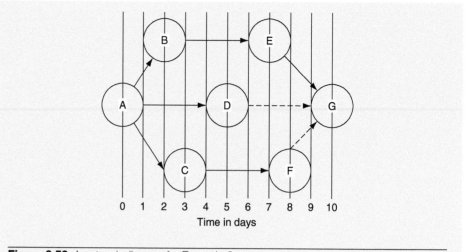

Figure 6.50 A network diagram for Example 2.

Constructing a network

The first step in constructing a network diagram is to identify and draw up a list of all the individual jobs, or activities, which require time for their completion and which must be completed to advance the contract from start to completion.

The next step is to build up the arrow network showing schematically the precise relationship of the various activities between the start and end event. The designer of the network must ask these questions:

1 Which activities must be completed before others can commence? These activities are then drawn in a similar way to a series circuit but with event circles instead of resistor symbols.
2 Which activities can proceed at the same time? These can be drawn in a similar way to parallel circuits but with event circles instead of resistor symbols.

Commencing with the start event at the left-hand side of a sheet of paper, the arrows representing the various activities are built up step by step until the final event is reached. A number of attempts may be necessary to achieve a well-balanced and symmetrical network diagram showing the best possible flow of work and information, but this time is well spent when it produces a diagram which can be easily understood by those involved in the management of the particular contract.

Example 3

A particular electrical contract is made up of activities A–F as described below:

A = an activity taking 2 weeks commencing in week 1

B = an activity taking 3 weeks commencing in week 1

C = an activity taking 3 weeks commencing in week 4

D = an activity taking 4 weeks commencing in week 7

E = an activity taking 6 weeks commencing in week 3

F = an activity taking 4 weeks commencing in week 1

(Continued)

Example 3 (Continued)

Certain constraints are placed on some activities because of the availability of men and materials and because some work must be completed before other work can commence as follows:

Activity C can only commence when B is completed

Activity D can only commence when C is completed

Activity E can only commence when A is completed

Activity F does not restrict any other activity

(a) Produce a simple bar chart to display the activities of this particular contract.

(b) Produce a network diagram of the programme and describe each event.

(c) Identify the critical path and the total contract time.

(d) State the maximum delay which would be possible on activity E without delaying the completion of the contract.

(e) State the float time in activity F.

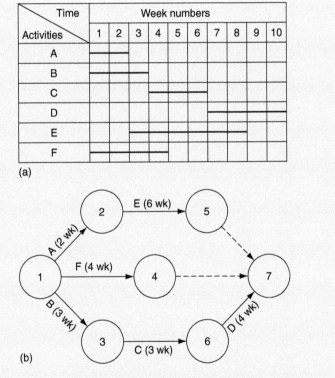

(a)

(b)

Figure 6.51 (a) Bar chart and (b) network diagram for Example 3.

(a) A simple bar chart for this contract is shown in Fig. 6.51(a).

(b) The network diagram is shown in Fig. 6.51(b). The events may be described as follows:

Event 1 = the commencement of the contract

Event 2 = the completion of activity A and the commencement of activity E

Event 3 = the completion of activity B and the commencement of activity C

(Continued)

Example 3 (Continued)

Event 4 = the completion of activity F
Event 5 = the completion of activity E
Event 6 = the completion of activity C
Event 7 = the completion of activity D and the whole contract.

(c) There are three possible paths:

1 via events 1−2−5−7
2 via events 1−4−7
3 via events 1−3−6−7.

The time taken for each path is:

path 1 = 2 weeks + 6 weeks = 8 weeks

path 2 = 4 weeks = 4 weeks

path 3 = 3 weeks + 3 weeks + 4 weeks = 10 weeks.

The critical path is therefore path 3, via events 1−3−6−7, and the total contract time is 10 weeks.

(d) We have that:

$$\text{Float time} = \text{Critical path time} - \text{Activity time}$$

Activity E is on path 1 via events 1−2−5−7 having a total activity time of 8 weeks

$$\text{Float time} = 10 \text{ weeks} - 8 \text{ weeks} = 2 \text{ weeks}.$$

Activity E could be delayed for a maximum of 2 weeks without delaying the completion date of the whole contract.

(e) Activity F is on path 2 via events 1−4−7 having a total activity time of 4 weeks

$$\text{Float time} = 10 \text{ weeks} - 4 \text{ weeks} = 6 \text{ weeks}.$$

Personal communications

Remember that it is the customers who actually pay the wages of everyone employed in your company. You should always be polite and listen carefully to their wishes. They may be elderly or of a different religion or cultural background than you. In a domestic situation, the playing of loud music on a radio may not be approved of. Treat the property in which you are working with the utmost care. When working in houses, shops and offices use dust-sheets to protect floor coverings and furnishings. Clean up periodically and make a special effort when the job is completed.

Dress appropriately: an unkempt or untidy appearance will encourage the customer to think that your work will be of poor quality.

The electrical installation in a building is often carried out alongside other trades. It makes good sense to help other trades where possible and to develop good working relationships with other employees. The customer will be most happy if the workers give an impression of working together as a team for the successful completion of the project.

Finally, remember that the customer will probably see more of the electrician and the electrical trainee than the managing director of your firm and, therefore, the image presented by you will be assumed to reflect the policy of the company.

You are, therefore, your company's most important representative. Always give the impression of being capable and in command of the situation, because this gives customers confidence in the company's ability to meet their needs. However, if a problem does occur which is outside your previous experience and you do not feel confident to solve it successfully, then contact your supervisor for professional help and guidance. It is not unreasonable for a young member of the company's team to seek help and guidance from those employees with more experience. This approach would be preferred by most companies rather than having to meet the cost of an expensive blunder.

Construction site: Safe working practice

In Chapter 1 we looked at some of the laws and regulations that affect our working environment. We looked at safety signs and personal protective equipment (PPE), and how to recognize and use different types of fire extinguishers.

If your career in the electrical industry is to be a long, happy and safe one, you must always wear appropriate PPE such as footwear and head protection, and behave responsibly and sensibly in order to maintain a safe working environment. Before starting work, make a safety assessment; what is going to be hazardous, will you require PPE, do you need any special access equipment?

Construction sites can be hazardous because of the temporary nature of the construction process. The surroundings and systems are always changing as the construction process moves to its completion date when everything is finally in place.

Safe methods of working must be demonstrated by everyone at every stage. 'Employees have a duty of care to protect their own health and safety and that of others who might be affected by their work activities.'

To make the work area safe before starting work and during work activities, it may be necessary to:

- use barriers or tapes to screen off potential hazards;
- place warning signs as appropriate;
- inform those who may be affected by any potential hazard;
- use a safe isolation procedure before working on live equipment or circuits;
- obtain any necessary 'permits-to-work' before work begins.

Get into the habit of always working safely and being aware of the potential hazards around you when you are working.

Having chosen an appropriate wiring system which meets the intended use and structure of the building and satisfies the environmental conditions of the installation, you must install the system conductors, accessories and equipment in a safe and competent manner.

The structure of the building must be made good if it is damaged during the installation of the wiring system; for example, where conduits and trunking pass through walls and floors.

All connections in the wiring system must be both electrically and mechanically sound. All conductors must be chosen so that they will carry the design current under the installed conditions.

If the wiring system is damaged during installation it must be made good to prevent future corrosion. For example, where galvanized conduit trunking or tray is cut or damaged by pipe vices, it must be made good to prevent localized corrosion.

All tools must be used safely and sensibly. Cutting tools should be sharpened and screwdrivers ground to a sharp square end on a grindstone.

It is particularly important to check that the plug and cables of handheld electrically powered tools and extension leads are in good condition. Damaged plugs and cables must be repaired before you use them. All electrical power tools of 110 and 230 V must be tested with a portable appliance tester (PAT) in accordance with the company's health and safety procedures, but probably at least once each year.

Tools and equipment that are left lying about in the workplace can become damaged or stolen and may also be the cause of people slipping, tripping or falling. Tidy up regularly and put power tools back in their boxes. This is called 'good housekeeping' and is an important part of Appendix F of the Electrician's Guide to the Building Regulations. You personally may have no control over the condition of the workplace in general, but keeping your own work area clean and tidy is the mark of a skilled and conscientious craftsman.

The electrical team must always be respectful of the customers property and contents when carrying out their electrical activities. This might involve laying dust sheets over some items, or moving others to a safe area while work is carried out, or wearing plastic protectors over your shoes so that carpets. are not damaged. Always be polite to the customer. The customer or their representative will probably see more of the electrician and the electrical trainee than the managing director of the company, and therefore the image presented by them to the customer will be assumed to reflect the company policy. You are therefore your company's most important representative at that time, and indirectly, it is the customer who pays your wages.

Finally, when the job is finished, clean up and dispose of all waste material responsibly. This is an important part of your company's 'good customer relationships' with the client. We also know that we have a 'duty of care' for the waste that we produce as an electrical company.

We have also said many times in this book that having a good attitude to health and safety, working conscientiously and neatly, keeping passageways clear and regularly tidying up the workplace is the sign of a good and competent craftsman. But what do you do with the rubbish that the working environment produces? Well:

- All the packaging material for electrical fittings and accessories usually goes into either your employer's skip or the skip on-site designated for that purpose.
- All the offcuts of conduit, trunking and tray also go into the skip.
- In fact, most of the general site debris will probably go into the skip and the waste disposal company will take the skip contents to a designated local council landfill area for safe disposal.
- The part coils of cable and any other reusable leftover lengths of conduit, trunking or tray will be taken back to your employer's stores area. Here it will be stored for future use and the returned quantities deducted from the costs allocated to that job.
- What goes into the skip for normal disposal into a landfill site is usually a matter of common sense. However, some substances require special consideration and disposal. We will now look at asbestos and large quantities of used fluorescent tubes which are classified as 'special waste' or 'hazardous waste'.

Asbestos is a mineral found in many rock formations. When separated it becomes a fluffy, fibrous material with many uses. It was used extensively in the construction industry during the 1960s and 1970s for roofing material, ceiling and floor tiles, fire-resistant board for doors and partitions, for thermal insulation and commercial and industrial pipe lagging.

In the buildings where it was installed some 40 years ago, when left alone, it does not represent a health hazard, but those buildings are increasingly becoming in need of renovation and modernization. It is in the dismantling and breaking up of these asbestos materials that the health hazard increases. Asbestos is a serious health hazard if the dust is inhaled. The tiny asbestos particles find their way into delicate lung tissue and remain embedded for life, causing constant irritation and eventually, serious lung disease.

Working with asbestos materials is not a job for anyone in the electrical industry. If asbestos is present in situations or buildings where you are expected to work, it should be removed by a specialist contractor before your work commences. Specialist contractors, who will wear fully protective suits and use breathing apparatus, are the only people who can safely and responsibly carry out the removal of asbestos. They will wrap the asbestos in thick plastic bags and store them temporarily in a covered and locked skip. This material is then disposed of in a special landfill site with other toxic industrial waste materials and the site monitored by the Local Authority for the foreseeable future. See Control of Asbestos at Work Ragulations in Chapter 1 of this book.

There is a lot of work for electrical contractors in many parts of the country updating and improving the lighting in government buildings and schools. This work often involves removing the old fluorescent fittings hanging on chains or fixed to beams, and installing a suspended ceiling and an appropriate number of recessed modular fluorescent or LED fittings. So what do we do with the old fittings? Well, the fittings are made of sheet steel, a couple of plastic lamp holders, a little cable, a starter and ballast. All of these materials can go into the ordinary skip. However, the fluorescent tubes contain a little mercury and fluorescent powder with toxic elements, which cannot be disposed of in the normal landfill sites. New Hazardous Waste Regulations were introduced in July 2005 and under these regulations lamps and tubes are classified as hazardous. While each lamp contains only a small amount of mercury, vast numbers of lamps and tubes are disposed of in the United Kingdom every year, resulting in a significant environmental threat.

The environmentally responsible way to dispose of fluorescent lamps and tubes is to recycle them. See Hazardous Waste Regulations: 2005 in Chapter 1 of this book.

Secure site storage

The ability to resolve a technical practical problem using tools and equipment is a part of what those of us in the electrical industry are about, and so tools are very important to us. As a safety precaution it makes sense not to leave them unattended on-site and to keep them in the back of your locked vehicle when not being used. Power tools are expensive, make your working life less difficult and are very easy to be carried away, so they too should leave with you at the end of the working day. However, some pieces of equipment are too big to take home such as access equipment, generators and other electrical equipment brought in or hired specifically for a particular job. At the end of the working day these larger pieces of equipment must be made secure. On large construction sites the main contractors often make secure storage in the shape of metal containers available to their subcontractors and this may be a good solution in some situations.

However, the security of a store can be put at risk if there is more than one keyholder. One of the other keyholders might leave the store unlocked and your material and equipment could be stolen through no fault of your own. So what are the alternatives? You could put your materials and equipment inside a large locked box that has been bolted to the wall or floor inside the main contractor's locked store, or you could use motorcycle chains and locks to secure your plant inside the main contractor's locked store. Materials which are to be installed at some later date also require secure storage. Electrical contractors these days often work out of an industrial unit which may also be their head office incorporating facilities for 'office work' and will therefore be occupied during the working day. Materials could probably be delivered and stored at this industrial unit and only small quantities taken to site each day by the installing electricians. Alternatively, if you have a good relationship with your local wholesalers, they too will deliver smaller quantities to site so that there is no requirement for storage. You may think that I am painting a rather bleak picture of secure site storage but electrical goods and equipment are very expensive and most desirable to a thief, and you must, therefore, give serious consideration to how you will securely store goods on-site if you have no other alternative.

Commissioning electrical systems

The commissioning of the electrical and mechanical systems within a building is a part of the 'handing-over' process of the new building by the architect and main contractor to the client or customer in readiness for its occupation and intended use. To 'commission' means to give authority to someone to check that everything is in working order. If it is out of commission, it is not in working order.

Following the completion, inspection and testing of the new electrical installation, the functional operation of all the electrical systems must be tested before they are handed over to the customer. It is during the commissioning period that any design or equipment failures become apparent, and these must be corrected before the installation is handed over to the client or 'signed off'. We call this process 'snagging'.

This is the role of the commissioning engineer, who must assure him-or herself that all the systems are in working order and that they work as they were designed to work. The engineer must also instruct the client's representative, or the staff who will use the equipment, in the correct operation of the systems, as part of the handover arrangements.

The commissioning engineer must test the operation of all the electrical systems, including the motor controls, the fan and air-conditioning systems, the fire alarm and emergency lighting systems. However, before testing the emergency systems, everyone in the building must be made aware of the test taking place, so that alarms may be ignored during the period of testing.

Commissioning has become one of the most important functions within the building project's completion sequence. The commissioning engineer will therefore have access to all relevant contract documents, including the building specifications and the electrical installation certificates as required by the IET Regulations (BS 7671), and have a knowledge of the requirements of the Electricity at Work Regulations and the Health and Safety at Work Act.

The building will only be handed over to the client if the commissioning engineer is satisfied that all the building services meet the design specification in the contract documents.

Check your understanding

When you have completed the questions, check out the answers at the back of the book.

Note: more than one multiple-choice answer may be correct.

1 'The conductive mass of the earth' is one definition of:
   a. earth
   b. earthing
   c. bonding conductor
   d. circuit protective conductor.

2 A protective conductor connecting exposed conductive parts of equipment to the main earthing terminal is one definition of:
   a. earth
   b. earthing
   c. bonding conductor
   d. circuit protective conductor.

3 A protective conductor connecting exposed and extraneous parts together is one definition of:
   a. earth
   b. earthing
   c. bonding conductor
   d. circuit protective conductor.

4 The act of connecting the exposed conductive parts of the installation to the main earthing terminal is called:
   a. earth
   b. earthing
   c. bonding conductor
   d. circuit protective conductor.

5 The act of linking together the exposed and extraneous metal parts is called:
   a. earthing
   b. bonding
   c. basic protection
   d. fault protection.

6   The protection provided by insulating live parts is called:
    a.   extraneous conductive parts
    b.   basic protection
    c.   exposed conductive parts
    d.   fault protection.

7   A current which exceeds the rated value in an otherwise healthy circuit is one definition of:
    a.   earthing
    b.   bonding
    c.   overload
    d.   short-circuit.

8   An overcurrent resulting from a fault of negligible impedance is one definition of:
    a.   earthing
    b.   bonding
    c.   overload
    d.   short-circuit.

9   The linking together of the exposed or extraneous conductive parts of an installation for the purpose of safety is one definition of:
    a.   earthing
    b.   protective bonding
    c.   exposed conductive parts
    d.   extraneous conductive parts.

10 The conduit and trunking parts of the electrical installation are:
    a.   earth conductors
    b.   bonding conductors
    c.   exposed conductive parts
    d.   extraneous conductive parts.

11 The gas, water and central heating pipes of the building not forming a part of the electrical installation are called:
    a.   earthing conductors
    b.   bonding conductors
    c.   exposed conductive parts
    d.   extraneous conductive parts.

12 The protection provided by insulating the live parts of the electrical installation is called:
    a.   overload protection
    b.   short-circuit protection
    c.   basic protection
    d.   fault protection.

13 The protection provided by protective bonding and automatic disconnection of the supply is called:

   a.   overload protection

   b.   short-circuit protection

   c.   basic protection

   d.   fault protection.

14. The metalwork of the electrical installation is called:

   a.   extraneous conductive parts

   b.   basic protection

   c.   exposed conductive parts

   d.   fault protection.

15 The metalwork of the building and other service pipes is called:

   a.   extraneous conductive parts

   b.   basic protection

   c.   exposed conductive parts

   d.   fault protection.

16 The protection provided by protective bonding and automatic disconnection of the supply is called:

   a.   extraneous conductive parts

   b.   basic protection

   c.   exposed conductive parts

   d.   fault protection.

17 Cutting off the electrical supply in order to ensure the safety of those working on the equipment is one definition of:

   a.   basic protection

   b.   fault protection

   c.   protective bonding

   d.   isolation switching.

18 A current which exceeds the rated current in an otherwise healthy circuit is one definition of:

   a.   fault protection

   b.   an overcurrent

   c.   an overload current

   d.   a short-circuit current.

19 The weakest link in the circuit designed to melt when an overcurrent flows is one definition of:

   a.   fault protection

   b.   a circuit protective conductor

   c.   a fuse

   d.   a consumer unit.

20 According to IET Regulation 411.3.2.2 all final circuits not exceeding 32 A in a building supplied with a 230 V TN supply shall have a maximum disconnection time not exceeding:
    a. 0.2 s
    b. 0.4 s
    c. 5.0 s
    d. unlimited.

21 To ensure the effective operation of the overcurrent protective devices, the earth fault loop path must have:
    a. a 230 V supply
    b. a very low resistance
    c. fuses or MCBs in the live conductor
    d. a very high resistance.

22 A load connected to the three phases of a star-connected three-phase four-wire supply system from the local substation would have a voltage of:
    a. 230 V
    b. 400 V
    c. 25 kV
    d. 132 kV.

23 A load connected to phase and neutral of a star-connected three-phase four-wire supply system from the local substation would have a voltage of:
    a. 230 V
    b. 400 V
    c. 25 kV
    d. 132 kV.

24 An electrical cable is made up of three parts which are:
    a. conduction, convection and radiation
    b. conductor, insulation and outer sheath
    c. heating, magnetic and chemical
    d. conductors and insulators.

25 An appropriate wiring method for a domestic installation would be a:
    a. metal conduit installation
    b. trunking and tray installation
    c. PVC cables
    d. PVC/SWA cables.

26 An appropriate wiring method for an underground feed to a remote building would be a:
    a. metal conduit installation
    b. trunking and tray installation
    c. PVC cables
    d. PVC/SWA cables.

27 An appropriate wiring method for a high-temperature installation in a boiler house is:
   a. metal conduit installation
   b. trunking and tray installation
   c. FP 200 cables
   d. MI cables.

28 An appropriate wiring system for a three-phase industrial installation would be:
   a. PVC cables
   b. PVC conduit
   c. one which meets the requirements of Part 2 of the IET Regulations
   d. one which meets the requirements of Part 5 of the IET Regulations.

29 A standard form completed by every employee to inform the employer of the time spent working on a particular site is called:
   a. job sheet
   b. time sheet
   c. delivery note
   d. daywork sheet.

30 A record which confirms that materials ordered have been delivered to site is called:
   a. job sheet
   b. time sheet
   c. delivery note
   d. daywork sheet.

31 A standard form containing information about work to be done usually distributed by a manager to an electrician is called:
   a. job sheet
   b. time sheet
   c. delivery note
   d. daywork sheet.

32 A standard form which records changes or extra work on a large project is called a:
   a. job sheet
   b. time sheet
   c. delivery note
   d. daywork sheet.

33 A graph which shows the sequence or time to be taken on various electrical activities within the contract as a whole is called a:
   a. network analysis
   b. variation order
   c. bar chart
   d. tender.

34   A formal contract to supply goods or carry out work at a stated price is called a:
a.   network analysis
b.   variation order
c.   bar chart
d.   tender.

35   A record of the work done which is outside or in addition to the original electrical contract is called a:
a.   network analysis
b.   variation order
c.   bar chart
d.   tender.

36   A method of showing the separate activities or tasks which make up a large or complex electrical contract is called a:
a.   network analysis
b.   variation order
c.   bar chart
d.   tender.

37   We know that we have a 'duty of care' to dispose responsibly of all waste material which the working environment creates. So what is the 'responsible' way to dispose of the packaging from boxes of electrical equipment and fittings?
a.   place it in the general waste skip which will go to landfill
b.   place it in the skip designated for recycling materials
c.   have it removed by a specialist waste contractor before your electrical work begins
d.   recycle through a specialist hazardous waste company.

38   We know that we have a 'duty of care' to dispose responsibly of all waste material which the working environment creates. So what is the 'responsible' way to dispose of asbestos material?
a.   place it in the general waste skip which will go to landfill
b.   place it in the skip designated for recycling materials
c.   have it removed by a specialist waste contractor before your electrical work begins
d.   recycle through a specialist hazardous waste company.

39   We know that we have a 'duty of care' to dispose responsibly of all waste material which the working environment creates. So what is the 'responsible' way to dispose of the offcuts of conduit and trunking and old metal fluorescent fittings?
a.   place it in the general waste skip which will go to landfill
b.   place it in the skip designated for recycling materials
c.   have it removed by a specialist waste contractor before your electrical work begins
d.   recycle through a specialist hazardous waste company.

40 We know that we have a 'duty of care' to dispose responsibly of all waste material which the working environment creates. So what is the 'responsible' way to dispose of dozens of old fluorescent tubes and SON lamps?
    a. place it in the general waste skip which will go to landfill
    b. place it in the skip designated for recycling materials
    c. have it removed by a specialist waste contractor before your electrical work begins
    d. recycle through a specialist hazardous waste company.

41 The current demand of 10-65W fluorescent tubes connected to a 230V supply is:
    a. 283mA
    b. 5.07A
    c. 150W
    d. 650W.

42 Ten lighting points are to be wired in a domestic 230V home. Calculate the assumed current demand of the 10 lighting points which are to be connected to fittings of unknown style or rating at a later date.
    a. 43.5mA
    b. 4.35A
    c. 7.83A
    d. 1000W.

43 Calculate the assumed current demand of a circuit comprising three shaver units, 1 bell chime transformer, and five 2 amp socket outlets to provide power for table lamps in the sitting area of the property. The final circuit is connected to a consumer unit connected to the 230V mains supply.
    a. 23mA
    b. 43mA
    c. 2.5A
    d. 10A.

44 Calculate the assumed current demand of a cooker having four 3kW boiling rings and a 2kW oven connected to a 230V supply.
    a. 14kW
    b. 15.26A
    c. 25.25A
    d. 10A.

45 The total current demand of a lighting circuit in an individual household is calculated to be 5A. The current demand applying the diversity factor given in Table A2 of the *On Site Guide* will be:
    a. 3.3A
    b. 3.75A
    c. 4.5A
    d. 5.0A.

46   The total current demand of a lighting circuit in a small shop is calculated to be 5A. The current demand applying the diversity factor given in Table A2 of the *On Site Guide* will be:

    a. 3.3A

    b. 3.75A

    c. 4.5A

    d. 5A.

47   The total current demand of a lighting circuit in a small hotel is calculated to be 5A. The current demand applying the diversity factor given in Table A2 of the *On Site Guide* will be:

    a. 3.3A

    b. 3.75A

    c. 4.5A

    d. 5A.

48   The allowance for diversity on a lighting circuit in an individual household given in Table A2 of the *On Site Guide* is:

    a. none

    b. 66%

    c. 75%

    d. 90%.

49   The allowance for diversity on a lighting circuit in business premises given in Table A2 of the *On Site Guide* is:

    a. none

    b. 66%

    c. 75%

    d. 90%.

50   The allowance for diversity on a thermostatically controlled water heater in a small hotel given in Table A2 of the *On Site Guide* is:

    a. none

    b. 66%

    c. 75%

    d. 90%.

51   The allowance for diversity for 13A socket outlets installed according to the standard arrangements of ring and radial circuits given in Table H2.1 of the *On Site Guide* when installed in individual households is:

    a. 100% of all circuits

    b. 100% of the largest circuit + 75% of every other circuit

    c. 100% of the smallest circuit + 50% of every other circuit

    d. 100% of the largest circuit + 40% of every other circuit.

52 The allowance for diversity for 13A socket outlets installed according to the standard arrangements of ring and radial circuits given in Table H2.1 of the *On Site Guide* when installed in offices and business premises is:

a. 100% of all circuits

b. 100% of the largest circuit + 75% of every other circuit

c. 100% of the largest circuit + 50% of every other circuit

d. 100% of the largest circuit + 40% of every other circuit.

53 The current carrying-capacity of a two-core single-phase 2.5mm non-armoured PVC cable enclosed in trunking as given in Table F5 (i) of the *On Site Guide* or Table 6.6 of this book is:

a. 18A

b. 23A

c. 27A

d. 30A.

54 The current-carrying capacity of a two-core single-phase 2.5mm non-armoured PVC cable clipped direct to a wall as given in Table F5 (i) of the *On Site Guide* or Table 6.6 of this book is:

a. 18A

b. 23A

c. 27A

d. 30A.

55 The current-carrying capacity of a two-core single-phase 2.5mm non-armoured PVC cable laid on a perforated cable tray or cable basket as given in Table F5 (i) of the *On Site Guide* or Table 6.6 of this book is:

a. 18A

b. 23A

c. 27A

d. 30A.

56 The volt drop in mV per ampere per metre of a two-core single-phase 2.5mm cable as given in Table F5 (ii) of the *On Site Guide* or Table 6.7 of this book is:

a. 18

b. 11

c. 7.3

d. 4.4.

57 The volt drop in mV per ampere per metre of a two-core single-phase 4.0mm cable as given in Table F5 (ii) of the *On Site Guide* or Table 6.7 of this book is:

a. 18

b. 11

c. 7.3

d. 4.4.

58 The volt drop in mV per ampere per metre of a two-core single-phase 6.0mm cable as given in Table F5 (ii) of the *On Site Guide* or Table 6.7 of this book is:

a.   18

b.   11

c.   7.3

d.   4.4.

59 The volt drop in mV per ampere per metre of a two-core single-phase 10.0mm cable as given in Table F5 (ii) of the *On Site Guide* or Table 6.7 of this book is:

a.   18

b.   11

c.   7.3

d.   4.4.

60 Why does the volt drop in the last four questions, 56 to 59, decrease as the cable size increases? Remember that $V = I \times R$.

a.   volt drop is proportional to resistance

b.   volt drop is proportional to temperature

c.   as the cable size increases, the resistance of the conductors increases

d.   as the cable size increases, the resistance of the conductors decreases.

61 Ten 1.0mm single conductors are to be drawn into a conduit. Use Table E3 of the *On Site Guide* or Table 6.10 of this book to find the sum of the cable factors for these cables.

a.   16

b.   30

c.   160

d.   300.

62 A 3-metre length of conduit with two bends is to contain the ten 1.0mm cables. Determine the size of conduit required using Table E4 of the *On Site Guide* or Table 6.11 of this book.

a.   16mm

b.   20mm

c.   25mm

d.   32mm.

63 Ten 2.5mm single conductors are to be drawn into a conduit. Use Table E3 of the *On Site Guide* or Table 6.10 of this book to find the sum of the cable factors for these cables.

a.   16

b.   30

c.   160

d.   300.

64 A 3-metre length of conduit with two bends is to contain the ten 2.5mm cables. Determine the size of conduit required using Table E4 of the *On Site Guide* or Table 6.11 of this book.

    a.   16mm

    b.   20mm

    c.   25mm

    d.   32mm.

65 Ten 4.0mm single conductors are to be drawn into a conduit. Use Table E3 of the *On Site Guide* or Table 6.10 of this book to find the sum of the cable factors for these cables.

    a.   43

    b.   160

    c.   300

    d.   430.

66 A 3-metre length of conduit with two bends is to contain the ten 4.0mm cables. Determine the size of conduit required using Table E4 of the *On Site Guide* or Table 6.11 of this book.

    a.   16mm

    b.   20mm

    c.   25mm

    d.   32mm.

67 Ten 1.5mm solid PVC cables are to be drawn into trunking. Use Table E5 of the *On Site Guide* or Table 6.12 of this book to find the sum of the cable factors for these cables.

    a.   80

    b.   60

    c.   40

    d.   20.

68 Ten 2.5mm solid PVC cables are to be drawn into trunking. Use Table E5 of the *On Site Guide* or Table 6.12 of this book to find the sum of the cable factors for these cables.

    a.   116

    b.   117

    c.   118

    d.   119.

69 Ten 4.0mm stranded PVC cables are to be drawn into trunking. Use Table E5 of the *On Site Guide* or Table 6.12 of this book to find the sum of the cable factors for these cables.

    a.   80

    b.   119

    c.   166

    d.   180.

70 Determine the total trunking factor for all 30 cables described in questions 67 to 69 above.
   a. 85
   b. 155
   c. 265
   d. 365.

71 Determine the factor for a 75mm × 75mm trunking.
   a. 738
   b. 1146
   c. 1555
   d. 2371.

72 What is the maximum number of 4.0mm PVC conductors which may be drawn into a 75mm × 50mm trunking?
   a. 52
   b. 93
   c. 110
   d. 120.

73 Cutting off the electrical supply to a piece of equipment in order to ensure the safety of those working on it is one definition of:
   a. emergency switching
   b. functional switching
   c. switching for isolation
   d. switching for mechanical maintenance.

74 Cutting off the full load electrical supply to a piece of equipment in order to ensure the safety of those working on it is one definition of:
   a. emergency switching
   b. functional switching
   c. switching for isolation
   d. switching for mechanical maintenance.

75 The rapid disconnection of the electrical supply to remove or prevent danger is one definition of:
   a. emergency switching
   b. functional switching
   c. switching for isolation
   d. switching for mechanical maintenance.

76 Switching electrical equipment in normal service is:
   a. emergency switching
   b. functional switching
   c. switching for isolation
   d. switching for mechanical maintenance.

77 Briefly explain why an electrical installation needs protective devices.

78 List the four factors on which the selection of a protective device depends.

79 List the five essential requirements for a device designed to protect against overcurrent.

80 Briefly describe the action of a fuse under fault conditions.

81 State the meaning of 'discrimination' or 'coordination' as applied to circuit protective devices.

82 Use a sketch to show how 'discrimination' or 'coordination' can be applied to a piece of equipment connected to a final circuit.

83 List typical 'exposed parts' of an installation.

84 List typical 'extraneous parts' of a building.

85 Use a sketch to show the path taken by an earth fault current.

86 Use bullet points and a simple sketch to briefly describe the operation of an RCD.

87 State the need for RCDs in an electrical installation:
   a. supplying socket outlets with a rated current not exceeding 20A and
   b. for use by mobile equipment out of doors as required by IET Regulation 411.3.3.

88 Briefly describe an application for RCBOs.

89 In your own words, state the meaning of circuit overload and short-circuit protection. What will provide this type of protection?

90 State the purpose of earthing and earth protection. What do we do to achieve it and why do we do it?

91 In your own words, state the meaning of earthing and bonding. What types of cables and equipment would an electrician use to achieve earthing and bonding on an electrical installation?

92 In your own words, state what we mean by 'basic protection' and how it is achieved.

93 In your own words, state what we mean by 'fault protection' and how it is achieved.

94 How does a bar chart help with the organization of a work programme?

95 State five methods of making your work area safe on a construction site.

96 Why are good relationships important between yourself and the customer and other trades on-site when carrying out work activities?

97 Briefly state why time sheets, fully and accurately completed, are important to:
   a. an employer
   b. an employee.

98 State the reasons why you should always present the right image to a client, customer or his representative.

99 Briefly describe what we mean by a schedule of work. Who would use a bar chart or schedule of work in your company and why?

100 Changes to the Building Regulations now require an electrician to make 'reasonable provision' for energy-efficient lamps and controls. Use five bullet points to describe what we must now do.

101 Briefly describe what the 'code for sustainable homes' aims to do between now and 2016.

102 Briefly describe the importance of 'sustainable design' for buildings and the services within buildings.

103 Use four bullet points to state how the government will achieve level 6 of the code for sustainable homes by 2016.

104 Describe the procedure for making safe terminations and connections in your electrical work.

105 Describe what we mean by a 'maintenance-free' junction box.

106 Describe a resin compound joint and give a typical application for such a joint.

107 Draw a quick sketch to show the complete earth fault loop path Z_s. Fig 6.11 may give guidance.

108 State the meaning of 'diversity' when applied to current demand.

109 When signing a contract to carry out work described in the contract, what is the electrical contractor actually agreeing to do?

110 How would you secure each of the following pieces of equipment on-site to prevent theft? Electrical power tools in individual boxes, hand tools in a toolbox, a small aluminium tower scaffold, a bundle of conduit, 20 3-metre lengths of trunking and 20 cardboard boxes about 500mm square by 50mm deep containing LED lighting modules, each retailing at about £90.

111 Use bullet points to describe the 'commissioning procedure'.

112 What do we mean by the 'handing-over' process?

113 What do we mean when we say the installation has been 'signed off'?

114 What do we mean by 'snagging' during the commissioning process?

115 Waste can be described as ordinary landfill waste or recyclable waste or hazardous waste. Describe each type of waste and give typical examples of each type that an electrical contracting company may encounter when working on-site.

Answers to check your understanding questions

Chapter 1

1. D	2. A	3. A, B
4. B, C	5. D	6. D
7. D	8. C	9. A, B
10. B, C, D	11. A, B	12. C, D
13. B, C, D	14. C, D	15. C
16. C	17. A, C	18. B, D
19. D	20. B	21. C
22. B, C, D	23. C	24. D
25. C	26. A	27. C
28. C	29. A, B, D	

30 to 48 Answers in text of Chapter 1.

Chapter 2

1. C	2. A, B	3. B, C	4. A, D	5. B, D
6. C	7. C	8. D	9. B	10. C
11. D	12. C	13. A		

14 to 28 Answers in text of Chapter 2.

Chapter 3

1. A	2. C	3. D	4. C	5. A
6. B	7. C	8. C	9. C	10. A, D
11. C	12. B	13. C	14. C	15. C
16. D	17. B	18. A	19. B	20. C, D
21. A	22. C	23. D	24. B	25. B

26 to 45 Answers in text of Chapter 3.

Chapter 4

1. B, C, D	2. A, C	3. B, C, D	4. D	5. C
6. B, D	7. A, B	8. C, D	9. C	10. A
11. D	12. B, C			

13 to 33 Answers in text of Chapter 4.

Chapter 5

1. B	2. C	3. C, D	4. B, C	5. C
6. D	7. B	8. D	9. B	10. D
11. A, D	12. C			

13 to 25 Answers in text of Chapter 5.

Chapter 6

1. A	2. B	3. C	4. B	5. B
6. B	7. C	8. D	9. B	10. C
11. D	12. C	13. D	14. C	15. A
16. D	17. D	18. C	19. C	20. B
21. B	22. B	23. A	24. B	25. C
26. D	27. D	28. D	29. B	30. C
31. A	32. D	33. C	34. D	35. B
36. A	37. B	38. C	39. A	40. D
41. B	42. B	43. C	44. C	45. A
46. C	47. B	48. B	49. D	50. A
51. D	52. C	53. B	54. C	55. D
56. A	57. B	58. C	59. D	60. A, D
61. C	62. B	63. D	64. C	65. D
66. D	67. A	68. D	69. C	70. D
71. D	72. B	73. C	74. D	75. A
76. B				

77 to 115 Answers in text of Chapter 6.

Preparing for assessment

City & Guilds will assess your knowledge and understanding of the topics covered in this electrical installation course in two ways:

1. Internally and externally set written assignments, internally marked, externally verified.

2. Online multiple-choice questions similar to the questions at the end of each chapter in this book. Each question will consist of a statement with four possible answers. **Only one answer will be correct in the online examination.**

Preparing for your online assessment examination

Before the exam:

- You should have been revising throughout the course – revision is not just something you start with a day or two before the exam.
- Don't cram all the subject areas into one revision session – narrow areas down into bite-sized chunks and tackle them one at a time.
- Don't try to revise or learn new things at the last minute.
- Do lots of sample papers before the exam.
- Look online for sample questions and answers.
- Make sure that you are used to the style, type and wording of the questions that you will have to answer in your exam.
- Make sure you know what documents are required and expected in the exam (e.g. the *On Site Guide* or Wiring Regulations).
- Go to bed early the night before your exams (no partying until they're all finished, not just one exam).
- If your exam is in the morning make sure you get up early enough to eat a decent breakfast, and don't eat just before exams, as this can make you sleepy.
- Drink water well before exams in order to ensure that you are fully hydrated, which in turn makes you more alert.
- Sip water during exams (if allowed).
- Go to the toilet before the exam starts – it's too late when you are in the room.
- Make sure you know where and when the exam is taking place, and how long it will take you to get there.
- Check that your method of transportation to the exam centre is going to work on the day of the exams (e.g. timetables for buses and trains, etc.).
- Arrive early at the exam centre, but not so early that you're hanging around outside for hours.
- Try to relax in the minutes leading up to the exam.
- Check all your personal stationery before going into the exam, ensuring that you've got everything you need (e.g. pens, a calculator, etc.).

Taking the online assessment:

- Listen carefully to the exam instructions that the invigilator gives you;
- You may not know the invigilator but don't be frightened to ask questions if you don't understand the instructions;
- Check the exam title details and codes to make sure that you're in the right exam;
- Make sure you know how many questions there are and the time limit of the exam. Use this information to allocate an appropriate amount of time for each question;
- Make sure you do the tutorial session before the real exam. This will ensure that you become familiar with the techniques required to successfully complete the online exams;
- Questions are not always clear, but should contain clues to the answer. Take your time reading each question when you get to it, and read questions more than once before answering them;
- Look at the four possible answers and read them more than once before picking an answer to run with; don't tick the first thing that you recognize in a list of answers – use a process of elimination to rule out wrong answers before settling on a final answer;
- If you are unsure about an answer put a flag on it and leave it until the end, when you can go back and spend more time on it;
- Answer the questions that you find easy first, which will leave you more time to spend on the questions that you find difficult;
- Check any answers that need a calculation at least twice;
- Never leave the exam with questions unanswered – guess if you have to, but use the process of elimination mentioned earlier first, as this will give you more chance of getting it right;
- Don't try to be first out of the exam, take your time – there are no prizes for finishing first!;
- Don't press the finish button until you are totally sure that you can't do any more. Once the finish button is pressed you cannot restart the test or go back to check your answers.

APPENDIX
Environmental organizations

A

The Department of Energy and Climate Change (DECC) for grants

BS 7671:2008 Engineering Recommendations G 83/1 and G 59/1 published by the Energy Network Association and the Department for Business, Enterprise and Regulatory Reform (BERR) for technical specifications

Building Regulations England and Wales: the Department of Communities and Local Government at www.communities.gov.uk and, for Scotland, The Scottish Building Standards Agency at www.sbsa.gov.uk

For information on the Feed-in Tariff Scheme see the Office of the Gas and Electricity Markets (OFGEM) website at www.ofgem.gov.uk/fits

Energy Saving Trust at 020 7222 0101 and www.energysavingtrust.org.uk for advice on grants and products

Microgeneration product advice and their own certification scheme at 01752 823 600 and www.microgeneration.com

The Carbon Trust at www.carbontrust.co.uk/energy offers free advice on loans to businesses which are upgrading to more energy-efficient equipment. The size of the loan will depend upon the CO_2 savings

Planning Guide for solar, PV and wind turbine installations can be found at www.planningportal.gov.uk/uploads/hhg/houseguide.html

Best practice guide for installing microgeneration systems can be found on the Electrical Safety Council website at www.esc.org.uk/bestpracticeguides.html

Rainwater harvesting guidance and products can be found in abundance by Googling 'rainwater harvesting'

APPENDIX
Health and Safety Executive (HSE) publications and information

B

HSE Books, Information Leaflets and Guides may be obtained from
HSE Books, P.O. Box 1999, Sudbury, Suffolk CO10 6FS

HSE Infoline – Telephone: 0845 3450055 or write to
HSE Information Centre, Broad Lane, Sheffield S3 7HO

HSE home page on the World Wide Web
http:/www.hse.gov.uk

The Health and Safety Poster and other HSE publications are available from
www.hsebooks.com

Environmental Health Department of the Local Authority
Look in the local telephone directory under the name of the authority

Glossary of terms

Activities Activities are represented by an arrow, the tail of which indicates the commencement, and the head the completion of the activity.

Advantages of a d.c. machine One of the advantages of a d.c. machine is the ease with which the speed may be controlled.

Appointed person An appointed person is someone who is nominated to take charge when someone is injured or becomes ill, including calling an ambulance if required. The appointed person will also look after the first aid equipment, including restocking the first aid box.

As-fitted drawings When the installation is completed a set of drawings should be produced which indicate the final positions of all the electrical equipment.

Asphyxiation Asphyxiation is a condition caused by lack of air in the lungs leading to suffocation. Suffocation may cause discomfort by making breathing difficult or it may kill by stopping the breathing.

Assembly point The purpose of an assembly point is to get you away from danger to a place of safety where you will not be in the way of the emergency services.

Bar chart There are many different types of bar chart used by companies but the object of any bar chart is to establish the sequence and timing of the various activities involved in the contract as a whole.

Basic protection Basic protection is provided by the insulation of live parts in accordance with Section 416 of the IET Regulations.

Bill of quantities The size and quantity of all the materials, cables, control equipment and accessories. This is called a 'bill of quantities'.

Block diagram A block diagram is a very simple diagram in which the various items or pieces of equipment are represented by a square or rectangular box.

Bonding The linking together of the exposed or extraneous metal parts of an electrical installation.

Bonding conductor A conductor providing protective bonding.

BS 5750/ISO 9000 certificate A BS 5750/ISO 9000 certificate provides a framework for a company to establish quality procedures and identify ways of improving its particular product or service. An essential part of any quality system is accurate record-keeping and detailed documentation which ensures that procedures are being followed and producing the desired results.

Cables Most cables can be considered to be constructed in three parts: the conductor, which must be of a suitable cross-section to carry the load current; the insulation, which has a colour or number code for identification; and the outer sheath, which may contain some means of providing protection from mechanical damage.

Cable tray Cable tray is a sheet-steel channel with multiple holes. The most common finish is hot-dipped galvanized but PVC-coated tray is also available. It is used extensively on large industrial and commercial installations for supporting MI and SWA cables which are laid on the cable tray and secured with cable ties through the tray holes.

Cage rotor — The solid construction of the cage rotor used in many a.c. machines makes them almost indestructible.

Calibration certificates — Calibration certificates usually last for a year. Test instruments must, therefore, be tested and recalibrated each year by an approved supplier.

Capacitance — The property of a pair of plates to store an electric charge is called its capacitance.

Capacitive reactance (X_C) — Capacitive reactance (X_C) is the opposition to an a.c. current in a capacitive circuit. It causes the current in the circuit to lead ahead of the voltage.

Capacitor — By definition, a capacitor has a capacitance (C) of one farad (symbol F) when a p.d. of one volt maintains a charge of one coulomb on that capacitor.

Cartridge fuse — The cartridge fuse breaks a faulty circuit in the same way as a semi-enclosed fuse, but its construction eliminates some of the disadvantages experienced with an open-fuse element.

CFLs — CFLs (compact fluorescent lamps) are miniature fluorescent lamps designed to replace ordinary GLS lamps.

Circuit diagram — A circuit diagram shows most clearly how a circuit works.

Circuit Protective Conductor (CPC) — A protective conductor connecting exposed conductive parts of equipment to the main earthing terminal.

Clean Air Act — The Clean Air Act applies to all small and medium-sized companies operating furnaces, boilers or incinerators.

Competent person — A competent person is anyone who has the necessary technical skills, training and expertise to safely carry out the particular activity.

Conductor — A conductor is a material in which the electrons are loosely bound to the central nucleus and are, therefore, free to drift around the material at random from one atom to another.

Conduit — A conduit is a tube, channel or pipe in which insulated conductors are contained.

Copper losses — Copper losses occur because of the small internal resistance of the windings.

Copper losses and iron losses — As they have no moving parts causing frictional losses, most transformers have a very high efficiency, usually better than 90%. However, the losses which do occur in a transformer can be grouped under two general headings: copper losses and iron losses.

Critical path — Critical path is the path taken from the start event to the end event which takes the longest time.

Dangerous occurrence — Dangerous occurrence is a 'near miss' that could easily have led to serious injury or loss of life. Near-miss accidents occur much more frequently than injury accidents and are, therefore, a good indicator of hazard, which is why the HSE collects this data.

Daywork Daywork is one way of recording variations to a contract; that is, work done which is outside the scope of the original contract. It is extra work.

Delivery note By signing the delivery note the person is saying 'yes, these items were delivered to me as my company's representative on that date and in good condition and I am now responsible for these goods'.

Designer The designer of any electrical installation is the person who interprets the electrical requirements of the customer within the regulations.

Detail drawings and assembly drawings These are additional drawings produced by the architect to clarify some point of detail.

Direct current motors Direct current motors are classified by the way in which the field and armature windings are connected, which may be in series or in parallel.

Discharge lamps Discharge lamps do not produce light by means of an incandescent filament but by the excitation of a gas or metallic vapour contained within a glass envelope.

Disconnection and separation We must ensure the disconnection and separation of electrical equipment from every source of supply and that this disconnection and separation is secure.

Dummy activities Dummy activities are represented by an arrow with a dashed line.

Duty holder Duty holder is someone who has a duty of care for health, safety and welfare matters on-site.

Duty of care Everyone has a duty of care but not everyone is a duty holder. The person who exercises 'control over the whole system, equipment and conductors' and is the electrical company's representative on-site is the duty holder.

Earth The conductive mass of the earth whose electrical potential is taken as zero.

Earthing The act of connecting the exposed conductive parts of an installation to the main protective earthing terminal of the installation.

Eddy currents Eddy currents are circulating currents created in the core material by the changing magnetic flux. These are reduced by building up the core of thin slices or laminations of iron and insulating the separate laminations from each other.

Efficacy The performance of a lamp is quoted as a ratio of the number of lumens of light flux which it emits to the electrical energy input which it consumes. Thus efficacy is measured in lumens per watt; the greater the efficacy the better is the lamp's performance in converting electrical energy into light energy.

Electricity At Work Act The Electricity at Work Act tells us that it is 'preferable' that supplies be made dead before work commences (IET Regulation 4(3)).

Emergency lighting Since an emergency occurring in a building may cause the mains supply to fail, the emergency lighting should be supplied from a source which is independent from the main supply.

Emergency switching Emergency switching involves the rapid disconnection of the electrical supply by a single action to remove or prevent danger.

Employees   Employees have a duty to care for their own health and safety and that of others who may be affected by their actions.

Employer   Under the Health and Safety at Work Act an employer has a duty to care for the health and safety of employees.

Environmental conditions   Environmental conditions include unguarded or faulty machinery.

Event   An event is a point in time, a milestone or stage in the contract when the preceding activities are finished.

Exit routes   Exit routes are usually indicated by a green and white 'running man' symbol. Evacuation should be orderly; do not run but walk purposefully to your designated assembly point.

Exposed conductive parts   This is the metalwork of an electrical appliance or the trunking and conduit of an electrical system.

Extraneous conductive parts   This is the structural steelwork of a building and other service pipes such as gas, water, radiators and sinks.

Fault   A fault is not a natural occurrence; it is an unplanned event which occurs unexpectedly.

Fault protection   Fault protection *is provided by protective bonding and automatic disconnection of the supply (by a fuse or MCB) in accordance with IET Regulations 411.3 to 6.*

Fire   Fire is a chemical reaction which will continue if fuel, oxygen and heat are present.

Fire extinguishers   Fire extinguishers remove heat from a fire and are a first response for small fires.

First aid   First aid is the initial assistance or treatment given to a casualty for any injury or sudden illness before the arrival of an ambulance, doctor or other medically qualified person.

First aider   A first aider is someone who has undergone a training course to administer first aid at work and holds a current first aid certificate.

Flexible cable   A flexible cable is a cable whose structure and material make it suitable to be flexed while in service. Previously called flexible cord.

Flexible conduit   Flexible conduit is made of interlinked metal spirals often covered with a PVC sleeving.

Float time, slack time or time in hand   Float time, slack time or time in hand is the time remaining to complete the contract after completion of a particular activity.

Fluorescent lamp   A fluorescent lamp is a linear arc tube, internally coated with a fluorescent powder, containing a low-pressure mercury vapour discharge.

FP 200 cable   FP 200 cable is similar in appearance to an MI cable in that it is a circular tube, or the shape of a pencil, and is available with a red or white sheath. However, it is much simpler to use and terminate than an MI cable.

Functional switching   Functional switching involves the switching on or off or varying the supply of electrically operated equipment in normal service.

Fuse element   The fuse element is encased in a glass or ceramic tube and secured to end-caps which are firmly attached to the body of the fuse so that they do not blow off when the fuse operates.

GLS lamps   GLS lamps produce light as a result of the heating effect of an electrical current. Most of the electricity goes to producing heat and a little to producing light. A fine tungsten wire is first coiled and coiled again to form the incandescent filament of the GLS lamp.

GU10   GU10 describes a bi-pin lamp having a 10mm gap between the pins.

Hazardous malfunction   If a piece of equipment was to fail in its function, that is fail to do what it is supposed to do and, as a result of this failure has the potential to cause harm, then this would be defined as a hazardous malfunction.

HBC cartridge fuses   As the name might imply, these HBC (high breaking capacity) cartridge fuses are for protecting circuits where extremely high-fault currents may develop such as on industrial installations or distribution systems.

Human errors   Human errors include behaving badly or foolishly, being careless and not paying attention to what you should be doing at work.

Hysteresis loops   Some materials magnetize easily, and some are difficult to magnetize. Some materials retain their magnetism, while others lose it.

Impedance   The total opposition to current flow in an a.c. circuit is called impedance and given the symbol Z.

Improvement notice   An improvement notice identifies a contravention of the law and specifies a date by which the situation is to be put right.

Induction heating processes   Induction heating processes use high-frequency power to provide very focused heating in industrial processes.

Induction motor rotor   There are two types of induction motor rotor: the wound rotor and the cage rotor.

Inductive reactance (X_L)   Inductive reactance (X_L) is the opposition to an a.c. current in an inductive circuit. It causes the current in the circuit to lag behind the applied voltage.

Instructed person (electrically)   An instructed person (electrically) is a person adequately advised or supervised by a skilled person (electrically) to be able to perceive risks and avoid the hazards which electricity can create.

Insulator   An insulator is a material in which the outer electrons are tightly bound to the nucleus and so there are no free electrons to move around the material.

Inverse square law   The illumination of a surface follows the inverse square law,

where $E = \dfrac{I}{d^2}$ (lx)

Investors in People   Investors in People is a national quality standard that focuses on the needs of the people working within an organization. It recognizes that a company or business is investing some of its profits in its workforce in order to improve the efficiency and performance of the organization. The objective is to create an environment where what people can do and are motivated to do, matches what the company needs them to do to improve.

Iron losses   Iron losses are made up of hysteresis loss and eddy current loss. The hysteresis loss depends upon the type of iron used to construct the core and consequently core materials are carefully chosen.

Isolation   Isolation means the disconnection and separation of the electrical equipment from every source of electrical energy in such a way that this disconnection and separation is secure.

Isolator   An isolator is a mechanical device that is operated manually and is provided so that the whole of the installation, one circuit or one piece of equipment, may be cut off from the live supply.

Job sheet or job card   A job sheet or job card carries information about a job which needs to be done, usually a small job.

Layout drawings or site plan   These are scale drawings based upon the architect's site plan of the building and show the positions of the electrical equipment which is to be installed.

Line conductor   a conductor in an ac system for the transmission of electrical energy other than a neutral or protective conductor. Previously called a phase conductor.

Low smoke and fume cables   Low smoke and fume cables give off very low smoke and fumes if they are burned in a burning building. Most standard cable types are available as LSF cables.

Lumen method   When designing interior lighting schemes the method most frequently used depends upon a determination of the total flux required to provide a given value of illuminance at the working place. This method is generally known as the lumen method.

Manual handling   Manual handling is lifting, transporting or supporting loads by hand or by bodily force.

Metallic trunking   Metallic trunking is formed from mild steel sheet, coated with grey or silver enamel paint for internal use or a hot-dipped galvanized coating where damp conditions might be encountered.

Michael Faraday   Michael Faraday demonstrated on 29 August 1831 that electricity could be produced by magnetism. He stated that: 'When a conductor cuts or is cut by a magnetic field an e.m.f. is induced in that conductor. The amount of induced e.m.f. is proportional to the rate or speed at which the magnetic field cuts the conductor.'

MI cable   An MI cable has a seamless copper sheath which makes it waterproof and fire- and corrosion-resistant. These characteristics often make it the only cable choice for hazardous or high-temperature installations.

Mini-trunking   Mini-trunking is very small PVC trunking, ideal for surface wiring in domestic and commercial installations such as offices.

Mobile equipment   Electrical equipment which is moved while in operation or can be moved while connected to the supply. Previously called portable equipment.

Motor starter   The purpose of the motor starter is not to start the machine, as the name implies, but to reduce heavy starting currents and provide overload and no-volt protection in accordance with the requirements of Regulation 552.

Mutual inductance   A mutual inductance of 1H exists between two coils when a uniformly varying current of 1A/s in one coil produces an e.m.f. of 1V in the other coil.

Network diagram   A network diagram can be used to coordinate all the interrelated activities of the most complex project.

Ohm's law   Ohm's law, which says that the current passing through a conductor under constant temperature conditions, is proportional to the potential difference across the conductor.

Ordinary person   A person who is neither a skilled person nor an instructed person.

Overload current   An overload current can be defined as a current which exceeds the rated value in an otherwise healthy circuit.

Paper insulated lead covered steel wire armour cables   Paper insulated lead-covered steel wire armour cables are only used in systems above 11 kV. Very high-voltage cables are only buried underground in special circumstances when overhead cables would be unsuitable, for example, because they might spoil a view of natural beauty.

Permit-to-work procedure   The permit-to-work procedure is a type of 'safe system to work' procedure used in specialized and potentially dangerous plant process situations.

Power factor   The power factor of the consumer is governed entirely by the electrical plant and equipment that is installed and operated within the consumer's buildings.

Power factor improvement   Power factor improvement of most industrial loads is achieved by connecting capacitors to either:
- individual items of equipment, or
- banks of capacitors.

PPE   PPE is defined as all equipment designed to be worn, or held, to protect against a risk to health and safety.

Prohibition notice   A prohibition notice is used to stop an activity which the inspector feels may lead to serious injury.

Protective bonding   This is protective bonding for the purpose of safety.

Public nuisance   A public nuisance is 'an act unwarranted by law or an omission to discharge a legal duty which materially affects the life, health, property, morals or reasonable comfort or convenience of Her Majesty's subjects'.

Pulsating field   Once rotation is established, the pulsating field in the run winding is sufficient to maintain rotation and the start winding is disconnected by a centrifugal switch which operates when the motor has reached about 80% of the full load speed.

PVC insulated steel wire armour cables   PVC insulated steel wire armour cables are used for wiring underground between buildings, for main supplies to dwellings, rising sub-mains and industrial installations. They are used where some mechanical protection of the cable conductors is required.

Quality   Quality generally refers to the level of excellence, but in the business sense it means meeting the customer's expectations regarding performance, reliability and durability.

RCD   An RCD is a type of circuit-breaker that continuously compares the current in the line and neutral conductors of the circuit.

Rectification	Rectification is the conversion of an a.c. supply into a unidirectional or d.c. supply.
Resistance	In any circuit, resistance is defined as opposition to current flow.
Resistivity	The resistivity (symbol ρ – the Greek letter 'rho') of a material is defined as the resistance of a sample of unit length and unit cross-section.
Resistor	All materials have some resistance to the flow of an electric current but, in general, the term resistor describes a conductor specially chosen for its resistive properties.
Safety Officer	The Safety Officer will be the specialist member of staff, having responsibility for health and safety within the company. He or she will report to the senior manager responsible for health and safety.
Schematic diagrams	A schematic diagram is a diagram in outline of, for example, a motor starter circuit.
Semi-enclosed fuse	The semi-enclosed fuse consists of a fuse wire, called the fuse element, secured between two screw terminals in a fuse carrier.
Shock protection	Protection from electric shock is provided by basic protection and fault protection.
Short-circuit	A short-circuit is an overcurrent resulting from a fault of negligible impedance connected between conductors.
Single PVC insulated conductors	Single PVC insulated conductors are usually drawn into the installed conduit to complete the installation.
Skilled person (electrically)	A skilled person (electrically) is a person with relevant education, training and practical skills and sufficient experience to be able to perceive risks and to avoid the hazards which electricity can create.
Skirting trunking	A trunking manufactured from PVC or steel and in the shape of a skirting board is frequently used in commercial buildings such as hospitals, laboratories and offices.
Socket outlets	Socket outlets provide an easy and convenient method of connecting portable electrical appliances to a source of supply.
Space factor	The ratio of the space occupied by all the cables in a conduit or trunking to the whole space enclosed by the conduit or trunking is known as the space factor.
Special waste	Special waste is covered by the Special Waste Regulations 1996 and is waste that is potentially hazardous or dangerous and which may, therefore, require special precautions during handling, storage, treatment or disposal. Examples of special waste are asbestos, lead-acid batteries, used engine oil, solvent-based paint, solvents, chemical waste and pesticides.
Static charge	Static charge builds up between any two insulating surfaces or between an insulating surface and a conducting surface, but it is not apparent between two conducting surfaces.
Static electricity	Static electricity is a voltage charge which builds up to many thousands of volts between two surfaces when they rub together.
Statutory nuisance	'A statutory nuisance must materially interfere with the enjoyment of one's dwelling. It is more than just irritating or annoying and does not take account of the undue sensitivity of the receiver.'

Steel wire armoured	Steel wire armoured PVC insulated cables are now extensively used on industrial installations and often laid on cable tray.
Switching for mechanical maintenance	The switching for mechanical maintenance requirements is similar to those for isolation except that the control switch must be capable of switching the full load current of the circuit or piece of equipment.
Team working	Team working is about working with other people.
Test instruments and test leads	The test instruments and test leads used by the electrician for testing an electrical installation must meet all the requirements of the relevant regulations.
The safety representative	The safety representative will be the person who represents a small section of the workforce on the Safety Committee. The role of the safety representative will be to bring to the Safety Committee the health and safety concerns of colleagues and to take back to colleagues information from the Committee.
Three-phase supply	If a three-phase supply is connected to three separate windings equally distributed around the stationary part or stator of an electrical machine, an alternating current circulates in the coils and establishes a magnetic flux.
Time sheet	A time sheet is a standard form completed by each employee to inform the employer of the actual time spent working on a particular contract or site.
Transformer	A transformer is an electrical machine which is used to change the value of an alternating voltage. Transformers vary in size from miniature units used in electronics to huge power transformers used in power stations.
Transformers rating	Transformers are rated in kVA (kilovolt-amps) rather than power in watts because the output current and power factor will be affected by the load connected to the transformer.
Trunking	Trunking is an enclosure provided for the protection of cables and is normally square or rectangular in cross-section, having one removable side. Trunking may be thought of as a more accessible conduit system.
Universal motor	A series motor will run on both a.c. or d.c. and is, therefore, sometimes referred to as a 'universal' motor.
UPS	A UPS is essentially a battery supply, electronically modified to provide a clean and secure a.c. supply.
Visual inspection	The aim of the visual inspection is to confirm that all equipment and accessories are undamaged and comply with the relevant British and European standards, and also that the installation has been securely and correctly erected.
Wiring diagram	A wiring diagram or connection diagram shows the detailed connections between components or items of equipment.
Written messages	A lot of communications between and within larger organizations take place by completing standard forms or sending internal memos by e-mail or text.

Index